4000000444757

College & Career Readiness

Texas Edition

Steve Mariotti

Neelam Patel Chowdhary

Suzanne Weixel

Boston • Columbus • Indianapolis • New York • San Francisco
Amsterdam • Cape Town • Dubai • London • Madrid • Milan • Munich • Paris • Montréal • Toronto
Delhi • Mexico City • São Paulo • Sydney • Hong Kong • Seoul • Singapore • Taipei • Tokyo

Pearson

© 2017. Pearson Education, Inc. All Rights Reserved.

Printed in the United States of America. This publication is protected by copyright, and permission should be obtained from the publisher prior to any prohibited reproduction, storage in a retrieval system, or transmission in any form or by any means, electronic, mechanical, photocopying, recording, or otherwise. For information regarding permissions, request forms and the appropriate contacts, please visit www.pearsoned.com/permissions to contact the Pearson Education Rights and Permissions Department.

Unless otherwise indicated herein, any third party trademarks that may appear in this work are the property of their respective owners and any references to third party trademarks, logos or other trade dress are for demonstrative or descriptive purposes only. Such references are not intended to imply any sponsorship, endorsement, authorization, or promotion of Pearson Education Inc. products by the owners of such marks, or any relationship between the owner and Pearson Education Inc. or its affiliates, authors, licensees or distributors.

330 Hudson Street, New York, NY 10013

Hardcover ISBN 10: 0-13-442803-X
Hardcover ISBN 13: 978-0-13-442803-1

Contents

Introduction   ix

Part I   Careers and You

1 ■ Your Personal Strengths   3
Analyzing Career Requirements and Rewards   4
  Career Requirements   4
  Career Rewards   5
  Planning for College and Career   5
Recognizing Your Values   6
  Types of Values   6
  Changing Values   7
  Are Your Standards Flexible?   8
  Understanding Ethics   9
Identifying Your Strengths and Interests   9
  What Are Strengths?   9
  Discovering Your Interests   10
Recognizing Character Qualities   11
  Personal Qualities and Your Career   11
  Developing Your Character   12
  Recognizing Your Character   12
  Benefits of Character Qualities   13
  A Positive Attitude   14
Analyzing Employability Skills   15
  Employability Characteristics   15
  Transferable Skills   16
  How Can I Build Employability Skills?   16

2 ■ The Roles You Play   21
Identifying Roles and Responsibilities   22
  Different Roles = Different Responsibilities   22
  Assigned Roles vs. Acquired Roles   23
  Roles and Responsibilities Change   24
  Fulfilling Your Responsibilities   24
  Rules and Responsibilities   25
  Different Rules for Different People   25
Analyzing Well-Being   26
  Well-Being: Six Ways   26
  Promote Your Own Well-Being   27
  Roadblocks to Well-Being   27
Recognizing Influences   28
  Understanding Influence   28
  Good Influence or Bad Influence?   29
Identifying Resources   29
  Types of Resources   30
  Available Resources   30
  Managing Resources   31
  Making the Most of What You Have   32

3 ■ Why We Work   37
Examining Lifestyle Goals and Factors   38
  What Are Lifestyle Factors?   38
  Setting Lifestyle Goals   39
  Making Trade-Offs   39
Recognizing Needs vs. Wants   40
  Why Do Needs and Wants Matter?   40
  Do You Want It or Need It?   41
Measuring the Value of Work   42
  The Economic Value of Work   42
  The Emotional Value of Work   43
  The Educational Value of Work   43
Examining Career Trends   44
  What Are Employment Trends?   45
  Economic Trends   45
  Impact of Technology   46
  Changing Lifestyles   46
  Flexible Lifestyles   47
  Entrepreneurship   48
  Nontraditional Occupations   48
  Diversity in the Workforce   49

4 ■ Exploring the Career Clusters   53
Analyzing the Career Clusters   54
  Agriculture, Food & Natural Resources   55
  Architecture & Construction   56
  Arts, Audio-Video Technology & Communications   57
  Business Management & Administration   58

Education & Training   59
Finance   60
Government & Public Administration   61
Health Science   62
Hospitality & Tourism   64
Human Services   65
Information Technology   66
Law, Public Safety, Corrections & Security   67
Manufacturing   68
Marketing   69
Science, Technology, Engineering & Mathematics   70
Transportation, Distribution & Logistics   71

Exploring Other Resources for Identifying Careers   72
Resources at School   72
The Bureau of Labor Statistics   72
Resources in Your Community   73

5 ■ Think Like an Entrepreneur   77

Exploring Entrepreneurial Thinking   78
What Is Entrepreneurial Thinking?   78
Thinking Like an Entrepreneur   79

Recognizing Entrepreneurial Risks and Rewards   80
What Are the Rewards?   81
What Are the Risks?   82

Developing the Characteristics of an Entrepreneur   83
Qualities of an Entrepreneur   83
Entrepreneurial Skills   84
Keys to Success   85

Identifying Entrepreneurial Opportunities   88
Finding Ideas   88
Evaluating an Opportunity   89
Turning an Opportunity into a Business   90

Part II Career Preparation

6 ■ Skills for Success   97

Making Decisions   98
Every Decision Is Unique   98
Six Steps to a Decision   98
Decision-Making for College and Career Readiness   99
Can Your Decisions Be Hurtful?   99
Overwhelmed by Options   100
More About Consequences   100
Did I Make the Best Choice?   100

Solving Problems   102
Owning the Problem   102
Six Steps to a Solution   103
Asking for and Giving Help   103

Setting Goals   105
Five Steps to a Goal   105
Short-Term Goals   106
Long-Term Goals   106
What Are Milestones?   106

Thinking Critically   107
Critical Thinking in Action   107
Are Emotions Always a Problem?   108

Using Management Skills   109
What a Manager Does   109
What Makes a Manager Effective?   109
What Does the Title "Manager" Mean?   110
Becoming a Leader   110
What Makes a Leader?   111

7 ■ Academic Planning   115

Recognizing the Value of School   116
Why School Matters   116
Why Dropping Out Is Not Cool   116
How Education Affects Employment   117

Analyzing Your Learning Style   118
Visual Learners   118
Auditory Learners   118
Tactile Learners   119
Study Strategies for Success in School   119

Transitioning to High School   120
How to Prepare for High School   120
Choosing a High School   121
What to Expect in High School   121

Developing an Academic Plan   122
The Purpose of the Plan   122
How Do I Create an Academic Plan?   123
Setting Long-Term and Short-Term Academic Goals   123

Developing Career-Related Skills in School   124
Building Math Skills   124
Developing Communication Skills   125
Choosing Electives   125
Choosing Advanced Courses   125
Learning Beyond High School   126

Contents

Planning for Postsecondary Education   127
   Researching Colleges   127
   Which College Is Right for Me?   128
   Applying to College   128
   College-Ready Assessments   128
   Paying for College   130
Identifying Career-Related Opportunities for Students   132
   What Is an Internship?   132
   Volunteer!   133
   Other Ways to Gain Experience   133
   Joining a Student Organization   134
   Common Career and Technical Student Organizations   134

8 ■ Communicating with Others   139

Using Written Communications   140
   Reading   140
   Writing   141
   Electronic Communication   142
Using Verbal Communications   143
   Using the Communication Process   143
   Different Types of Verbal Communication   144
   How Verbal Communication Helps You Think   146
   Delivering an Oral Presentation   146
   Steps to Becoming a Great Speaker   146
Developing Listening Skillss   148
   The Difference Between Hearing and Listening   148
   Listening on the Job   149
Identifying Nonverbal Communications   150
   Positive Body Language   151
   Negative Body Language   151
   Using Visual Aids   152
Recognizing Obstacles to Communications   153
   Recognizing Communication Barriers   154
   Overcoming Miscommunication   155

9 ■ Building Relationships   159

Relating to Others   160
   Relating to People at Work   160
   Qualities of a Healthy Relationship   161
   A Two-Way Street   161
   Communicate, Communicate, Communicate!   162
Building Team Relationships   163
   Challenges of Teamwork   163
   What Makes a Team Successful?   164
   Developing As a Leader   165
   What About Team Members?   166
   Recognizing Peer Pressure   166
   Building Ethical Relationships   167
Managing Conflict   168
   Types of Conflict   168
   Causes of Conflict   169
   Resolving Conflict   169
   Team Conflicts   170
   Asserting Yourself   170
   Unresolved Conflict   171

10 ■ Basic Math Skills   175

Adding, Subtracting, Multiplying, and Dividing   176
   What Good Is Addition?   176
   When Do I Subtract?   176
   Memorize Those Multiplication Tables!   177
   Do I Need Division?   177
Working with Percentages, Fractions, and Decimals   178
   What's a Fraction?   179
   What's a Decimal?   179
   What's a Percent?   180
Working with Ratios and Proportions   181
   What's a Ratio?   182
   What's a Proportion?   183
Analyzing Data and Probability   184
   Data Analysis   184
   What's the Chance of That?   185
Analyzing Charts   187
   Jobs That Use Charts   187

11 ■ Technology and Your Career   191

Recognizing the Impact of Technology   192
   How Technology Impacts Career Options   192
   New Careers Driven by Technology   193
   Technology Trends in the Workplace   193
High Technology in Industry   194
   Technology in Manufacturing   195
   Technology in E-Commerce   196
   Careers in Information Technology (IT) Systems   197
   Technology in Agriculture   199
Using Technology in Your Career   200
   Transferrable Computer Skills   201
   Using Communications Devices   202
   Using Computer Applications   203

Part III Career Development

12 ■ Career Planning 209

Planning a Career 210
- Steps for Planning a Career 210
- Mapping out a Career Plan 211
- What to Include in a Career Plan 211

Managing a Career Self-Assessment 213
- Developing a Career Self-Assessment Worksheet 213
- Identifying Your Abilities 215
- Identifying Your Interests 216
- Completing Your Career Self-Assessment Worksheet 217
- Using Your Career Self-Assessment Worksheet 217

Analyzing Career Planning Resources 218
- What Is Networking? 218
- How Do I Network? 219
- Organizing Your Contacts 220
- Using Online Resources 220
- Using a Career Center or Agency 221
- Career Fairs 222
- What About Want Advertisements? 222

13 ■ Managing a Job Search 227

Creating Job Search Materials 228
- Why Do I Need a Cover Letter? 228
- Listing References 229
- Do I Need a Portfolio? 230
- Managing Your Job Search Materials 231

Preparing a Resume 233
- How Should I Format My Resume? 233
- What Do I Include on My Resume? 234
- Use Action Words! 234

Applying for a Job 235
- Tips for Applying 235
- Filling out a Job Application 236
- Use a Personal Information Card 236

Interviewing for a Job 237
- Preparing for a Job Interview 237
- Practicing for a Job Interview 238
- Making the Most of the Job Interview 238
- Telephone Interviews 239
- Write a Thank-You Note 239
- Follow Up 239
- What If You Get the Job? 240
- What If You Do Not Get the Job? 240

Evaluating a Job Offer 242
- How Much Does It Pay? 242
- What Are Benefits? 243
- Analyzing Insurance 243
- How to Negotiate Your Compensation 244
- Do You Need Accommodations? 245

14 ■ Getting Started in Your Career 249

Beginning a New Job 250
- Filling out Forms 250
- Making the Most of Orientation 251
- Making a Good Impression 252

Building Work Relationships 253
- Types of Work Relationships 254
- Teamwork and Leadership 255

Benefiting from a Performance Review 256
- Knowing What to Expect 257
- What the Performance Review Means 258
- Requesting a Review 258

Requesting Additional Education and Training 259
- How Can You Develop New Skills? 260
- Finding a Mentor 261
- Becoming a Lifelong Learner 261

Obtaining a Raise or Promotion 263
- The Importance of Advancement 263
- How Can You Earn a Raise or a Promotion? 264

Making a Career Change 265
- How to Say Goodbye 266
- Coping with a Layoff 267
- Finding a New Job 268
- Taking Time Off 269

15 ■ Being Productive in Your Career 273

Being a Successful Employee 274
- Being Professional 274
- Being a Problem Solver 275
- Being Ethical at Work 276

Applying Time Management Techniques in the Workplace 278
- Tools for Managing Time at Work 279
- Tips for Managing Time at Work 280

Contents

Managing Workplace Conflict 282
   Valuing Differences 282
   Breaking Down Workplace Barriers 283
   Peer Pressure at Work 283
Recognizing Your Rights and Responsibilities 284
   Overcoming Discrimination and Harassment 284
   Understanding Unions and Trade Organizations 286
   Understanding Professional Organizations 287
   Staying Safe at Work 288

16 ■ Living a Healthy and Balanced Life 293

Analyzing Lifestyle Choices 294
   Balancing Roles 294
   Career Choices and a Balanced Lifestyle 296
   Balancing Work and Leisure 298
   Thinking About the Future 299
Recognizing the Importance of Lifelong Learning 301
   Staying Informed 302
   Learning Outside of School 303
Living in Your Community 304
   Being a Responsible Citizen 305
   Volunteering 305
   Taking Care of the Environment 306
Staying Healthy 307
   Stress Management 308
   Nutrition 310
   Physical Activity 311
   Avoiding Risky Behaviors 313

Part IV Financial Management

17 ■ Personal Money Management 365

Why We Use Money 320
   What Money Is Worth 320
   The Power to Purchase 321
Comparing Financial Needs and Wants 322
   Establishing Your Needs 322
   Changing Wants and Needs 323
   Making Financial Decisions 323
Setting Financial Goals 324
   Getting Your Priorities in Order 325
   Short-Term and Long-Term Financial Goals 325
   Spending Goals and Saving Goals 326

Managing a Budget 326
   What's in a Budget? 327
   Fixed or Flexible? 328
   Setting Up a Budget 328
   Make Your Budget Work for You 329
   Expecting the Unexpected 330
   Staying on Budget 331
   Running in the Red 332
Analyzing Your Paycheck 333
   How You Will Be Paid 333
   Understanding Your Paycheck 334
   Understanding Your Pay Stub 334
Choosing a Method of Payment 336
   When to Use Cash 337
   When to Write a Check 337
   When to Use a Debit Card 338
   When to Use a Credit Card 338
   Using an Installment Plan 338
   Using Electronic Funds 339

18 ■ Personal Financial Planning 343

Choosing a Bank and a Bank Account 344
   Types of Banks 344
   Which Bank Is Right for You? 345
   Types of Bank Accounts 346
   Opening a Banking Account 347
   Using Basic Transactions 347
   Using a Checking Account 348
   Can I Use a Debit Card? 349
Managing Your Bank Account 349
   Recording Transactions 350
   Balancing Your Account 350
   Using an ATM 351
Saving and Investing 352
   The Difference Between Saving and Investing 353
   Using Savings Accounts 353
   Opening a Savings Account 354
   Understanding Investments 354
   Planning for Retirement 355
   Using a Money Management Service 357
Comparing and Contrasting Forms of Credit 357
   Getting a Loan 359
   Using a Credit Card 359
   How Do I Establish Credit? 360
   What's a Credit Report? 361
   Managing Debt 362

Paying Taxes 364
   Where Do Your Tax Dollars Go? 365
   Paying Income Tax 365
   Filing Income Tax Returns 366

Keeping Your Personal and Financial Information Safe 367
   Protect Your Records 367
   Protect Your Wallet 368
   Protect Yourself Online 368

Analyzing Banking and Credit Regulations 369
   Regulating Banks 369
   Protecting Credit 369

Appendices

A ■ **Language Arts Review** 373

B ■ **Math Review** 389

Glossary 399

Index 415

Introduction

"What will I be when I grow up?" is an age-old question that, for most students, has no easy answer. *College & Career Readiness* is a broad exploration of educational and career opportunities and the rewards of higher education and different career pathways. It provides the essential tools that students need to set and achieve career and academic goals and make positive decisions that will affect their futures.

College & Career Readiness highlights the most promising careers and prepares students to connect academic achievement to real-world success. Students learn to identify their career interests, aptitudes, and learning styles. They discover that they can gain critical experience and develop job-related skills, including math, technology, teamwork, leadership, and effective communications. This book also helps students recognize and manage their resources and develop a career portfolio while practicing the principles of personal finance.

College & Career Readiness is a clear and concise text, designed to engage students by focusing on topics that are relevant to their lives today. It uses contemporary, real-world examples to encourage students to apply their skills in practical situations and to meet responsibilities by providing them with challenging learning experiences that require practical application of academic skills.

Content Overview

College & Career Readiness is organized into four parts.

In **Part I, Careers and You**, students are encouraged to identify how their skills and interests can help them succeed in the world of work. They learn about the different roles and responsibilities they have in the five critical areas of living: home, school, peers, community, and careers. In addition, they are introduced to the career clusters and pathways and to the risks and rewards of entrepreneurship.

Part II, Career Preparation, emphasizes the importance of using processes for decision making, goal setting, and problem solving. The relationship between academic planning, goal setting, and future success takes center stage; students are encouraged to develop an academic plan, and to take advantage of opportunities to develop their skills and interests from middle school through high school and beyond. Students have the opportunity to practice effective communication, management, leadership, and teamwork, and learn how math and technology are used in real-world situations.

Part III, Career Development, takes students through the process of identifying career opportunities, planning for a career, and conducting a career search. Students learn to perform a career self-assessment so they can match their interests and abilities with career opportunities. Students are given the opportunity to create career search documents and practice interviewing for a job. Key concepts include making lifestyle choices, balancing work and family life, and making use of lifelong learning opportunities.

In **Part IV, Financial Management**, students study personal finance. They are introduced to the concept of money, and how to differentiate between needs and wants. They are encouraged to set financial goals and manage a budget to achieve those goals. In addition, they identify the benefits and drawbacks of different types of payment methods. They learn about bank accounts and the difference between saving and investing. Responsible use of credit is emphasized, along with how to keep personal and financial information safe.

Two appendices and a glossary complete the text.

- **A** **Language Arts Review** provides a review of essential grammar and punctuation skills, including how to identify subjects and verbs, sentence structure, and the use of punctuation.
- **B** **Basic Math Review** provides a review of essential mathematics lessons, including place value, addition, subtraction, multiplication, and division, as well as working with decimals, fractions, and percents.

Glossary provides an alphabetical list of terms found throughout the text.

Introduction

21st Century Learning

As the future leaders of our families, communities, government, and workforce, it is imperative that today's students develop the skills they need to succeed in work and life. The Partnership for the 21st Century, an advocacy organization focused on infusing 21st Century Skills into education, has created the Framework for 21st Century Learning, which describes the skills, knowledge, and expertise students must master to succeed in work and life (www.p21.org).

Among other things, the Framework for 21st Century Learning identifies the following:

- *Core subjects.* English, reading or language arts, world languages, arts, mathematics, economics, science, geography, history, government, and civics.
- *21st Century interdisciplinary themes.* Global awareness; financial, economic, business, and entrepreneurial literacy; civic literacy; health literacy.
- *Learning and innovation skills.* Creativity and innovation; critical thinking and problem solving; communication and collaboration.
- *Information, media and technology skills.* Information literacy; media literacy; information, communications and technology literacy.
- *Life and career skills.* Flexibility and adaptability; initiative and self-direction; social and cross-cultural skills; productivity and accountability; leadership and responsibility.

Preview of the Textbook

Parts

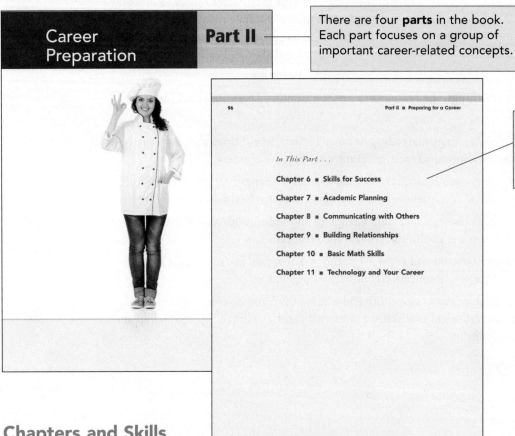

There are four **parts** in the book. Each part focuses on a group of important career-related concepts.

Parts are organized into **chapters**. There are between three and six chapters in each part.

Chapters and Skills

Each chapter has between two and seven objectives that identify the key **skills** explained in the chapter.

Each chapter begins with an introduction, called **Getting Started**, that highlights the concepts covered.

The introduction includes a **"bellringer"** activity designed to focus the class on the current concept while encouraging discussion and collaboration.

Introduction

xiii

Working within a Chapter

Main headings correspond to the objectives listed at the beginning of each chapter.

Each objective has at least two sections, which are identified by a **subheading**.

After important figures, an open-ended **figure question** asks the students to consider relevant issues related to the figure and text.

Vocabulary terms are in bold and are defined in the text and in the glossary at the end of the book.

Career Facts, **Tips**, and **Trends** throughout the chapter provide useful information to help the students put the concepts into real-world context.

Margin features enhance and support the text, and prompt the students to apply critical thinking and 21st Century Skills to the covered concepts.

Features and Activities

Integrated throughout *Exploring Careers for the 21st Century* are opportunities for the student to use critical 21st Century Skills to solve problems, make decisions, and develop and present information individually and cooperatively with their classmates.

Career Profile is a career exploration activity. It highlights a career within one of the 16 career clusters, and prompts the student to use available resources to investigate the skills, qualities, education, and abilities that might be necessary to succeed in that career.

Tech Connect is a technology awareness feature that introduces a technology concept relating to the current topic. The student is asked to research the concept and use available technology to illustrate and present the information they discover to their teacher or classmates.

Money Madness is a financial literacy activity that invites the student to use their problem solving, goal setting, and decision-making skills along with math to solve a problem involving money.

Introduction

Global Awareness feature illustrates how the current topic relates to the world outside of students' own community. Each feature includes a critical-thinking prompt or activity that encourages students to investigate the topic to learn more.

Career Counsel is a 21st Century Learning activity. It presents a story or scenario about a real-world career situation that has no conclusion. Students are asked to use 21st Century Skills such as decision making, critical thinking, and problem solving to think of an ending and discuss, write, or perform it for their peers.

Margin Features

Margin features within each chapter are designed to stand out visually on the page. They support and reinforce the chapter content.

Numbers Game introduces a math concept. It provides instruction on how to apply the concept in a real world situation, and then prompts the student to do the math.

Check-Off lists present content in easy-to-read lists.

What If? lists present the student with a problem or decision in the form of a What If question, followed by a list of possible solutions or results. A critical-thinking prompt encourages the student to consider possibilities or situations when the information might be useful in their own lives.

Introduction

xvii

No Excuses is a list of possible pitfalls or problems relating to the topic, marked with thumbs-down icons, followed by a list of positive actions, marked with thumbs-up icons. A critical-thinking prompt encourages students to consider how you might act in a similar situation.

Myth/Truth presents a common misconception and the facts behind the truth.

End-of-Chapter Activities

At the end of every chapter is a series of questions and activities.

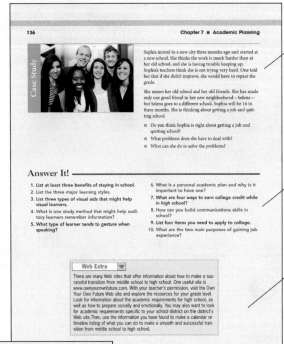

Case Study presents a real-life career-related scenario, followed by a series of open-ended questions designed to provide students with the opportunity to use 21st Century Skills.

Answer It! is a set of ten questions designed to assess reading comprehension.

Web Extra provides suggestions for using the Internet to locate additional information about the chapter topics.

Career-Ready Practices is a group activity designed to encourage teamwork and to support career-ready practices. It provides the opportunity for students to work collaboratively on projects ranging from research presentations to debates and competitive games.

Write Now is a language arts activity that encourages students to use language to answer questions, research and present information, or respond to critical-thinking prompts.

Career Portfolio provides students with the opportunity to develop a collection of documents and other items that they can use to illustrate their accomplishments, skills, and abilities to potential employers.

College-Ready Practices is an individual activity designed to encourage students to think about what it means to be college-ready.

Careers and You — Part I

In This Part...

Chapter 1 ■ **Your Personal Strengths**

Chapter 2 ■ **The Roles You Play**

Chapter 3 ■ **Why We Work**

Chapter 4 ■ **Exploring the Career Clusters**

Chapter 5 ■ **Think Like an Entrepreneur**

1 Your Personal Strengths

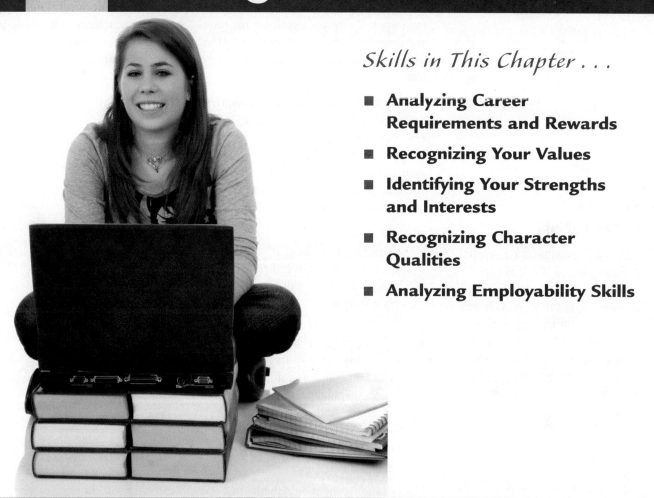

Skills in This Chapter . . .

- **Analyzing Career Requirements and Rewards**
- **Recognizing Your Values**
- **Identifying Your Strengths and Interests**
- **Recognizing Character Qualities**
- **Analyzing Employability Skills**

GETTING STARTED

Has anyone ever asked you what you want to be when you grow up? Did you answer firefighter or doctor? Did you answer artist or president of the United States? These are all good careers. However, they are not the right careers for everyone. If you choose a career that interests you and that makes use of your skills and abilities, you are more likely to be happy and satisfied with your work.

Why do people work? Ask ten people that question, and you are likely to get ten different answers! Yes, people work to make money. They also work to feel pride and satisfaction, to stay busy, and because it can be fun. The average person spends between 40 and 50 hours a week at work. If that work is enjoyable and rewarding, that person is going to be happier in all areas of life.

▶ In two groups, take a few minutes to select and write down a career without the other group knowing what you have picked. Take turns asking the other group "Yes" or "No" questions about their career, until you can guess what it is. As a class, discuss the skills and abilities you might need for the careers each group picked, and what it would be like to work in those careers.

You Want Me to Do What?

When you consider a career, it is important to think about whether or not you are willing to meet its requirements. If you do not live up to your responsibilities at work, you will soon be looking for a new job. What are some ways you show that you can or cannot meet responsibilities?

- 👎 Showing up late and leaving early
- 👎 Letting co-workers do your job
- 👎 Skipping meetings
- 👍 Cooperating with co-workers
- 👍 Always trying your best
- 👍 Making sure you have the training you need to complete your work

How can researching the requirements of a job help you avoid problems at work?

Analyzing Career Requirements and Rewards

A **career** is a chosen field of work in which you try to advance over time by gaining responsibility and earning more money. A **job** is any activity that you do in exchange for money or other payment. A job does not necessarily lead to advancement. **Occupation** is a word that means career or job.

Every career comes with its own **requirements**—things you must do or have—and **rewards**—things you receive in return. The specific requirements and rewards of the career you choose have a major impact on the kind of life you will lead.

- A doctor spends years in school and in training. It is expensive and leaves little time for extra activities. But, doctors help people and save lives, which can be very rewarding.

- A restaurant owner invests his own money in the business. He works long hours and must interact with temperamental chefs, kitchen staff, and customers. However, he is his own boss and can take pride in starting and running his own exciting business.

Understanding the requirements and rewards can help you choose a career that you will enjoy for many years.

Career Requirements

Career requirements are the responsibilities that you must perform in order to succeed in the career. A **responsibility** is something people expect you to do, or something you must accomplish. As a student, your teachers expect you to come to class and do your homework. As an employee—worker—your supervisor expects certain things from you, too.

Education is a requirement for all careers. Most employers expect employees to have diplomas from high school and college. For many careers, additional training and education is required.

The requirements of a career determine things such as the type of training and education you will need. Career requirements might also determine where you will live, how you will spend your time, and who you will be working with.

- If you choose to become an electrician, you will need education and training different from that of someone who chooses to become a chef.

Your Personal Strengths ■ Chapter 1

- A kindergarten teacher spends most of the day with 5-year-olds.
- A lawyer must read and prepare legal documents.
- A software programmer is alone in front of a computer for hours at a time.

Career Rewards

Recall that the average person spends between 40 to 50 hours per week at work. Most people work because they require money to buy things they need, such as food, clothes, and shelter. Some people choose their careers based on how much money they can earn on the job. The money that you receive in exchange for work performed is called **wages** or a **salary**.

What are the requirements and rewards of a career as a firefighter?

The word *career* first appeared in the middle 1500s. Its original meaning was "road" or "race course."

Money is a good reason to choose a career, but it is not the only reason. People who are happy with their careers usually do something they enjoy. Even if the job does not pay very well, people might choose a career because it gives them satisfaction, fits their values, uses their skills and abilities, and gives them a role in the community.

- A woman who starts a small jewelry business may enjoy being her own boss and setting her own hours.
- A social worker who helps homeless people may believe she is making the world a better place.
- An artist may feel he is making the world more beautiful by creating paintings or sculptures.

Planning for College and Career

It is never to soon to start planning for what you will do after high school. When you identify a career you want, you can set goals for achieving the education and experience you will need. Effective planning for college and career will set you on a track for success. It helps you:

- Focus on what you have to do to get to college.
- Understand what will be expected of you once you are in college.
- Gain the skills and education you need to achieve college and career success.

What are some of the rewards of being a health care professional?

Identifying Work Values

Identifying your work values can help you choose a career that you will enjoy. Some common work values include:

- ✔ Creativity: Having a job that allows you to express yourself or use your imagination
- ✔ Physical activity: Having a job that allows you to move around and be active
- ✔ Independence: Being your own boss, an entrepreneur
- ✔ Good salary: Being paid well for your work
- ✔ Job security: Having a steady, long-term job
- ✔ Work environment: Being able to work in a space where you are comfortable, such as outside, or in an office
- ✔ Leadership opportunities: Directing or managing the work of others
- ✔ Prestige: Having a job that commands the respect and admiration of others
- ✔ Challenge: Having the opportunity to perform difficult or important tasks
- ✔ Work safety: Being able to work in a predictable job, or in a job where there is no risk
- ✔ Variety: Having the opportunity to do different things
- ✔ Working with people: Being able to work closely with others
- ✔ Thinking: Having a job that requires you to use your brain and intellectual abilities

Recognizing Your Values

Values are the thoughts, ideas, and actions that are important to you. You use them to gauge or evaluate the people and things in your life. Knowing what you value is key to choosing an occupation that is right for you.

You develop values by figuring out what is important to you. When you are young, your values might reflect the values of the people in your life. For example, you might value the same things your parents and friends value.

As you grow and learn more about the world around you, you can develop your own values. Your parents might value job security more than you do, or your friends might not value protecting the environment as much as you do.

Types of Values

Something has **instrumental value** if it is important for acquiring something else. For example, money has instrumental value because you use it to buy other things. A car has instrumental value because you use it to drive to school or work. Clothes and an education also have instrumental value.

Something has **intrinsic value** if it is important in and of itself. Things with intrinsic value may be hard to explain. They might include emotions or feelings. For example, happiness, love, respect, justice, and beauty have intrinsic value. If you do something because it makes you feel good about yourself and not because someone is going to give you credit for it, then it has intrinsic value.

As soon as you are old enough to tell right from wrong, you develop **moral values**. Moral values help us judge behavior based on what we think is good compared to what we think is wrong. Moral values usually include positive character traits such as honesty, respect, justice, and responsibility.

Your Personal Strengths ■ Chapter 1

Work-related values refer to how you like to work and the results that you produce. Your work-related values can help you make career decisions.

- If you value adventure, you might consider a career in tourism or recreation management.
- If you value job security—knowing you will have your job for a long time—you might consider a career in government.
- If you value time with your family, then you may not want a job that requires you to travel a lot.

Changing Values

Of course, values can change. When you are young, you might value the opportunity to travel and meet new people. You might consider a career as a flight attendant.

As you get older, you might value job security and a steady salary. You might consider a career in hotel management.

Not everyone values the same things. Your mother might value a flexible work schedule so she can spend more time at home. Your cousin might value expensive things, so he works a lot of hours to earn more money. It is important to be able to respect other people's values even if you don't agree with them.

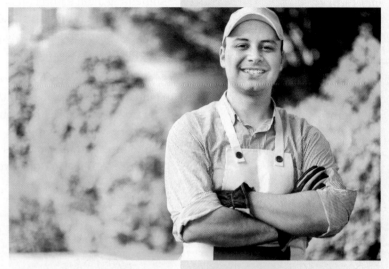

What types of careers might suit someone who values working outdoors?

Money Madne$$

You have been offered two jobs for the summer. They are both in food service, which is a field that interests you, and they have similar responsibilities. You have decided to choose the one that offers the best wages.

The first position is as a bus person at a restaurant. A bus person supports the wait staff by setting up and clearing tables, and performing tasks such as filling water glasses. It pays $6.15 per hour, and you are expected to work 30 hours a week for 8 weeks. The second position is as a caterer's assistant. A caterer's assistant also sets up and clears, and may also help prepare and transport food. It pays $1,500.00 for the summer. Which position should you accept?

NUMBERS GAME

Employees earning an hourly rate are paid for the number of hours they work. You can calculate the number of hours you work by measuring the **elapsed time**, which is the amount of time that has passed from one time to another. You find the amount of elapsed time by subtracting the earlier time from the later time.

For example, if you start working at 3:00 and finish at 5:00, how long did you work? 5:00 − 3:00 = 2:00, so you worked 2 hours.

If you start working at 3:30 and finish at 6:00, how long did you work?

Myth Once you develop a value, it stays with you for life.

Truth Values change over time. Things that are important to you when you are young might become less important as you grow older.

Are Your Standards Flexible?

We use values to set standards. Setting a high standard shows that you value something a lot; a lower standard shows that you don't value it so much. **Standards** are really just guidelines for whether or not something meets expectations. We use standards to measure performance on everything from how many miles per gallon a car gets, to how well students do in a class.

You probably know when someone is judging you to see if you meet his or her standards: a teacher gives you a test or a parent comments on your table manners. You might not realize that you set standards, too, and you use them to measure whether people and things meet your expectations. At work, an employer expects you to meet certain standards such as dressing appropriately and respecting your co-workers. You also expect your employer to meet your standards.

Why might someone who values the environment be unhappy working in a factory that creates pollution?

- If you value the environment, you will probably be happier working for an employer that supports environmentally-friendly practices, such as recycling.
- If you value making money, you might prefer an employer that rewards hard work with high pay.

Sometimes, having different values can cause conflicts—disagreements. One way to avoid conflict is to have **flexible standards**—standards that you can adapt to different situations. For example, you might lower your standard for high pay if your employer provides rewards such as more vacation time. Having flexible standards shows you are **tolerant** or willing to consider the opinions of others.

Tech Connect

Technology has made the option of working at home a reality for many employees.

- E-mail, instant messaging, and voice over Internet protocol (VOIP) make it easy to communicate.
- Applications and documents stored on Internet servers let workers collaborate on projects no matter where their desks are. Likewise, employees can sync multiple devices, allowing them to access the same information from any location on whichever device they choose.
- Video conferences make it possible to meet face to face—even when you are miles apart.

What qualities and values might a person have who enjoys working at home? What might be some benefits and drawbacks? What technology do you think would be most useful? Use a word processing program to write an essay that answers these questions.

Understanding Ethics

Your values and standards are also influenced by your **ethics**. Ethics are a set of beliefs about what is right and what is wrong.

- Some ethics are established by society. They determine how people are supposed to behave, usually in terms of human rights, responsibilities, and justice. They may be based on customs or on law.
- Some ethics are established by groups of people, such as the members of certain professions. For example, doctors and lawyers must abide by strict ethical standards, or they can lose their licenses to practice.
- Some ethics are personal and usually measure **virtues**—or positive character traits—such as honesty, compassion, and loyalty.

People don't always agree on what is ethical. One employee may think it is ethical to make personal phone calls while at work, while another employee may think it is unethical.

Even when there are laws defining what is ethical, some people still don't agree. For example, one person might think it is ethical to take paper or staples from an office, even though it is illegal.

Identifying Your Strengths and Interests

If you are like most people, you will spend half your life or more working. So, choosing a career that will bring you satisfaction makes good sense. How can you know what will bring satisfaction? One way is to choose a career that gives you a chance to use and develop your strengths and interests.

Being able to identify your strengths and interests will help you choose a career that is right for you. It will also help you set realistic educational and career goals.

What Are Strengths?

Your **strengths** are your positive qualities and skills. Your personal qualities are the characteristics and traits that make you unique. A **skill** is an ability or talent—something you do well.

- You might have a talent for growing flowers.
- You might understand how machines work.
- You might speak multiple languages.

Showing Your Values

People have different ways of showing what they value. What if you value animals? Does everyone who values animals show it in the same way?

★ You might volunteer at an animal shelter.
★ You might take care of your own pet.
★ You might walk your neighbor's dog.
★ You might eat a vegetarian diet.
★ You might not wear clothes made from animal products.

Showing values in different ways doesn't mean one person is right and the other is wrong. Do you think two people with different values can be friends?

Interests and Careers

You don't have to be an expert in something to have an interest in it. An interest might be a favorite activity or something you just like to do. For example, you might not be a star basketball player, but you can still seek a career that satisfies your interest in basketball and also uses your other skills and abilities. What if you are interested in basketball but know you do not have the talent to be a professional player?

★ If you have an ability for communications, you might consider a career as a sports journalist.

★ If you have an ability for food and nutrition, you might consider a career as a sports dietitian.

★ If you have an ability for marketing, you might consider a career in public relations, sales, or marketing for a sports team.

What interests and abilities do you have that you think might relate to a career?

Sometimes it is hard to recognize your own strengths. You must be very honest about what you do well, and what you do not do very well. Sometimes other people can help. You can ask family members and friends who you trust, or a teacher or counselor at school, to help you identify your strengths.

Strengths also include **accomplishments**, which are things you have achieved at home, in school, or in the community. Accomplishments might include specific skills, such as knowing how to create a chart using a spreadsheet program or how to change a flat tire on a car. Accomplishments might also include achievements such as being the treasurer of a club or scoring a goal in a soccer game.

Discovering Your Interests

Often, people have strengths that relate to their interests. **Interests** are subjects or activities that attract your attention and that you enjoy doing or learning about. You can have interests in the different areas of your life.

- At home, you might have interests in cooking, gardening, sewing, or design.
- At school, you might have interests related to your class subjects, music, theater, athletics, and clubs.
- With peers, you might share interests in movies, music, and sports.
- In your community, you might have interests in volunteer activities, politics, and public service.

Many of your interests are probably related to activities such as sports, arts, and entertainment. These interests may be useful in your career, or they may be things you do outside of work for fun.

For example, if you are interested in music, you might consider a career such as a music teacher, a disc jockey, an agent for a musician, an owner of a club that hosts performers, or a radio host. You might consider performing with a band for fun.

What interests do you have that might be useful in a career?

Multilingual Employees

Are you multilingual? Multilingual means that you can communicate in more than one language. It is an ability that is an advantage to anyone looking for a career.

Many businesses have offices in countries all around the world. Even those that have offices in only one country have customers who speak different languages. Employers are eager to hire qualified candidates who can speak multiple languages. If you understand different cultures and are interested in living internationally, that is an advantage as well.

What languages do you think are spoken by the most people around the world? Use the Internet or the library to find the answer. Make a poster showing how you might use multiple languages to succeed in a career.

Recognizing Character Qualities

Your **character** is the personal qualities or traits that make you unique. You show your character qualities by the way you act and the things you say. Most people have both positive and not-so-positive character traits.

These character traits combine to make you the type of person you are. You might be kind, helpful, or considerate. You might be stubborn, sad, or mean.

Your character traits influence the way other people see you and the way you see yourself. They can contribute to your self-esteem and your well-being. You are not stuck with the same character traits for your whole life. You might change some traits, and you might gain new ones.

Myth If a person has one bad quality, he is bad through and through.

Truth Most people have both positive and negative character qualities. To be a good person, you might want to try to have more positive qualities than negative ones.

Personal Qualities and Your Career

Everyone has positive and negative personal qualities. When you start looking for a career, it helps to understand the situations that bring out both types of qualities.

- Are you happy working alone, or would you get lonely and sad?
- Do you enjoy talking on the phone, or would you prefer to meet with people face to face?

- Are you relaxed sitting at a desk for a long time, or would you rather be outdoors and active?
- Do you enjoy the excitement of having to meet a deadline, or does it cause you to feel stress?

You will spend a lot of time and energy at work. The best work environment will bring out your positive personal qualities.

Developing Your Character

You develop character qualities over time as you come to understand what is important to you. You can also work to develop character qualities. For example, you can decide you want to be more **compassionate**—caring about the suffering of others.

- Can you show compassion by walking dogs at an animal shelter?
- Can you show compassion by stocking shelves at the food pantry?
- Can you show compassion by speaking up when someone is mean to another person?

Another way to develop character qualities is to find someone who has the character you admire and then try to **emulate**—or copy in a respectful manner—his or her behavior.

Sometimes, your character traits show up only in certain situations. You might be shy in class but outgoing when you are with your family and friends. You might be lazy when you have homework to do but enthusiastic when you play soccer.

Recognizing Your Character

Sometimes it is hard to recognize your own character qualities. You might be embarrassed to think about your positive traits. People might think you are bragging or conceited.

You might be unwilling to think about your negative traits. You might not want to admit you can sometimes be mean or stubborn.

The more you understand about yourself, the easier it is to identify all of your traits. You can accept both the good and the bad, and understand how you might try to improve.

One way to tell a positive character trait from a negative character trait is to ask yourself whether you would like that trait in someone else. For example, do you like it when someone else is rude? Do you like it when someone else is bossy? If not, you might not want to show those character traits yourself, either.

What character qualities do you think are important for a paramedic?

Your Personal Strengths ■ Chapter 1

J.T. is looking for a summer job. His friend Alex is a lifeguard and works at the town pool. He tells J.T. that lifeguarding is fun. The pay is good, he gets to be outside all day, the other guards are nice, and the girls all think the lifeguard is cool.

J.T. applies for the job. The town's recreation manager explains that the town will pay for the lifeguard training and certification as long as J.T. works all summer. If he quits, he will have to repay the cost of the training.

The training is harder than J.T expected. He does not enjoy learning first aid and CPR. The job is not much fun, either. Little kids are constantly asking for bandages. J.T. has to clean the bathrooms at the end of the day. After one week, J.T. is regretting his decision to become a lifeguard. He is thinking about quitting.

What problems are facing J.T.? Can he find solutions? Did he make the best decisions while looking for a summer job? What could he have done differently? Use your 21st Century Skills such as decision making, problem solving, and critical thinking to help J.T. Write an ending to the story. Read it to the class, or form a small group and present it as a skit.

Benefits of Character Qualities

Positive character qualities—or virtues—make you feel good about yourself and contribute to your well-being.

- People recognize that your character is positive, and they respect you for it.
- Positive character qualities also make you feel proud of yourself and confident in your abilities.
- When people think your character is negative, they don't trust or respect you.

Positive character qualities contribute to your well-being. Some common positive traits include enthusiasm, thoughtfulness, compassion, trustworthiness, tolerance, dedication, perseverance, and kindness.

Negative character qualities interfere with your well-being. Some common negative traits include laziness, indifference, dishonesty, intolerance, meanness, and stubbornness.

21st Century Learning

Character Qualities of an Entrepreneur

An entrepreneur is a person who organizes and runs his or her own business. Being your own boss has advantages, such as independence, flexibility, and the pride and excitement that come with ownership. It also has disadvantages, such as no regular salary, an unpredictable work schedule, and the risk of losing everything if the business fails.

What are some of the character qualities you think would be useful for someone who wants to be an entrepreneur? Make a list of at least five qualities and write a brief explanation of how each quality would help an entrepreneur succeed.

A Positive Attitude

Attitude is the way you think, feel, or behave, particularly when you are with other people. If you have a **positive attitude**, you are happy, you think life is good, and you have confidence in yourself and the people around you. If you have a **negative attitude**, you are unhappy, you think life is unfair, and you have little confidence in yourself and others. When you have a positive attitude, people want to be around you.

Attitude changes depending on your situation. You might have a positive attitude about school and a negative attitude about your family. You might have a positive attitude about language arts class and a negative attitude about social studies.

Most employers want to hire people who have a positive attitude. You can develop a positive attitude by developing positive character traits. You can pay attention to the good things in your life and learn how to change the bad things, or let them go, so they don't bother you or make you sad.

Why do you think an employer would prefer an employee with a positive attitude?

Career Trend

According to the Bureau of Labor Statistics, occupations related to healthcare are projected to add the most new jobs between now and 2022. Total employment in that sector is projected to increase 10.8 percent, or by 15.6 million. The healthcare occupation that will have the highest number of new jobs is that of personal care aide. The projected number of new personal care aide jobs between 2012 and 2022 is 581,000.

Appearance and Attitude

Do people judge your attitude based on your appearance? It might be a bias or a prejudice, but it happens all the time. Can you convey an attitude by the way you look or dress?

- 👎 Scowling and slouching
- 👎 Wearing clothing that doesn't fit
- 👎 Wearing heavy dark make-up and black nail polish
- 👍 Smiling
- 👍 Wearing neat, well-fitting outfits
- 👍 Making eye contact

Do you think people should judge you by the way you look and dress? How can you convey a positive attitude no matter what you look like?

Analyzing Employability Skills

You might graduate from college with a bachelor's degree in design, but if clients think you are unfriendly, or if they question your confidence, they are unlikely to hire you. Attitude and confidence are part of employability. **Employability** means having and using your life skills and abilities to be hired and stay hired. It is more than just meeting the qualifications for a position. It also means knowing how to:

- Present your positive qualities to an employer
- Communicate effectively with employers, co-workers, and customers
- Meet your responsibilities at work

Having employability skills will give you an advantage when you are ready to apply and interview for a position, no matter what career you choose.

Employability Characteristics

Many of the personal qualities that make you successful in other areas of your life are necessary for professional success. Characteristics such as integrity, patience, dedication, perseverance, a willingness to work with many types of people, and responsibility show an employer that you are someone who will not only get the work done, but be good to have around the workplace. You can develop other positive characteristics to improve your employability.

- A positive attitude shows that you enjoy your work. You do not complain or criticize others.
- Being cooperative shows that you respect your employer, co-workers, and customers. You meet your responsibilities, value other team members, and help others when it is necessary.
- Accepting and using **criticism**—advice about how to make positive changes in your actions or behavior—show that you want to improve and are willing to learn.
- Flexibility shows that you can accept change and adapt to new situations.
- Leadership shows that you are capable of making decisions and solving problems. It shows that others respect you and want to work with you.

Employers want to see that you are generally happy and easy to work with. Developing positive personal qualities and showing these qualities to an employer improves your chances of finding and keeping a job.

Myth Once I choose a career, I am stuck with it.

Truth Most people change careers at least three times in their lives, and many change more than that. Skills that you develop in one job can be used in a different line of work. Attending school or training programs at night or on the job may lead you in a different direction. Your goals are likely to change over time. You can change careers, too.

What Skills Are Transferable?

Here is a list of some transferable skills.

- ✔ Negotiating
- ✔ Problem solving
- ✔ Management
- ✔ Leadership
- ✔ Accepting responsibility
- ✔ Honesty
- ✔ Decision making
- ✔ Cooperating
- ✔ Respect
- ✔ Creativity
- ✔ Independence
- ✔ Communications

Can you think of skills you have that you can transfer to many different situations?

How can joining a career and technical student organization help you build employability skills?

Transferable Skills

A skill is the ability to do a task well. Skills can generally be placed into two groups: hard skills and **transferable skills**.

Hard skills relate to specific jobs or tasks. For example, operating a cash register is a hard skill. Baking a cake is a hard skill. Sewing a seam is a hard skill. Employers often look for people with hard skills to fill specific jobs. For example, an auto repair shop looks to hire people who are skilled at repairing automobiles. A florist looks to hire people skilled at arranging flowers.

Transferable skills can be used on almost any job. They are called transferable skills because you can transfer them from one situation or career to another. Being able to communicate effectively with other people is a transferable skill. So is being able to manage time and other resources. Employers are always looking for employees who have transferable skills.

How Can I Build Employability Skills?

You develop character qualities over time as you come to understand what is important to you. You can also work to develop character qualities by taking active steps. For example, you can make an effort to be more cooperative in school, or more flexible with your family. You can also emulate—copy—someone who shows qualities that you admire.

You can build employability skills that will appeal to an employer by practicing them in all areas of your life.

- At home, you can practice conflict resolution. Look for ways to fight less with your siblings and to communicate more effectively with your parents.
- At school, you can set academic goals and work with a teacher or counselor to create a plan for achieving them.

- With your peers, you can practice leadership by standing up against bullies or becoming active in a club or career technical student organization such as Family, Career and Community Leaders of America (FCCLA), or Health Occupations Students of America (HOSA).
- In your community, you can practice problem solving by looking for ways to support public resources. You might look for ways to reduce trash or to improve safety.

You can demonstrate employability skills by developing a personal career portfolio. A **portfolio** is a collection of information and documents that illustrate and support the skills and abilities you will need for college and a career. For example, a school report card or a resume might be in your portfolio.

Your Personal Strengths ■ **Chapter 1**

Career Profile: Employment and Placement Manager

Business Management & Administration Career Cluster

Job Summary

An employment and placement manager is a professional who advises other people on the process of choosing a career. Other job titles for the same position are career counselor, vocational counselor, and employment counselor.

Employment and placement managers meet with clients who are looking for jobs and help them assess their interests, skills, and abilities in order to identify suitable jobs. They may also work to help people develop job search skills as well as to locate and apply for jobs. Some employment and placement managers help clients cope with career-related issues such as job loss or stress.

Employment and placement managers may work in schools, for private companies, or for government agencies. Managers should be trustworthy and honest and enjoy helping others. Education requirements vary, but in most states, counselors must earn a master's degree in order to receive a license or become certified.

Use the Internet, a library, or career center to research the responsibilities, education and training requirements, and salary range of an employment and placement manager in your state. Write a job description for the position.

What might be the qualities and interests of a person who would be successful as an employment and placement manager?

Case Study

Casey is preparing to attend a job fair to find part-time work. He is not sure what type of job he wants. His mother suggests he consider a position as a bank teller, because banking careers are generally secure and pay well. His father—a landscape architect—thinks Casey might want to be a landscape assistant.

Casey has a list of the companies that will be at the fair. There are more than 50! He is not sure if he will be able to visit every booth, and he is worried he might miss the one that has the best opportunity for him.

- What can Casey do to narrow the list of companies he should visit at the job fair?
- What else can Casey do to prepare for the job fair?
- Do you think Casey's parents are being helpful with their suggestions?

Answer It!

1. What is a career?
2. What is a job?
3. List three things that might be determined by career requirements.
4. List four reasons people might choose a career.
5. What are values?
6. What are interests?
7. What is the impact of effective college and career planning?
8. List at least five characteristics that are necessary for professional success.
9. List five negative character qualities.
10. What are transferable skills? List at least ten skills that can be transferable among a variety of careers.

Web Extra

Many job Web sites include sections specifically for teens, such as groovejob.com and myfirstpaycheck.com.

Use the Internet to identify the sites that you think might be the most useful for your peers. Make a directory of the sites, including the Web site address and a brief description. Post the directory on your school Web site or in your school career center or library.

Your Personal Strengths ■ Chapter 1

Write Now

Sometimes it is difficult to identify your own interests, skills, and abilities. Working in pairs, spend 15 minutes talking to your partner about the things you like to do. Discuss the things you think you are good at, your favorite subjects in school, and the clubs or organizations you have joined. Take notes so you will remember what your partner tells you.

When you have finished talking, look over your notes and select a career that you think suits your partner's interests, skills, and abilities. Write a letter of recommendation to a potential employer on behalf of your partner, explaining why you think this would be the right career match.

COLLEGE-READY PRACTICES

Extra-curricular activities—activities you do out of class—are a great way to explore and develop your interests and abilities and to build characteristics that will help you achieve academic and professional success. In addition, most colleges consider extra-curricular activities as part of your application.

Conduct research into the extra-curricular activities available at your school and in your community. When you have completed your research, make a chart that lists at least three activities that you might like to participate in. In the chart, describe how each selected activity will help you develop your interests and abilities and build characteristics that will help you prepare for college. Present your chart to a partner or to the class.

Career-Ready Practices

Participating in school clubs and organizations can provide valuable opportunities to gain career-related experience. As a class or in small groups, brainstorm ways you can inform students and families about the different clubs and organizations at your school. For example, you might publish a booklet, design a Web site, or organize an Activity Fair. Make sure you consider the importance of using reliable and thorough research strategies to brainstorm ideas and gather information.

Working together, select one idea that will let you educate students and families about the clubs and organizations available at school. Make a plan for completing the project, and then put the plan into action. Keep in mind that career-ready individuals work productively in teams to evaluate the methods for sharing and presenting information to others. They find ways to encourage all team members to participate and contribute to the team's goals.

When the project is complete, evaluate your success. Did you achieve your goal? Did you create a useful product? Did the students and families find it useful and informative? Did you run into any obstacles? If so, how did you overcome them? Did you have to make any trade-offs? Write a report explaining what you learned, whether you were successful or not, and what you would do differently if you were going to start the project all over again.

Career Portfolio

A **portfolio** is a collection of information and documents. A personal career portfolio can help you stay focused and organized as you explore college and career options and develop a career plan. A career portfolio includes information and documents that show the progress you make in school and in your career plan. It helps you stay on track to achieve your educational and career goals. You can also take it to college and job interviews and job fairs, so you have the information you need to fill out applications and the documents you want to show to potential employers.

As a class, discuss the types of information and documents you might include in a personal career portfolio. For example, you might include a list of skills, interests, and abilities; a personal academic plan; a resume; and a sample cover letter. You might also include examples of achievement, such as a brochure you designed, artwork, or writing samples, as well as report cards or other assessments.

On your own, set up a career portfolio. It might be a paper folder or an electronic folder on your computer. For the first document, make a list of the documents and information you plan to include in the portfolio. As you collect the items, you can cross them off the list.

2 The Roles You Play

Skills in This Chapter . . .

- **Identifying Roles and Responsibilities**
- **Analyzing Well-Being**
- **Recognizing Influences**
- **Identifying Resources**

GETTING STARTED

How well do you know yourself? You know the facts: your birthday, your address, what grade you are in. But do you know what makes you happy? Do you know why you choose one flavor of ice cream over another, or why you would rather play soccer than basketball—or basketball instead of volleyball? These are hard questions to answer. Thinking about them might even make you uncomfortable. But understanding yourself is an important step in understanding the world around you, how you fit in, and why you might find one career more satisfying than another.

➤ Make a list of the things that are important to you, such as people, pets, sports, places, and even things. Pick the one that you think best defines you, write it on a slip of paper, and then give it to your teacher. As a class, read the slips of paper and try to figure out who wrote each one.

Identifying Roles and Responsibilities

An actor might play one **role** in a movie and a different role on television. In real life, you play different roles, too. That doesn't mean you pretend to be someone you are not—it just means you act differently depending on who you are with and what you are doing.

A role is the way you behave in a specific situation. As a student, you are expected to behave a certain way in school, but that's not the same way a coach expects you to behave as an athlete in practice, or an employer expects you to behave on the job.

A **responsibility** is something people expect you to do, or something you must accomplish. Responsibilities help define the way people interact and communicate with each other in different circumstances. When you live up to your responsibilities, you are successful in your role.

Different Roles = Different Responsibilities

You have roles in five main areas of your life: at home, at work, in school, with people your own age, or **peers**, and in your community. You also have a role as an individual—a **unique** (or one-of-a-kind) person.

You relate to people in each area in different ways. For example, you probably use different words when you talk to your friends than you do when you talk to your parents or teachers. You might dress differently, too. At home, you might wear sweats, at school you might want to express your style, and at work you might wear a uniform.

Responsibilities are different in each role, too.

- As an employee, you are responsible for showing up on time.
- As a family member, you are responsible for making your bed.
- As a friend, you are responsible for being a good listener.
- As a student, you are responsible for completing your assignments.
- As a neighbor, you are responsible for picking up trash in front of your house.

The Roles You Play ■ Chapter 2

Career FACT: Working Americans spend an average of 8.8 hours a day at work or on work-related activities. High school students spend an average of 7.5 hours a day attending class and doing homework.

Assigned Roles vs. Acquired Roles

Many of our roles happen because of who we are. Take the role of family member, for example. We don't choose to be a brother or sister or son or daughter. These roles are assigned to us.

Some roles we choose for ourselves.

- You might want to be a musician, so you learn to play an instrument and audition for a band.
- You might want to be a member of the student government, so you campaign for office.
- You might want to work in a restaurant, so you apply for a job.
- You might want to contribute to your community, so you volunteer at a senior center.

Not everyone chooses the same roles. We choose certain roles because they are important to us. The roles we choose are the ones that make us unique—different from other people.

A person who starts a small business is said to "wear many hats." That means he or she has to play many different roles at work. For example, you might start a company to sell hand-printed T-shirts because you are skilled at designing the artwork. But you must also manage the other aspects of the business if you want to succeed, including advertising and sales, filling orders, purchasing supplies, and handling customer complaints.

Who Am I Now?

Your roles vary depending on the five main areas of your life:

✔ Family:
- Son or daughter
- Brother or sister
- Niece or nephew
- Grandchild

✔ Work:
- Employee
- Co-worker
- Manager
- Owner

✔ Peers:
- Friend
- Rival
- Teammate

✔ School:
- Student
- Classmate

✔ Community:
- Neighbor
- Volunteer

Think about a small business you might like to operate. What different roles would you play in the business? What responsibilities might you have in each role?

Roles and Responsibilities Change

Roles change depending on your **stage of life**, or how old you are. Right now, you are a student. You might not have a paying job; in a few years, you might work part-time after school or in the summer. One day, you will start a career and work full time.

Like roles, responsibilities change over time or in different situations.

- Now, you might be responsible for watching your little brother in the afternoon. Someday, you might be a paid babysitter for your neighbor's children, care for your own children, or run a daycare center.
- Now, you might be responsible for making your own lunch. Someday, you might be responsible for preparing dinner for your whole family, managing a kitchen as a head chef, or owning your own restaurant.

Just because the role and responsibility are the same for you as for someone else doesn't mean you will both behave the same way. Everyone in your math class is responsible for doing the homework, but one person might do it every morning on the bus, while you meet with a classmate to do it after school.

The way we meet our responsibilities is unique, because we are unique. When you understand your roles and responsibilities, you understand more about yourself and why you behave the way you do in a certain situation.

Fulfilling Your Responsibilities

You might think responsibilities are a burden. After all, why should you care what other people expect from you? But people treat you with more respect if you fulfill your responsibilities.

Why do you think an employer would prefer to hire someone who fulfills his or her responsibilities?

- They trust you.
- They take you seriously.
- They enjoy being with you.

If you do not fulfill your responsibilities, people stop expecting very much from you. They think you are immature.

- People feel you let them down.
- They see you as a slacker.
- They find someone else who will meet their expectations.

Rules and Responsibilities

Living up to your responsibilities means playing by the rules. A **rule** is a written or unwritten statement about how something is supposed to be done. Rules are meant to protect people and to help people live together in communities. When you follow the rules, you show people that you are responsible.

There are rules in every area of your life—home, at work, in school, in the community, and even when you are with your peers.

- At home, you may have to be in bed at a certain time. You may not be able to watch television or use your phone until your homework is complete.
- At work, you may have to arrive at a specific time, or wear your hair a certain way.
- In school, you may have to sit in assigned seats. You may have to raise your hand to speak in class.
- In your community, you may have to turn off your phone in a movie. You may have to wait for the signal to cross a street.

When you break the rules, you show that you are not responsible. There may be consequences for breaking the rules.

- If you arrive late for work, your supervisor may fire you.
- If you use your phone during a movie, the usher may ask you to leave.

Different Rules for Different People

If you think that there are different rules for different people, you are right—there are. Many rules are meant to protect people, such as rules for wearing a seatbelt or a motorcycle helmet. Some rules vary depending on factors such as your age or your occupation.

- Health care workers are not allowed to discuss their client's health with others.
- Politicians are not allowed to keep secrets from the public.

If you think a rule is unreasonable, try to find out why the rule is in place. Once you understand the reason for the rule, you might understand why it is important not to break it.

You may be able to change some rules. For example, if you show your employer you are responsible and can be trusted to complete your assigned tasks, he or she might be more flexible about the time you are expected to start working.

NUMBERS GAME

You can use a circle—or pie—chart to show the relationship of parts to a whole. The entire circle represents the whole—or 100%. Each part represents a portion—or percentage—of the whole.

For example, you can use a circle chart to show the relationship among the times you spend in the different roles in your life.

The entire circle represents 24 hours, or one day. Sections of the circle represent the number of hours you spend in each role.

Say you spend 7 hours a day as a student in school, 2 hours as a teammate in practice, 1 hour a day as a friend, and 14 hours as a family member. First, calculate the percentage of time spent in each role relative to one day. (*Hint:* Divide the number of hours in each role by the total number of hours in a day. Round the answer to the nearest whole number.)

- 29%—student
- 8%—teammate
- 5%—friend
- 58%—family member

Next, draw the sections to represent each role. Your circle chart would look like this:

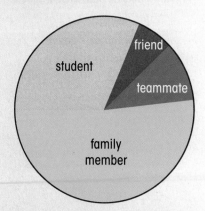

What would the chart look like if you spent 3 hours at work, no time as a teammate, and 13 hours at home? What about a chart representing the roles you play?

Analyzing Well-Being

Well-being is the feeling and understanding that everything is going right in your life. It is when you have a positive outlook about yourself and the people and things around you. You are happy, healthy, confident, and satisfied. Often, well-being has a lot to do with how well you meet your responsibilities.

Having a career that brings you satisfaction is important for your well-being. When you enjoy your work, you are more likely to be happy and have a positive attitude. If you do not enjoy your work, you are more likely to feel angry, sad, and frustrated. Recognizing the things that contribute to your well-being can help you choose a satisfying career path.

Well-Being: Six Ways

Six areas of your life contribute to your well-being, and when all six work together successfully, you experience an overall feeling of well-being, or **wellness**.

- **Emotional well-being** depends on your ability to deal with problems and stress.
- **Physical well-being** depends on your health.
- **Social well-being** depends on how you get along with other people.
- **Personal well-being** depends on how satisfied and confident you are with yourself.
- **Intellectual well-being** depends on your ability to think and learn new things.
- **Environmental well-being** depends on your comfort and satisfaction with the environment in which you live and work.

Wellness is not an accident. It happens when you make healthy choices about the way you live your life. When you make wellness part of your **lifestyle**—the way you think and behave every day—you take control of your health and well-being. You accept responsibility for your health and happiness instead of letting other people determine what is best for you.

Whose Well-Being Is More Important?

Sometimes, looking out for your own well-being could actually interfere with the well-being of others. While some people, such as your parents, might be willing to put your well-being first, you cannot expect that from everyone. What if you want to take a long lunch in order to attend an exercise class?

★ Your co-workers must cover for you, and they may miss their own lunch time.

★ Your customers must wait until you are back at work.

★ Your supervisor may turn to someone else for assistance.

★ You may lose pay, which affects your family.

Is your action fair to everyone else? What could you do so that your well-being does not interfere with the well-being of others?

Promote Your Own Well-Being

You are responsible for your own well-being. It is not up to other people to make sure you are happy and confident. To promote your own well-being, you can be aware of what makes you happy and satisfied, and then take the action you need to make it happen. You can take action in one particular area, such as attending a class to learn a new skill, or you can make general changes that will promote overall wellness and well-being, such as getting more exercise or eating a healthy diet.

- To promote emotional well-being, you can start writing your thoughts, fears, worries, and concerns in a journal.
- To promote physical well-being, you can make sure you get enough sleep.
- To promote social well-being, you can join a club or student organization.
- To promote personal well-being, you can be honest with yourself and the people around you.

Why do you think problems in one area of well-being affect you in other areas?

Roadblocks to Well-Being

A **roadblock** is something that gets in the way and interferes with your progress. Lots of things get in the way of well-being. You might feel tired or sick. You might be frustrated with a classmate who has trouble understanding a new subject, or angry because your brother or sister complains.

Problems in one area of well-being can affect the other areas. If you're tired at work, you might have less patience with a co-worker. If you're angry at your supervisor, you might be rude to a customer.

It is important to recognize roadblocks that are interfering with your well-being so you can take action. For example, you might make time to help your classmate learn a new skill.

21st Century Learning

Malik recently started working after school as a cashier at the grocery store. He earns a good hourly wage and is gaining work experience. Most of his friends do not have jobs. They spend their afternoons hanging around and playing video games.

One day, Malik's friends show up at the grocery store. They follow him around while he stocks the shelves. They make a lot of noise, knock things over, and move items from one shelf to another, where they don't belong. Malik asks them to cut it out, but they just laugh. Finally, Malik's supervisor asks them to leave. The next day, Malik's friends come back.

What challenges are facing Malik? Use 21st Century Skills such as problem solving, decision making, and critical thinking to help Malik recognize the obstacles in front of him, and to make and implement a plan of action. Write an ending to the story and read it to the class, or form small groups and present it as a skit.

Promoting Your Well-Being

Wanting something that will promote your well-being shows that you are responsible. Interfering with the well-being of others in order to get what you want is not. What are some of the positive and negative ways you might promote your own well-being?

- 👎 Taking credit for someone else's achievement
- 👎 Stealing someone else's belongings
- 👎 Putting someone down to make you look better
- 👍 Recognizing roadblocks to well-being and finding ways to overcome them
- 👍 Being aware of the way your actions and words affect others

Can you think of responsible ways to promote your own well-being?

Recognizing Influences

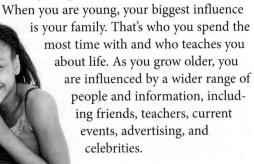

An **influence** is something that affects the way you think and act. You are influenced by pretty much everything in your life.

When you are young, your biggest influence is your family. That's who you spend the most time with and who teaches you about life. As you grow older, you are influenced by a wider range of people and information, including friends, teachers, current events, advertising, and celebrities.

Recognizing the things that influence you can help you understand yourself better. When you know where influence comes from, you can make up your own mind about whether the influence would be positive—a good influence—or negative—a bad influence. This knowledge can help you make decisions about which career will be right for you.

Understanding Influence

There are major influences in all areas of your life. You are influenced by your family and friends, by your culture and community. You may be influenced by things you read or see on television. You may be influenced by the behavior of athletes and celebrities you don't even know. A lot of influence happens without your even knowing it.

- For example, just living in a certain neighborhood can influence you and your opinions of others. In a friendly neighborhood, you might feel safe and spend a lot of time outside, but in a less friendly neighborhood, you might be wary and stay indoors.

- The media—television, radio, magazines, and the Internet—can influence you. For example, you see pictures of attractive, happy people in an ad for a particular store, and you want to shop in that store so you can be pretty and happy, too.

Good Influence or Bad Influence?

Sometimes it's pretty easy to tell whether an influence is positive or negative.

- If something influences you in a way that promotes your well-being, it's a good bet the influence is positive.
- If something influences you to break the law, or risk your life or health—or someone else's life or health—it's a good bet the influence is bad.

Sometimes it's not so easy to tell a good influence from a bad influence. A friend might want you to join a club or organization—usually that's a positive influence. But it's still a good idea to check out the club first and make sure it is right for you.

Some influences might create roadblocks to your success. For example, the behavior of the people around you can be a roadblock. You might see your co-workers making personal telephone calls or playing games online and be influenced to do the same.

Can you think of a situation when someone asked you to do something you knew was not good for your health or wellness? What did you do? How did you feel?

Identifying Resources

Resources are things you can use to get something else. We need resources to help us achieve our goals, which are the plans we make to obtain something. For example, if you set a career goal to become a kindergarten teacher, you might use resources such as high school classes in child development, volunteer opportunities at a day care center, and your ability to communicate with small children to achieve your goal.

What Influences You?

Your thoughts, beliefs, and actions are influenced by the people and things in your life. Major influences include:

- ✔ Family
- ✔ Friends
- ✔ Culture
- ✔ Community
- ✔ Current events
- ✔ Media

Sharing Resources

Lots of resources can be used by more than one person at a time. A family shares a home. In school, a whole class shares one teacher. A whole community shares the public library. Everyone shares the national parks. What if people don't manage shared resources very well?

★ Some resources might run out, such as coal and other fossil fuels.

★ Some resources might become unusable, such as parks overrun by criminals or destroyed by vandals.

★ Access to resources might cause conflicts, such as fighting over land or water rights.

★ Access to resources might be restricted to a few, such as beaches owned by the wealthy.

How might you be affected by mismanagement of shared resources?

Types of Resources

You can categorize resources into seven main groups.

- **Human resources** are the resources people provide that things cannot. Human resources include knowledge, talent, physical and intellectual abilities, time, energy, and even personal character.
- **Nonhuman** or **capital resources** include things such as money and cars.
- **Technological resources** have a great impact on your life. They include everything from computers to automated teller machines to medical equipment.
- **Community resources** are services that the government provides, such as public parks, public schools, libraries, and police and fire departments. Some businesses contribute to community resources as well. For example, a company might sponsor a softball team.
- **Natural resources** are things that exist in nature and are available for everyone. Natural resources include air, water, wildlife, minerals, and plants.
- **Renewable resources** are natural resources that can be recreated in unlimited quantities, such as air and sunlight.
- **Nonrenewable resources** are natural resources that are available in limited quantities and may one day be used up. Coal is a nonrenewable resource.

Available Resources

Some resources are available to everyone, such as air and water. Other resources may be available only to some people, such as artistic ability or talent. Some resources depend on your current role and your stage of life. For example, you might not have money if you're too young to have a job. You might have more physical strength and energy than a senior citizen.

You can develop your resources in order to make sure you have them available when you need them.

- You can eat well and exercise to develop your health.
- You can work and save money to develop financial security.
- You can practice to develop a skill or talent.

Whose Resources Are They?

Experts estimate that there are about 7 billion people living in about 195 countries on the planet Earth. Most people manage and use the resources available in their own communities. But who is responsible for managing the resources we all share?

One issue that affects the resources we all share is global warming. Scientists explain global warming as a gradual increase in the temperature of the Earth's atmosphere. Representatives from around the world have worked together to set rules and regulations to manage global warming, but they are not always successful.

Why is it important for people to work together to manage global resources? Can you think of reasons why it may be difficult to succeed? Individually or with a partner, research an issue affecting global resources. Write a letter to a politician explaining how the issue affects your community. Suggest ways to manage the resource for the benefit of all.

Managing Resources

Sometimes it's hard not to worry if you don't have as many resources as someone else, or if you have different resources. You might think someone who knows the president of a company will be hired for the job you want. Or you might worry that someone who has enough money to pay for a private college will have an advantage if you can only afford a public university.

Why is it important to manage the resources you have available?

You can make the most out of the resources you have by managing them. That means making sure you use your resources at the right time and for the right purpose.

For example, you and your parents may have been saving for your college education. You could spend your entire college fund on four years at a private college, or you could use it for four years at a public school and have enough left for graduate school as well. How will you manage your college fund resources?

To make that decision, you consider your values, standards, and ethics. You recognize the things that are influencing you. For example, are your friends going to public or private universities? Do you value the **prestige**—positive recognition and admiration—that comes with a degree from a private school? Finally, you ask yourself which option gives you the best opportunity to achieve your goals.

Making the Most of What You Have

Not everyone has access to the same resources. What if you want to play guitar in a band but cannot afford lessons?

- You could get a job and earn the money you need.
- You could offer to do chores for the guitar teacher in exchange for lessons.
- You could borrow a guitar lesson book from the library and teach yourself.

Can you think of creative ways you could manage your resources?

Myth Money is the most important resource.

Truth Money is important if you want to buy something. Other resources are important for achieving different goals. If you want to run a race, you need your health more than you need money. If you want to make a new friend, you need a sense of humor more than you need money.

Money Madne$$

Money is a resource we use to acquire things. For most people, money is a limited resource—we do not have an unending supply. That means we have to make decisions about how we use money. If you have $10.00 in your pocket, and you spend $6.75 to rent a video game, how much will you have left over to buy something to eat? What if you have $22.50 and rent two video games? (Hint: Multiply $6.75 times 2 to find out the cost of renting two video games.)

Tech Connect

Time is a valuable resource. An electronic day planner can help you manage your time effectively. It's a calendar you can use to enter appointments, activities, notes, and other information. Most let you view days, weeks, or even a month at a time so you can see what's coming in the days ahead and stay prepared and organized. Day planners are available for computers and cell phones.

Can you think of other technologies that can help you manage your resources? Research one and use a presentation graphics program to create a presentation about it.

The Roles You Play ■ Chapter 2

Career Profile: *Environmental Scientist*

Science, Technology, Engineering & Mathematics Career Cluster

Job Summary

Environmental scientists study ways to use science to protect the environment. They work to identify and fix problems with the air, water, and soil to make sure that the environment is safe.

Some environmental scientists work in laboratories, conducting experiments. Some work in offices, advising politicians and business owners. Some work outside—or in the field—collecting samples.

Environmental scientists study all fields of science, including biology and chemistry. They must know how to use math and computers. Some also study business, law, or engineering. They must have at least a college degree, and most need a master's or a doctoral degree.

Use the Internet, a library, or career center to research the responsibilities, education and training requirements, and salary range of an environmental scientist. Prepare a presentation about the career.

What skills and interests do you think might lead someone to pursue a career as an environmental scientist?

Career Trend

According to jobmonkey.com, jobs in the environmental science industry are expected to grow in the next decade. Find out more at jobmonkey.com/greenjobs/environmental-science.html.

Case Study

Melissa has an interest in finance and does well in math class. She applies for an after-school job as a bank teller. The bank manager, Karen Johnson, calls Melissa and asks her to come in for an interview.

Melissa's friend Jenna suggests that Melissa try appearing very grown up at the interview. Melissa borrows clothes from her older sister, including shoes with high heels, and calls the bank manager by her first name throughout the meeting. Melissa does not get the job.

- Did Melissa behave appropriately at the interview?
- What influenced Melissa's behavior?
- What could Melissa have done differently?
- Do you think Melissa understands the roles she plays in the different areas of her life?

Answer It!

1. What are the five main areas of your life that define your roles and responsibilities?
2. What is the purpose of rules?
3. What are the six areas of well-being?
4. Define the term roadblock.
5. Define the term influence.
6. What are six major influence factors?
7. Define the term resources.
8. List the seven main groups of resources.
9. List four examples of human resources.
10. How can you make the most of the resources you have available?

Web Extra

Use the Internet to locate information about community resources in your area. Are there state parks? National parks? How about libraries or museums? What types of career opportunities are available in these resources? Select one, and write a job description for it.

The Roles You Play ■ Chapter 2

Write Now

Think about your roles and responsibilities. What roles do you play in your family, with your peers, at school, at work, and in your community? Is one role more important to you?

Write a personal essay describing the role you think is most important and why. Be sure to contrast it to some of your other roles that you think are less important.

COLLEGE-READY PRACTICES

College-ready individuals know that it is never to soon to start identifying the resources needed to pay for college. According to the College Board, the average cost of tuition and fees for the 2014–2015 school year was $31,231 at private colleges, $9,139 for state residents at public colleges, and $22,958 for out-of-state residents attending public universities. And the cost goes up every year.

Conduct research to learn how much higher education in your area costs and what resources are available to help you pay. Do you have a college savings account? Are there scholarships you might qualify for? What kind of education loans might be available? Collect the information and enter it into a table or spreadsheet file. Discuss your findings with your parents, guardian, or school counselor. Together, develop a plan so that you will be on track to have the resources you need when it is time for college.

Career-Ready Practices

The purpose of advertising is to influence people to buy a certain product, use a certain service, or even think a certain way. In a small group, select an advertisement from a magazine, the Internet, or television and look at it closely to figure out how it influences people. Answer the following questions:

- What is being advertised?
- Who is doing the advertising?
- What methods does the ad use to influence you?
- Is the ad effective?

Then, pick a product, service, or idea and create your own advertisement. It might be a poster, video, presentation, or even a song. Use your communication skills, both written and verbal, to present your idea. Make sure you consider the audience you want to target with your advertisement. Career-ready individuals understand that knowing your audience is critical to communicating clearly and effectively. Work together as a team to complete the project, using cooperation, compromise, and teamwork to achieve your goal.

Career Portfolio

Resources are things you use to accomplish something else. Career resources are the things you will use to identify, get, and succeed in the career of your choice.

Make a table with three columns. In the left column, list career resources you think would be helpful. The list might include a career counselor or teacher, books about careers, computers and the Internet, and your parents. In the middle column, indicate whether you have the resource available to you or not. In the right column, if you have the resource, describe how you will use it. If you do not have the resource, describe how you might get it, or how you will make do without it.

Add the document to your career portfolio.

3 Why We Work

Skills in This Chapter . . .

- **Examining Lifestyle Goals and Factors**
- **Recognizing Needs vs. Wants**
- **Measuring the Value of Work**
- **Examining Career Trends**

GETTING STARTED

People work to achieve the things they want and need in life, from shelter and clothing to respect and pride. Do you daydream about things you want in life when you are an adult, such as the type of home you will live in, the car you will drive, or the number of children you will have? If you start planning now, you can identify a career path that will help make those daydreams become reality.

➤ Draw a picture of yourself twenty years in the future. What will you be doing? Where will you be? Who will you be with? What can you do now to make that vision of yourself come true?

Examining Lifestyle Goals and Factors

Does a clinical laboratory technician earn enough money to afford a vacation home? How many hours a week does an agricultural inspector spend on the job? Is a hotel clerk expected to work nights and weekends? Does a sales representative travel a lot?

The career you choose helps determine the lifestyle you will achieve, and your lifestyle might impact the career you choose. Recall that your lifestyle is the way you think and behave every day. Your lifestyle is the way you reflect personal values, standards, and attitudes in your life.

- If you value expensive items such as designer clothes, you might choose a career because it pays a high salary.
- If you value time with your family, you might choose a career with a flexible schedule, even if it pays less than other jobs.

You can consider lifestyle factors such as location, salary, education, time, and environment when you choose a career. You can also set goals for the way you want to live your life and the things you want to have in life as part of your career plan.

What Are Lifestyle Factors?

The things about your job that affect the way you live your life are **lifestyle factors**. There are five basic lifestyle factors you might want to consider when you are making decisions about the career that will be right for you.

1. *Location.* Where do you want to live? Do you want to be near your family? Do you want to live in the mountains or near the ocean? Do you prefer a city, the country, or the suburbs?

2. *Salary.* How much money do you think you must earn to pay for your needs and wants? Are you willing to work for less money to achieve other goals, such as more free time or less work-related stress?

3. *Education.* How many years are you willing to spend in school? Would you be happy taking classes at night while you work during the day? Would you be willing to earn less money while you train on the job?

4. *Time.* How many hours a week do you want to spend at work? Would you work at night or on weekends? Would you mind being **on-call**, which means being available at any time to cope with emergencies? Do you want to work part-time, even if it means earning less money?

5. *Environment.* What type of space do you want to be in? Do you like sitting at a desk? Do you want to be outdoors? What type of people do you want to work with? Do you need windows in your office? Do you like to dress in a suit or do you prefer casual clothes?

Career FACT About 80% of the U.S. population lives in urban areas (cities) while about 20% lives in rural areas (the countryside).

Setting Lifestyle Goals

Lifestyle goals are the things you want in life such as a family, where you want to live, how much money you have, and how much free time you have for friends, sports, and other activities that you enjoy.

- You might want to live in the mountains and spend as much time as possible out of doors. You might consider a career in recreation.
- You might want to have children and be available in the afternoons when they get home from school. You might consider a career in part-time sales.
- You might want to have enough money to travel, even if it means working many hours between trips. You might consider a career such as corporate lawyer or surgeon.

Selecting a career path that gives you the opportunity to achieve your lifestyle goals will make you a happier, more satisfied person.

Why do you think a career that matches your lifestyle goals will be more enjoyable and satisfying?

Making Trade-Offs

When you make decisions about your lifestyle and career plans, you may have to make trade-offs. A **trade-off** is a compromise, or giving up one thing in order to get something else.

You might want to be a social worker helping young adults in New York City. You might also want to live in a house with a backyard. Houses in New York City are very expensive, and social workers typically do not earn very high salaries. What trade-off could you make?

- You could choose to rent an apartment in the city, and put off the goal of living in a house with a backyard.
- Or, you could choose to live in a less expensive community outside the city. You would have a long commute to work, but you would be able to afford a house with a backyard.

Focusing on the things that are most important to you and setting priorities can help you make the best decisions. You can use the decision-making process described in Chapter 6 to consider all possible options and the consequences of each option. Then, choose the option that helps you achieve your career goals and your lifestyle goals.

Are Your Lifestyle Goals Realistic?

Like career-related goals, lifestyle goals should be realistic. While you might choose to aim high, it is important that you believe your goals are attainable, which means that you have the skills and resources to achieve them. Are the following goals realistic and attainable, or not?

- 👎 Work 10 hours a week and earn a $1,000,000.00 annual salary
- 👎 Live in a mansion and be named the NBA's most valuable player for three years in a row
- 👍 Study hard to become a doctor and open your own medical practice
- 👍 Enlist in the United States Air Force, learn to fly jet planes, and then travel the world as a commercial pilot
- 👍 Attend cosmetology school and get a job as a stylist at a salon in a city

Knowing you can achieve your lifestyle goals will help you work hard and stay on track. What lifestyle goals do you have that might influence your career plans?

How Long Is Your Work Week?

A **work week** is the number of hours you work per week. How do you think the length of a work week affects your lifestyle? How might it affect the employer's business, as well as the economy?

In the United States, the number of hours an employee works is set by the employer and agreed to by the employee. Typically, a full-time employee works 8 hours a day, 5 days a week, which equals a 40-hour work week.

In France, most employees work a 35-hour work week. In South Korea, they work 44 hours per week.

Make a chart comparing the typical work-week hours in three or four countries around the world. Use the Internet or the library to find the information.

Recognizing Needs vs. Wants

When you say that you absolutely, positively need a new video game, you're wrong. You might *want* a new video game. But you don't *need* it.

- A **need** is a something you can't live without.
- A **want** is something you desire.

Humans *need* certain basic items in order to survive. We *want* things to improve our quality of life, or to achieve a certain lifestyle. Our wants and needs and how we achieve them influence the career decisions that we make.

Why Do Needs and Wants Matter?

Wanting or needing something can be a powerful **motivator**, which means it can encourage you to set goals and make decisions. Recall that a goal is a plan to obtain something. Goals help you focus on what is really important to you and what you are willing to work for. When you know what you need and what you want, you can use goals to get those things.

For example, wanting a red convertible can motivate you to set a goal.

- You study hard in school so you become qualified for a career.
- The career brings you personal satisfaction as well as a good salary.
- You earn enough to pay for your needs and save to buy the car you want.

Do You Want It or Need It?

There are two categories of human needs: physical needs and psychological needs. Physical needs are basic items you need to survive, such as food, water, shelter, and clothing. Psychological needs are things that affect the way you think and feel, such as safety, security, love, acceptance, and respect.

Wants are things that are not necessary for survival. They are things that you think will make you happy and contribute to your well-being—and they might. But you can make do without them.

There are some things most people want, such as entertainment and dessert. Some wants are very personal and help define us as individuals. They are influenced by our values, standards, and goals. For example, you might value creativity and independence and want a career as a musician. Your friend might value job security and adventure and want a career in the military.

People want different things in life. What do you want that's different from what your friends want?

Money Madne$$

You work eight hours a week at the public library shelving books. You earn $7.50 an hour. Your Spanish teacher is organizing a class trip to Mexico during spring break. It costs $589.00. How many weeks must you work to earn enough to pay for the trip? Do you think a trip to Mexico with your Spanish class is a good thing to spend your money on? Why or why not?

Measuring the Value of Work

Value is a measure of the quality or importance of something. So, how do you measure the importance of work? Start by defining the word *work*. Work has different meanings in the different areas of your life.

In terms of a career, work refers to the way you spend your time in order to earn an income.

- At home, it might mean the chores you do around the house.
- At school, it might mean homework, tests, or class projects.
- With your peers, it might mean the effort you put into friendships.
- In your community, it might mean the time you spend volunteering.

Recognizing the value of work in school will help you achieve your education goals. Recognizing the value of work in a career will help you choose a career that is right for you.

The Economic Value of Work

Economics is the study of the way people use resources to achieve their needs and wants. If you have a job now, you probably know the economic value of work. You work, and you are paid for it. When you receive a paycheck, you can use the money to buy things you want and need. You can contribute to your family's finances. You can save for the future.

The economic value of work is greater than just putting cash in your pocket. When you work, you contribute to the success of the economy. The **economy** is activity related to the production and distribution of goods and services.

- *Goods* are things you can see, touch, buy, and sell. Clothing, food, cell phones, and televisions are goods.
- *Services* are work that one person does for other people, in exchange for payment. Doctors, hairstylists, bus drivers, and lawyers are people who provide services.

Work produces the goods and services that people need and want. People who work are able to buy the goods and services. The stronger the economy, the more jobs there are, and the more people work.

Employers measure the value of work in terms of **productivity**—the amount of work an employee accomplishes. When workers are productive, the employer is successful.

The community measures the value of work in terms of taxes. When you earn an income, you pay taxes that support your local community, your state, and the federal government.

Benefits of Feeling Good About Work

A positive work environment contributes to your overall wellness. What if you successfully complete a complicated project?

★ Your employer appreciates your efforts.
★ Your customers respect your knowledge.
★ Your co-workers enjoy working with you.
★ Your family recognizes that you can achieve goals outside the home.

When you feel good about yourself, you put more effort into your work. How might feeling good about yourself help you achieve your career goals?

Why We Work ■ Chapter 3

Tech Connect

One way employers measure productivity is by keeping track of how many hours each employee works. Some ask employees to fill out a **time sheet**, or table with cells for entering the time in, time out, and total hours at work.

Some use time clocks to record employee hours. A **time clock** is a machine that automatically records the time you *clock-in*—arrive—and the time you *clock-out*—leave. With older, mechanical time clocks, you actually insert a card in the machine and the current time prints on the card.

Newer time clocks are computerized. The current time is recorded when you swipe your ID card or punch in a code. **Biometric time clocks** use a fingerprint or handprint to record when an employee comes and goes.

Why do you think employers want to track their employees' hours? Can you think of any advantages to using a biometric time clock? Use a spreadsheet program to make a sample time sheet for one employee for one week. If you know how, use a formula to total the number of hours for each day and for the whole week.

The Emotional Value of Work

Some days you might think your job is not worth your paycheck. Your friends might be playing games while you are stocking shelves at the market. Or they are watching videos while you are babysitting.

Recall from Chapter 1 that every job has rewards as well as responsibilities. Luckily, the benefits of work include more than just money. A career can meet emotional needs by providing respect, satisfaction, and pride. You also gain confidence knowing you are developing skills and experience you will be able to use later on in life. When you have a career that meets emotional needs, it helps you feel good about yourself. It contributes to emotional well-being and overall wellness.

The Educational Value of Work

Jobs you have when you are still in school provide the opportunity to gain knowledge that will be valuable as you begin to plan and search for a career. You can discover more about the workplace and about the types of jobs you enjoy.

- As a data entry clerk, you might realize that you have a knack for computer programming.
- As a landscaping assistant, you might realize that you get satisfaction from working outdoors.

Value of Work As an Entrepreneur

Entrepreneurs—people who start their own businesses—must work many hours to build a successful business. Sometimes it takes a long time before they earn a profit. How do you think being an entrepreneur might impact your lifestyle? How might you measure the value of work as an entrepreneur?

Can you think of skills you have developed while working or in school that will help you succeed in a future career?

Myth There is one perfect job for everyone.

Truth Trying to find one perfect job limits your opportunities. Many occupations can satisfy your career goals, and most people will have several jobs and careers throughout life.

Throughout your life, work is valuable because it provides an opportunity to learn new skills and information.

- A waiter learns how to take orders and serve food.
- A childcare assistant learns how to clean up after snack time and monitor children in a play area.
- An office assistant learns how to answer phones, manage inventory, and use the office computer system.
- A sale representative learns about new products and services.

Every job you have—for wages and as a volunteer—provides an opportunity to develop skills that will help you succeed in the present and in the future. Useful skills you can develop on the job include:

- Showing respect to employers, customers, and co-workers.
- Dressing appropriately for the workplace.
- Arriving on time.
- Making healthy decisions.
- Solving problems.
- Setting and achieving goals.
- Working as part of a team.

Examining Career Trends

Will there be jobs for construction workers when you get out of school? Will California be hiring music teachers? Is the salary for transportation managers increasing or decreasing? One way to learn the answers to these questions is to look at career, or employment, trends. A **trend** is a general move in a certain direction. An employment trend is one way the labor market is changing over time. The **labor market** refers to any situation in which workers compete for jobs and employers compete for workers. Knowing how to research and identify employment trends can help you choose a career with a positive outlook.

What Are Employment Trends?

Employment trends include things such as the number of jobs and workers in a certain industry. For example, experts see a growing need for healthcare workers and educators, but a shrinking need for assembly-line workers. Other employment trends include where the jobs are. For example, if businesses are opening offices in the Southwest and closing them in the Northeast, there will be more job openings in the Southwest.

Employment trends are influenced by trends in other areas of life. There are four major factors:

- The economy
- Technology
- Lifestyle trends
- Changes in population

Economic Trends

The economy impacts employment in many ways. The **unemployment rate** is the percentage of unemployed people looking for jobs. The economy affects the unemployment rate, and the unemployment rate affects the economy.

- When the economy is strong, there is demand for goods and services. Businesses hire workers, so the unemployment rate is low.
- When the economy is weak, there is less demand for goods and services. Businesses cannot afford to pay wages. The unemployment rate is high.
- When the unemployment rate is low, many people are working and earning income. They have money to spend, which increases the demand for goods and services and boosts the economy.
- When the unemployment rate is high, many people are out of work. They have little money to spend, which decreases the demand for goods and services.

Is This Industry Growing?

You can spot career trends by looking for statistics that show how industries are changing over time. How can you tell if an industry is growing—a positive sign for jobs—or shrinking—a negative sign?

- 👎 There are fewer jobs in this field than there were last year.
- 👎 The salaries in this industry are the same or lower than they were last year.
- 👎 Companies in this industry are closing.
- 👍 Companies in this industry have higher profits than they did last year.
- 👍 Companies in this industry are hiring more employees than they did last year.
- 👍 There are more training programs for jobs in this industry than there were last year.

How can identifying employment trends help you choose a career in a growing industry?

Why do you think high demand for goods and services leads to lower unemployment?

Impact of Technology

Technology has a strong influence on employment. It creates new jobs, replaces old jobs, and changes the way some people perform their existing jobs.

- The development of new software and apps for mobile devices creates new jobs in areas such as application development, sales, and research and development.
- The trend toward smaller computers and mobile devices has shifted the manufacturing of systems from desktops to laptops, notebooks, and netbooks.
- Improvements in robotics have made it possible to use robots in positions that people once held, such as on automobile assembly lines.
- Electronic recordkeeping in fields such as healthcare has changed the way medical professionals enter patient information, order prescriptions, and access patient records.
- The trend toward storing information and applications on the Internet instead of on local computers has eliminated the need for some information technology managers at large companies.
- The trend toward using videoconferencing instead of traveling to meetings impacts travel agents, hotel workers, and people who work in restaurants where travelers might eat.

Changing Lifestyles

Lifestyle trends affect employment. For example, the trend toward **dual-income families**—both parents working to earn a salary—and the trend toward divorce both cause an increase in demand for childcare workers. Other lifestyle trends that affect employment include:

- The number of children in a family
- The number of adults in a family
- Where a family lives
- How a family chooses to spend and save money

Improved health also impacts employment. When people have an increased lifespan—when they live longer—jobs are affected. People may retire at an older age, so there are fewer jobs for younger workers. There is more demand for housing, care, and activities for senior citizens, and therefore more jobs in those areas. The government must pay more Social Security, which can reduce resources for other groups.

The Pros and Cons of Temp Work

Companies save money by hiring just-in-time employees, but how does it affect the worker?

Benefits

- ✔ Independence
- ✔ Flexible work schedules
- ✔ Opportunities to try different work experiences and gain new skills
- ✔ Development of a wide variety of work contacts

Drawbacks

- ✔ No guaranteed annual salary
- ✔ No work security
- ✔ No employee benefits
- ✔ The need to frequently look for work

Can you think of current lifestyle trends that might affect careers?

Why We Work ■ Chapter 3

Career Profile: Flight Attendant

Hospitality & Tourism Career Cluster

Job Summary

A flight attendant is a member of the cabin crew employed by an airline to ensure the safety and comfort of airplane passengers. Airlines are required by law to provide at least one flight attendant for every 50 passenger seats.

Flight attendants spend a lot of time away from home. They usually have to work some nights, weekends, and holidays. Their most important responsibility is to help passengers in case of an emergency. They communicate directly with many different people, and they must be friendly and patient but also assertive in order to command respect.

Use the Internet, a library, or career center to research the responsibilities, education and training requirements, and salary range of a flight attendant. Make a chart or table comparing the benefits and drawbacks of the position.

What lifestyle goals might influence a person's desire to become a flight attendant?

Flexible Lifestyles

A trend to a more flexible lifestyle is impacting employment. People in your grandparents' generation may have chosen a career early in life and then stayed in that business until retirement. **Retirement** is when someone leaves a job or career after many years and does not return to a paid job.

For some of these people, the biggest reward was **job security**—knowing that they would always have a job and receive a paycheck. They knew they would receive a pension when they retired. A **pension** is a regular payment given to a retired person by a former employer.

Career Trend

Some trends toward flexibility are the result of employers looking for ways to save money or increase productivity. Instead of hiring permanent full-time employees, companies may want to hire temporary workers. That means some industries are trending toward a just-in-time workforce, which is made up of part-time, temporary, and consulting positions.

Population Shifts

Where people live, the number of people in a community, and lifespan—how long people live—affect the number of jobs, the types of jobs, and the level of pay. What if people leave rural areas to live in cities? In the rural areas:

★ There are fewer people available to work in agricultural careers. Farmers might have to pay high wages to hire farm hands.

★ Businesses in rural towns might not have enough customers.

★ The towns will not collect enough taxes to support community programs and services.

★ In the cities, there will be more competition for existing jobs. Employers might be able to pay lower salaries.

★ New businesses will open in the cities to provide goods and services for the many residents.

★ City services such as trash collection and schools may be overwhelmed by demand.

Why do you think employers might have to pay higher wages when there are fewer job applicants and lower wages when there are more job applicants?

Today, workers are more flexible, or willing to make changes. People want careers that let them spend time with their families or doing leisure activities that they enjoy. They may want to work from a home office, have flexible hours, or even **job-share**, which means share the responsibilities for one job with another person. They change companies, locations, and even careers in order to find the job they like the best.

- A television news reporter might move to a station in a different city because the salary is higher, or because the station is watched by more viewers.
- A teacher may switch to selling real estate.
- A factory worker may retrain to become a nurse.

Entrepreneurship

Entrepreneurship is a trend resulting from a number of factors. A desire for independence, the possibility of great rewards, and even the lack of more traditional opportunities all encourage people to start their own businesses. Therefore, the number of entrepreneurs in the United States is rising, as people look for options that let them be their own boss and set their own working conditions.

Nontraditional Occupations

A **nontraditional occupation** is any job that a man or woman does that is usually done by someone of the other gender. For example, nurse is an occupation usually done by women, so it is nontraditional for men. Construction worker is an occupation usually done by men, so it is nontraditional for women. Some other nontraditional careers include:

Men
- Receptionist
- Librarian
- Hair stylist
- Childcare worker
- Elementary school teacher

Women
- Auto mechanic
- Detective
- Security guard
- Farmer
- Pilot

Try not to rule out a nontraditional career because you associate it with one gender or another; it might be a good match for your skills and abilities.

Career Tips

The Bureau of Labor Statistics (BLS) tracks the number of jobs in different careers and uses the data to develop the *Occupational Outlook Handbook*. You can use the OOH to identify careers with a positive outlook.

Diversity in the Workforce

Population is the number of people in a country or area. Facts about a population are called **demographics**. These facts include information such as age, sex, cultural background, and race. Trends in demographics impact trends in employment. For example, new businesses often open where there are a lot of people.

Increased cultural diversity in the population leads to increased cultural diversity in the workplace. For example, between 2000 and 2010, both the Hispanic/Latino and Asian populations in the United States grew by more than 40 percent. The change is reflected in the diversity of the workforce.

What might be some of the benefits and challenges of a diverse workforce?

21st Century Learning

Ari is 15. His father works as a financial analyst for a large company. His mother is a bank teller. Ari has one sister and two brothers. They live in a two-bedroom apartment in Pittsburgh, Pennsylvania.

One day Ari's parents say they have good news. The family is moving to Charlotte, North Carolina. The company has offered Ari's dad a promotion, with more responsibilities and a much larger salary. The family will be able to buy a house, and Ari's mom will be able to work part-time so she can be home more.

Ari's sister and brothers are excited, but Ari is angry. He does not think it is good news at all.

What are the benefits of making the move? What are the drawbacks? Use your 21st Century Skills such as problem solving, decision making, and critical thinking to help Ari and his family successfully meet the challenges to come. Write an ending to the story and read it to the class, or form small groups and present it as a skit.

Case Study

In clothing lab at school, Matthew discovers he has a talent for clothing construction. He understands the way the patterns work, and he has a good eye for selecting fabric and colors. Other students like his designs. One girl asked if she could buy a pillow he made.

Matthew tells his parents about it. He says he is thinking about buying a sewing machine. He thinks he can develop his skills and make some money selling his work. He thinks he might want to study textile and fashion design in college. His parents do not think it is a good idea. They tell him to look for a different hobby.

- Why do you think Matthew's parents don't want him to continue sewing?
- Do you agree that he should look for a different hobby?
- What can Matthew do to convince his parents that sewing is right for him?

Answer It!

1. List the five basic lifestyle factors.
2. What is a trade-off?
3. List four physical needs.
4. Why can wanting or needing something be a powerful motivator?
5. List four psychological needs.
6. List four examples of goods.
7. List four examples of people who provide services.
8. List three ways people measure the value of work.
9. What is the labor market?
10. List four types of demographic facts.

Web Extra

The Bureau of Labor Statistics publishes information and statistics about employment trends on its Web site: bls.gov.

Use the Web site to research trending fields relating to your career interest areas.

Write Now

Can you picture yourself in a nontraditional occupation? Many employers actively recruit males for traditionally female careers and females for traditionally male careers. Spend some time researching different types of nontraditional occupations, so you can understand the benefits or drawbacks they might provide.

Write a want ad for a nontraditional occupation that you think might appeal to the underrepresented gender—men for a traditionally female career or women for a traditionally male career. Share your ad with the class.

COLLEGE-READY PRACTICES

College-ready individuals understand that not every school will meet their needs and wants. For example, you might need a school that offers a specific course of study. You might need it to be within driving distance from your home, or offer online classes. You may want an award winning dining hall, new dormitories, or a championship football team.

Conduct research to learn about the types of opportunities available at colleges in your area. Then, make a 2-column chart listing the things you want in one column and the things you need in the other. Present your chart to a partner or to the class.

Career-Ready Practices

As a class, brainstorm different ways to measure the economic, emotional, and educational value of work. Divide into teams of four or five and select a career you all find interesting. Write and present a skit that demonstrates the different ways you can measure the value of work in your selected career.

Make sure you consider the impact your career has on the economy, the community, and the environment. Career-ready individuals are aware of how their roles and responsibilities on the job can have an effect—positive or negative—on other people, organizations, and the environment.

Career Portfolio

Do you have lifestyle goals that might influence the career path you choose? How are those goals affected by the five lifestyle factors? Make a list of the lifestyle factors, including an explanation of how each factor might impact your choice of career. Add the list to your career portfolio.

4 Exploring the Career Clusters

Skills in This Chapter . . .

- **Analyzing the Career Clusters**
- **Exploring Other Resources for Identifying Careers**

GETTING STARTED

There are an overwhelming number of available career opportunities. How can you possibly identify the ones that might be of interest to you? The 16 career clusters—developed by the National Association of State Directors of Career Technical Education Consortium (NASDCTEc)—are a good place to start. They can help you identify types of careers and specific jobs and help you pick out the ones that you might find interesting. They also show you the educational and skill requirements you might need and the path you might take to achieve the career of your dreams.

➤ Think of some of the activities that interest you. For example, do you like to write stories or climb trees? Do you like solving math problems or fixing broken toys or machines? Write a list of five things you like to do. As a class or in small groups, discuss possible career pathways that might match those interests and activities.

Analyzing the Career Clusters

The United States Department of Education and the NASDCTEc (National Association of State Directors of Career Technical Education Consortium) identifies 16 career clusters by classifying specific jobs and industries into similar categories. You can use the information on career clusters to learn about careers while you are in school or any time you are doing career research.

Within each of the 16 career clusters are related job, industry, and occupation types known as **pathways**. For example, the Architecture & Construction cluster has three pathways: Construction, Design/Pre-Construction, and Maintenance/Operations. The pathways are the building blocks of the 16 career clusters. Each pathway offers a variety of careers you might choose. There are also relationships between occupations in different clusters. For example, an information technology manager (Information Technology cluster) may work in almost any industry, including health care (Health Science cluster), legal services (Law, Public Safety, Corrections & Security cluster), and airline reservation systems (Hospitality & Tourism cluster).

The clusters and their pathways help job seekers and individuals interested in specific careers to identify professions that best suit their interests and abilities. Through the clusters and pathways, you can learn about the education and skills you will need to be effective in a specific job and career. They provide you with clear-cut choices and options as you begin your career exploration.

Tech Connect

Almost every job requires you to use technology in one way or another. For example, a bank teller uses technology to:

- Electronically process customer transactions
- Ensure the security of the bank through computerized code entries
- Create and generate financial reports

Select a profession that interests you. In your school library or on the Internet, research the types of technology someone in that profession might use. Then, use a presentation software program to create a presentation about it, and share your presentation with the class.

Agriculture, Food & Natural Resources

The Agriculture, Food & Natural Resources career cluster is for people interested in the production, processing, marketing, distribution, financing, and development of agricultural commodities—economic goods—and resources. Their interests in the field might include food, fuel, fiber, wood products, natural resources, horticulture, and other plant and animal products and resources. People in this field select jobs as diverse as farmer, food scientist, and meat inspector.

A fish and game warden are responsible for enforcing laws relating to the hunting, fishing, and trapping of wild animals. What skills and interests do you think might be necessary to become a fish and game warden?

Pathways

Agribusiness Systems
Animal Systems
Environmental Service Systems
Food Products and Processing Systems
Natural Resources Systems
Plant Systems
Power, Structural and Technical Systems

Sample Occupations

Agricultural Communications Specialist
Animal Scientist
Greenhouse Manager
Embryo Technologist
Feed Sales Representative
Fish and Game Warden
Food Scientist
Livestock Buyer
Microbiologist
Tree Trimmer and Pruner

Working Abroad

Of course you are not limited to working only in the United States to pursue a career. You might want to move to Germany, China, Nigeria, Australia, Panama, or Martinique to work.

Before you move to another country to work, there are things you should know and questions you should ask: Will you need a work permit? Are there language barriers? What will be the cost of living there? Who will be there for you if you get homesick or ill?

Select a country that interests you. Use your school library or the Internet to look for information about the rules and policies that govern foreign workers. Focus on finding answers to the questions on work permits and cost of living in an area. Share your findings with your classmates.

Architecture & Construction

A construction worker builds houses and buildings. What skills and interests do you think might be necessary to become a construction worker?

People in the Architecture & Construction career cluster design, plan, manage, build, and maintain the built environment. Take a look around you. The houses across the street, the roads that cars drive on, and the parks that you walk through every day were created or maintained by the people performing jobs in this cluster and its pathways.

Pathways

Construction

Design/Pre-Construction

Maintenance/Operations

Sample Occupations

Architect

Architectural and Civil Drafter

Carpenter

Civil Engineer

Civil Engineering Technician

Code Official

Computer Aided Drafter (CAD)

Concrete Finisher

Construction Worker

Cost Estimator

Drywall and Ceiling Tile Installer

Pipelayer

Entrepreneurial Opportunities

Many careers in the Architecture & Construction cluster are well-suited for someone interested in being an entrepreneur. An architect might work for a large firm for a while then start her own business when she feels that she has gained enough experience to succeed on her own. A drywall installer might prefer working independently instead of for a contractor. What are some of the benefits and drawbacks to working as an entrepreneur in the Architecture & Construction industry?

Exploring the Career Clusters ■ **Chapter 4**

Arts, Audio/Video Technology & Communications

Have you ever wanted to design, produce, exhibit, perform, write, and publish anything by making music or creating multimedia visual content? Do you love the performing arts? You could be dreaming of becoming an actress on the stage or television. Or perhaps you would rather be a world-renowned painter. Then again, your interests might direct you to journalism as the next star anchor of the evening news or as a respected blogger. You might want to be a public-relations specialist working behind the scenes to guide the careers of the rich and famous. If so, this cluster is for you.

Journalists report the news in newspapers and on television and radio stations. What skills and interests do you think might be necessary to become a journalist?

Pathways

Audio and Video Technology and Film
Journalism and Broadcasting
Performing Arts
Printing Technology
Telecommunications
Visual Arts

Sample Occupations

Animation Technician
Broadcast Technician
Cinematographer
Editor
Graphics and Printing Equipment Operator
Journalist
Publisher
Reporter
Special Effects Technician
Video Graphics Engineer
Video Systems Technician

Business Management & Administration

Every business hires individuals to manage its operations. Businesses need office managers and accountants to ensure that the calls are answered and that the books are in order. These businesses can range in size from small convenience stores to large hospitals. They all need employees who ensure that operations run smoothly within the organization. The employees who choose careers in the Business Management & Administration cluster play this role.

Pathways

Administrative Support

Business Information Management

General Management

Human Resources Management

Operations Management

A payroll manager ensures that all employees in a company receive accurate paychecks on time. What skills and interests do you think might be necessary to become a payroll manager?

Sample Occupations

Accountant

Accounting Manager

Accounts Payable Manager

Assistant Credit Manager

Billing Manager

Business and Development Manager

Chief Executive Officer

Compensation and Benefits Manager

Credit and Collections Manager

Entrepreneur

General Manager

Payroll Manager

Risk Manager

Career Trend

Accountants and auditors perform similar functions by tracking how companies spend their money. In 2010, there were more than 1.2 million accountants and auditors in the United States. According to the U.S. Bureau of Labor Statistics (www.bls.gov), employment of accountants and auditors is expected to grow 16 percent between 2010 and 2020.

Education & Training

Much thought and care go into planning, managing, and providing education as well as training and related learning-support services. People employed in this field educate children and adults. Workers in this cluster need patience and good customer service skills to be effective in their jobs.

Pathways

Administration and Administrative Support

Professional Support Services

Teaching/Training

A corporate trainer provides short, specialized courses to employees of large and small firms. What skills and interests does a corporate trainer need?

Sample Occupations

College President, Dean, Department Chair, Program Coordinator

Corporate Trainer

Curriculum Specialist

Education Researcher, Test Measurement Specialist/Assessment Specialist

Elementary and Secondary Superintendent, Principal, Administrator

Museum Coordinator

Post-Secondary Administrator

Supervisor and Instructional Coordinator

Teacher

Career Tips

Teaching is a key profession in the Education & Training cluster. If you are considering a teaching career, keep in mind that you must be certified to instruct in public schools.

There are different levels of teaching certification. Some teachers are licensed to teach through grade three. Others are licensed only for the middle and high school levels or for specialized subjects.

Check with your state or local government to find out the certification requirements in your community..

Finance

Finance is often linked to banking, but the industry has many facets because there are so many different types of financial institutions. Employees in this field provide services for financial and investment planning, banking, insurance products, business/financial management, and more. This field is directly connected to the stock market, which monitors the economic stability of the world's financial markets.

Pathways

Banking Services
Business Finance
Insurance
Securities and Investments

An insurance agent sells insurance to people who need it. What skills and interests do you think might be necessary to become an insurance agent?

Sample Occupations

Brokerage Representative
Development Officer
Insurance Agent
Investment Advisor
Personal Financial Advisor
Sales Agent, Securities and Commodities
Securities/Investments Analyst
Stock Broker
Tax Preparation Specialist

NUMBERS GAME

You are considering opening your own business but need $5,000.00 to start it. You have decided to borrow from your local bank. The bank charges 10% interest, so you will have to repay the loan plus the additional 10%. How much will you have to pay back?

First, calculate the 10% interest rate on your $5,000.00 loan:

1. Change the percentage to a decimal by replacing the percent sign with a decimal point and moving the decimal point two spaces to the left.

2. Multiply the decimal by the number. For example, to find 10% of $5,000.00:
   10% = .10
   5,000 × .10 = 500

So when you add the interest to the amount borrowed, how much will you repay the bank? What if the interest rate is 13%?

Government & Public Administration

Local, state, and federal governments need employees to perform functions that include **governance**—or making decisions about how government will act—national security, foreign service, planning, revenue flows and taxation, regulation, and management and administration. For example, the Secretary of State is the chief diplomat for the United States. As chief diplomat, the Secretary of State represents the United States at important meetings on the world stage. People employed in this field typically develop strong communication, problem-solving skills, and critical-thinking skills.

Pathways

Foreign Service

Governance

National Security

Planning

Public Management and Administration

Regulation

Revenue and Taxation

Sample Occupations

Abassador

Airborne Warning/Control Specialist

Assessor

Congressional Aide

Diplomatic Courier

Equal Opportunity Representative

Legislative Aide

National Security Advisor

President

Senator

Special Forces Officer

Tax Attorney

Vice President

Ambassadors represent their governments in foreign countries. What skills and interests do you think might be necessary to become an ambassador?

Chapter 4 ■ Exploring the Career Clusters

Doctors consult with their patients about their ailments and recommend treatments they need to recover from illnesses. What skills and interests do you think might be necessary to become a doctor?

Health Science

Health Science cluster employees plan, manage, and deliver therapeutic services, diagnostic services, health informatics, support services, and biotechnology research and development to individuals throughout the United States. They work in cities, suburbs, rural areas, and other communities to provide crucial services to a diverse client base. Individuals employed in these fields take on legal responsibilities, must have strong ethics, and often use their technical skills.

Pathways

Biotechnology Research and Development
Diagnostic Services
Health Informatics
Support Services
Therapeutic Services

Sample Occupations

Acupuncturist
Anesthesia Technologist/Technician
Anesthesiologist/Assistant
Art/Music/Dance Therapist
Athletic Trainer
Audiologist
Certified Nursing Assistant
Chiropractic Assistant
Chiropractor
Dental Assistant/Hygienist
Dental Lab Technician
Dietitian/Nutritionist
Doctor
Nurse

Exploring the Career Clusters ■ Chapter 4

Career Profile: Nurse

Health Science Career Cluster

Job Summary

One of the occupations within the Health Science cluster that appeals to a lot of individuals is nursing. Individuals interested in nursing typically have a passion for helping sick or injured people recuperate. They also participate in giving the preventive care that keep people healthy so they do not end up in hospitals at all.

Registered nurses (RNs) have direct interactions with patients. They take care of the sick and injured, give medicine, treat wounds, and often provide emotional support to patients and their families. Not all nurses are involved in providing direct care. Nurses also work in research or as administrators who manage clinics, nursing homes, or hospitals.

Use the Internet, a library, or career center to research the responsibilities, education and training requirements, and salary range of a nurse. Write a job description for the position.

Career Trend

The job outlook for the nursing field is very good. Employment of RNs is expected to grow 26 percent from 2010 to 2020. In 2010, the median pay for RNs was $64,690 per year.

Reasons for Choosing a Career

Sometimes people choose their careers for the wrong reasons. Someone interested in the creative arts might choose finance because he thinks he will earn more money. Someone interested in science might avoid the Health Science cluster because she thinks all health science jobs require direct contact with patients. How do you plan to choose a career?

- 👎 Choose a career because your parents say it is right for you
- 👎 Enter a field because it is popular and "hot"
- 👎 Select a job based only on the salary
- 👍 Match your career to your interests
- 👍 Do career research to find a career that appeals to you
- 👍 Dedicate yourself to getting the proper training for the career that most interests you

How might choosing a career for the wrong reasons lead to frustration and unhappiness?

Chapter 4 ■ Exploring the Career Clusters

Hospitality & Tourism

Hospitality and tourism careers comprise the world's largest industry. Its workers manage, market, and operate restaurants and other food services, lodging, attractions, recreation events, and travel-related services. They are very busy people who cater to the needs of individuals that visit your town, city, state, and other countries of the world as tourists. If you enjoy a vacation, a large part of the reason why is the services these employees perform.

Hotel managers are responsible for making sure guests are happy and satisfied. What skills and interests do you think might be necessary to become a hotel manager?

Pathways

Lodging

Recreation, Amusements and Attractions

Restaurants and Food/Beverage Services

Travel and Tourism

Sample Occupations

Bell Captain

Catering and Banquet Manager

Chef

Concierge

Convention Planner

Gaming and Casino Manager

Hotel Manager

Kitchen Manager

Restaurant Owner

Service Manager

Tour Guide

Travel Agent

Making Travel Fun

There are millions of people employed in the tourism and hospitality industries worldwide. What if there were no tourism and hospitality employees? Who would:

★ Call a cab to take vacationers to the beach?

★ Direct visitors from out of town to the best local restaurants?

★ Clean a traveler's hotel room?

★ Deliver room service to a hungry guest?

★ Serve meals in a resort dining room?

★ Sell souvenirs to remind people of the fun they had on vacation?

What problems might arise if you arrived at a hotel for a vacation and there were no hospitality workers?

Human Services

Human services careers prepare individuals for employment in pathways that relate to families and human needs. They work in daycare centers, drive the elderly and disabled to their appointments, provide massages in spas, and counsel those in need of emotional support. Their clients include students having problems in school, brides needing beautiful hairstyles for a wedding, congregants looking for guidance from their clergy, and homeless families trying to find housing. Working in this field can be very rewarding if you like to interact with people. The work that you do will help children and adults, but you must have a lot of compassion and patience to be effective in this field.

Social workers help people focus on improving their lives. What skills and interests do you think might be necessary to become a social worker?

Pathways

Consumer Services

Counseling and Mental Health Services

Early Childhood Development and Services

Family and Community Services

Personal Care Services

Sample Occupations

Barber

Clergy

Embalmer

Guidance Counselor

Hairdresser/Cosmetologist

Manicurist and Pedicurist

Nanny

Social Worker

Career Tips

If you need extra help in figuring out the right career for you, visit the Web site of the U.S. Bureau of Labor Statistics at www.bls.gov. It produces reports that outline the fastest growing industries, their salary range, and educational requirements.

Getting the Right Training for Jobs at School!

Just because you think a career cluster sounds interesting doesn't mean you have the skills or education you will need to land a job in that field. Use your time in school to make sure you are qualified for it.

✔ Stay on track to develop basic skills such as reading, writing, math, and languages.

✔ Work with a career counselor to select elective courses that match your interests and abilities.

✔ Spend time researching the career clusters and pathways.

✔ Build technology skills to help you with your job search.

You can use this approach to explore careers throughout your life until you find a good fit for you, your interests, and personality type.

Information Technology

Individuals employed in this cluster build the bridges between people and technology. They design, develop, support, and manage hardware, software, multimedia, and systems integration services that allow you to have access to the Internet, operate your remote control device, and use your mobile devices. Most people in this field have a strong mathematical and scientific academic foundation and are willing to learn information technology applications and systems. Its employees are constantly working on cutting-edge technology aimed at making life easier for everyone.

Computer technicians work to build or repair computer systems. What skills and interests do you think might be necessary to become a computer technician?

Pathways

Information Support and Services
Network Systems
Programming and Software Development
Web and Digital Communications

Sample Occupations

Analyst
Computer Programmer
Computer Security Specialist
Computer Technician
Database Developer
Modeler
Network Designer
Network Developer
Web Administrator

Myth Education beyond high school is unnecessary.

Truth Today, the careers that offer the best growth opportunities and pay the most require at least a two- or four-year degree. If you are considering a career path that will require a lot of education, begin to view graduating from high school as the start of your education. Embrace the idea of learning as a lifetime effort. Part of that lifelong process will include constantly training and updating your skills.

Exploring the Career Clusters ■ Chapter 4

Law, Public Safety, Corrections & Security

Individuals who plan, manage, and provide legal, public safety, protective services, and homeland security, including professional and technical support services, play important roles in your community. They are police officers, lawyers, security guards, and firefighters, and they deliver the services that make you feel safer in times of crisis.

Pathways

Correction Services
Emergency and Fire Management Services
Law Enforcement Services
Legal Services
Security and Protective Services

Sample Occupations

Arbitrator, Mediator and Conciliator
Correctional Officer and Jailer
Detective and Investigator
Dispatcher
Firefighter
Immigration and Custom Inspector
Judge
Lawyer
Paramedic
Police Officer

Police officers keep you safe, respond to emergencies, and issue citations to people who break the law. What skills and interests do you think might be necessary to become a police officer?

21st Century Learning

Sam has been having problems receiving and making calls on his cell phone. Finding solutions to the problem has been challenging yet interesting. The process has him curious about the clusters and pathways connected to the cell phone industry, such as Marketing, Manufacturing, and Transportation, Distribution & Logistics.

He is doing research to contact and meet with a career counselor at school to discuss the skills he will need to become employable in those clusters within cell phone companies. What are some of the ideas that the counselor might suggest to Sam? Do you think Sam will need to consult with professionals in those fields? What new information might the conversation with his counselor reveal? How would you advise Sam to use 21st Century Skills such as decision making, problem solving, and goal setting to learn more about the cell phone industry as a career path? Write a letter advising Sam, or form small groups and present your advice as a skit.

Manufacturing

Products ranging from plastic bottles to cars are built in factories and developed in manufacturing plants all over the world through specific processes. Those employed in the Manufacturing cluster perform the steps that go into intermediate or final products and related professional and technical support activities such as production planning and control, maintenance, and manufacturing/process engineering. They are in a field for individuals who love to use their hands to make things that are important to daily living such as cars, cabinets, or books.

Factory workers assemble a wide variety of products. What skills and interests do you think might be necessary to become a factory worker?

Pathways

Health, Safety and Environmental Assurance

Logistics and Inventory Control

Maintenance, Installation and Repair

Manufacturing Production Process Development

Production

Quality Assurance

Sample Occupations

Assembler

Automated Manufacturing Technician

Bookbinder

Calibration Technician

Electrical Installer and Repairer

Electromechanical Equipment Assembler

Logistical Engineer

Machinist

Energy Career Cluster

To meet the growing demand for skilled workers in the energy industry, a number of states have developed an Energy career cluster to identify energy-related occupations that are critical to their economic development.

Energy jobs fall within four of the 16 Career Clusters identified by NASDCTEc:

- ✔ *Agriculture, Food & Natural Resources*
- ✔ *Architecture & Construction*
- ✔ *Science, Technology, Engineering & Mathematics*
- ✔ *Manufacturing*

Investigate your state's career clusters to determine if an Energy cluster has been adopted.

Marketing

Marketing professionals plan, manage, and perform activities to reach organizational objectives. Their work can range from handing out flyers to publicizing an event or working in an office to researching consumers' spending habits. They work for small or large firms or independently as contractors in a number of roles where they develop their leadership and teamwork abilities as well as their analytical skills.

Pathways

Marketing Communications
Marketing Management
Marketing Research
Merchandising
Professional Sales

Sample Occupations

Account Executive
Creative Director
Key Account Manager
Market Research Analyst
Merchandise Displayer and Window Trimmer
Real Estate Agent
Sales Representative
Store Manager
Telemarketer
Wholesale and Retail Buyer

Store managers oversee the day-to-day operations of the stores, shops, and retail outlets in which you buy things. What skills and interests do you think might be necessary to become a store manager?

Money Madne$$

Esther is studying communications in college and is exploring careers in the Marketing career cluster. In the meantime, she is looking for a summer job. She is offered a position as a sales associate at a shoe store. It pays $10.25 per hour, for 25 hours per week. She will be able to work for 8 weeks until school starts again. She is also offered an internship at a public relations agency. The internship is full time, and it pays $2,500.00 for the entire summer. Which position do you recommend Esther accept? What factors should she consider? **Hint:** Use multiplication to calculate the amount Esther would earn.

Science, Technology, Engineering & Mathematics

Can you see yourself planning, managing, and providing scientific research as a career? Are you interested in laboratory and testing services or in research and development? If so, are you also willing to earn the certification or advanced education necessary to work in this career cluster? If your answer is a "yes" to any of these questions, then take a closer look at these pathways and occupations.

Pathways

Engineering and Technology

Science and Math

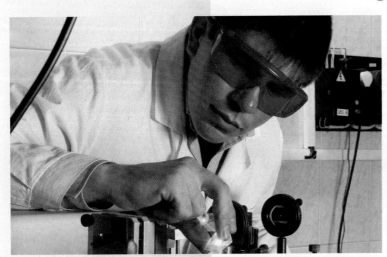

Academic institutions, government laboratories, and private industry hire physicists to study matter or energy within the universe. What skills and interests do you think might be necessary to become a physicist?

Sample Occupations

Anthropologist

Biologist

Geneticist

Mathematician

Nuclear Chemist

Paleontologist

Physicist

Statistician

Technical Writer

Transportation, Distribution & Logistics

How does food get to your supermarket shelves or fuel to your local gas station? A lot of planning goes into the transportation and shipping of products around the globe. Many people work behind the scenes to coordinate, manage, and move these goods to their final destinations by arranging for their transportation by road, pipeline, air, rail, and water. They also oversee the related professional and technical support services that are a part of the process. Beyond planning for the transportation of goods, there is another related sector in this field: urban planning. Planners decide where houses, parks, and stores should be built so that neighborhoods are organized in a way that is convenient for everyone living in them. Here are the pathways and some of the occupations that make up this cluster.

Urban planners oversee how land is used by and in communities. What skills and interests do you think might be necessary to become an urban planner?

Pathways

Facility and Mobile Equipment Maintenance

Health, Safety and Environmental Management

Logistics Planning and Management Services

Sales and Services

Transportation Operations

Transportation Systems/Infrastructure Planning, Management and Regulation

Warehousing and Distribution Center Operations

Sample Occupations

Air Traffic Controller

Driver

Industrial and Packaging Engineer

Logistician

Storage and Distribution Manager

Surveying and Mapping Technician

Traffic Manager

Traffic, Shipping, and Receiving Clerk

Urban and Regional Planner

Warehouse Manager

Exploring Other Resources for Identifying Careers

Identifying Careers through Job-Related Activities

Even if you do not hold a full- or part-time job, the tasks you do and activities in which you are involved on a daily basis can help you identify the career clusters that best match your interests, skills, aptitudes, and values. For example:

✔ Helping a younger sibling with his homework might lead you to a career in the Education & Training career cluster.

✔ Volunteering at a local park or zoo might mean you are cut out for a career in the Agriculture, Food & Natural Resources career cluster.

✔ Taking photos for your school newspaper could prepare you for a career in the Arts, Audio/Video Technology & Communications career cluster.

Think about the job-related activities you perform at home, in school, and in the community. How can they help you identify the career cluster(s) best suited to you?

In addition to the career clusters, you have many resources to help you identify the types of careers that are available. Recall that a resource is something you use to get or achieve something else. Some resources that will help you identify careers are available at school, and some you will find in your community. These and other resources will be discussed throughout this book.

You can use career search resources to learn about the types of jobs that are available, identify career requirements and responsibilities, and even to compare and contrast salaries. You can also use them to locate jobs that match your interests, skills, and abilities.

Resources at School

Some of the best resources for learning about career opportunities are available at school. Your guidance and career counselors have books, computer programs, videos, and lists of useful Web sites that provide information about types of careers. Your teachers can help you identify your strengths and abilities and match them with career requirements.

School clubs and organizations may also have information about types of careers. A business club might invite a representative from the community to speak about his or her occupation. An organization such as Family, Career and Community Leaders of America (FCCLA) or Health Occupations Students of America (HOSA) has career-related projects and activities.

The Bureau of Labor Statistics

The U.S. Bureau of Labor Statistics (BLS) is a government agency responsible for tracking information about jobs and workers. Its Web site has a section specifically for students (www.bls.gov/audience/students.htm). It includes information on more than 60 occupations, including descriptions of responsibilities, education requirements, salary ranges, and job outlook. A **job outlook** includes statistics and trends about whether the job is in an industry that is growing or shrinking.

The BLS also publishes the *Occupational Outlook Handbook* (OOH) in printed and online editions. The OOH describes more than 200 occupations, including responsibilities, working conditions, education requirements, salary ranges, and job outlook.

Exploring the Career Clusters ■ Chapter 4

The *Occupational Outlook Handbook* is massive. There may be a printed copy of the OOH in your local library or career center. You can start by looking through the table of contents to get an idea of how careers are organized. You may find it easier to use the online version at www.bls.gov/ooh. Online, you can search for specific careers, use the alphabetical listing to locate a career that interests you, or browse through the career clusters. As you look through the OOH, you might want to take notes on the information you find, and record the source—including the page number in the printed copy or the Web page address online.

Resources in Your Community

One of the best ways to learn about different careers is from someone who is employed. For example, you can job-shadow someone who has a job you find interesting. Job shadowing means to follow someone around at work for a day or part of a day. By job shadowing, you can see exactly what the responsibilities, tasks, and rewards are for a particular job.

Informational interviews are also a good way to explore careers. An informational interview gives you the opportunity to sit down with someone who is employed in a career or industry that interests you. You can ask specific questions about the job responsibilities, see the work environment, and maybe even meet other people in the industry.

Talking directly to someone in the job can teach you a lot more than you can learn from a book or online. For example, a roofer can tell you what it feels like to work on top of a house, 30 feet above the ground. Your career counselor may be able to help you arrange an informational interview.

Why do you think you should only request an informational interview when you are serious about a career?

Steps for Managing an Informational Interview

Follow these steps to plan, schedule, and conduct an informational interview:

✔ Find a person who has the job you are interested in.

✔ Call and make an appointment.

✔ Write down the questions you want to ask before the interview.

✔ Arrive on time for the interview.

✔ Write down the answers to your questions.

✔ Thank the person when you leave, and follow up by sending a written thank you note.

Career Trend

More informational interviews are being held over the telephone or by e-mail than in the past. People are busy and always on the go. Sometimes it is easier to make time for an interview using technology.

Career Tips

Do not ask for a job during an informational interview. If you ask for a job, the person you are interviewing might feel that you misled him or her about the purpose of the interview.

Case Study

Michael works as a full-time firefighter for his local municipality, making $35,000.00 a year. He has seen many of his colleagues laid off recently as a result of a downturn in the economy. In his firehouse of ten firefighters and two part-time clerical staff, half of the firefighters and one of the clerks lost their jobs.

He has been thinking about a career change because he wants more job and financial security. Many close friends have advised him not to change careers at this time, but to consider it in the future.

For now, Michael has decided to remain in his current position, but he will continue looking into other opportunities. He also decides to take classes in areas of interest as he waits to make a final decision about his future.

- What decisions did Michael face?
- Do you agree with his choices? Why or why not?
- How can the information on career clusters and their pathways be helpful to Michael?
- What other steps can Michael take now, while he waits for the right time to make a career change?

Answer It!

1. How many career clusters are defined by the National Association of State Directors of Career Technical Education Consortium?
2. List the pathways in the Architecture & Construction cluster.
3. What is the title of the chief diplomat for the United States?
4. List five sample occupations in the Hospitality & Tourism cluster.
5. What career cluster has a pathway for computer programmers?
6. In which cluster are firefighters and police officers employed?
7. What career cluster includes a pathway for urban planners?
8. List two resources available at school to help you identify career opportunities.
9. What government agency is responsible for tracking information about jobs and workers?
10. List five things you can learn about an occupation from the Occupational Outlook Handbook.

Web Extra

Many Web sites provide information on career clusters. One site is careertech.org.

Do you have effective research skills? Test them by using the Internet to locate resources that help you determine job and career opportunities related to the 16 career clusters defined by the U.S. Dept. of Education. Include a brief description of the information available through each resource.

Create a directory of all of the sites that you find. Post it on your school Web site, make it available in your school library, or add it to the school blog.

Exploring the Career Clusters ■ Chapter 4

Write Now

To attract new employees, companies post want ads on online job listings or in electronic or printed publications and newspapers announcing their vacancies. Now that you know about and understand the career clusters and their pathways, create ads for the occupations in the three career clusters that most interest you. Research the careers so you can include the skills and educational requirements and salary range. Look at examples of want ads online or in your local newspaper. Use them as models of how your ads should look. Show your ads to the class and discuss them.

COLLEGE-READY PRACTICES

College-ready individuals understand that education is the first step in achieving a rewarding career.

Conduct research to learn about the 16 career clusters defined by the U.S. Dept. of Education. Organize a list of all 16 based on your interest, putting the one that interests you most at the top and the one that interests you least at the bottom. From the cluster at the top of your list, select three occupations and research the educational options and requirements for each. From the list of three, select one, and write a paragraph explaining how you can start now to prepare for the educational options and requirements necessary to obtain that occupation. Read your paragraph out loud to a partner or to the class.

Career-Ready Practices

Divide into teams of four or five. Each team will select a career cluster and pathway. Use the Internet or other resources to research companies that work in your cluster and pathway. If possible, visit the Web sites of these companies and locate the employment opportunities section where job openings are listed. As a team, make a chart listing the job title, skills and educational requirements, and salary range of at least three available positions. As a class, compare and contrast the information in your charts. Discuss how skills and education impact the salary range for each position. How can you and your classmates prepare yourselves for the requirements of the positions that most interest you?

Career-ready individuals recognize that nearly all career paths require ongoing education and experience. They understand the time, effort, and experience that will be required to pursue the pathways that interest them. In addition, they seek counselors, mentors, and other experts to help them plan and meet their career and personal goals.

Career Portfolio

Matching your interests and abilities to a career is an important step in finding job satisfaction. After reviewing the 16 career clusters, make a list of the four that you find most interesting. Write a description of each, and include the reasons they appeal to you. You may list the clusters in a word processing document, or write them on a piece of paper. Include the title *Career Clusters that Interest Me* at the top of the page. Add the document to your career portfolio.

5 Think Like an Entrepreneur

Skills in This Chapter . . .

- **Exploring Entrepreneurial Thinking**
- **Recognizing Entrepreneurial Risks and Rewards**
- **Developing the Characteristics of an Entrepreneur**
- **Identifying Entrepreneurial Opportunities**

GETTING STARTED

Have you ever thought about starting your own business? Do you have ideas about what kind of business it would be? An entrepreneur is someone who creates and runs a business. If you have a business idea and can imagine a way to make it come true, you could be an entrepreneur. Even if you don't want to be an entrepreneur, thinking like one can help you develop a vision, or plan, of what you want your future to be.

▶ Make a list of the character qualities you think make someone a good boss. Put a star next to the ones that you think you possess. Would you be a good boss? Do you think it would be easier to work for yourself or for someone else?

Exploring Entrepreneurial Thinking

Most people earn a living by working in a **business**. A business is an organization that provides goods or services, usually to make money. A person who works in a business owned by someone else is an **employee** of that business. On the other hand, someone who creates and runs his or her own business is called an *entrepreneur*.

Owning a business isn't for everyone. But that's okay because both employees and entrepreneurs are needed in the world of work. Whether or not you become an entrepreneur, understanding entrepreneurial thinking can help you in any career you choose.

Do You Think Like an Entrepreneur?

Thinking like an entrepreneur can help you succeed in school as well as in your career.

- 👎 Arrive late for class
- 👎 Never participate in class discussions
- 👎 Turn in incomplete assignments
- 👍 Come to class prepared
- 👍 Listen to your teachers
- 👍 Respect your classmates
- 👍 See projects through to completion
- 👍 Contribute new ideas in class

Can you think of other ways you can use entrepreneurial skills and qualities to succeed in school?

What Is Entrepreneurial Thinking?

Entrepreneurial thinking is the process of starting a new business. It may mean thinking of, creating, and developing a brand-new product or service. For example, you might develop a new type of fuel for powering automobiles, or a new method of fitness training. It may also mean creating a new business to sell an existing product or service, such as opening a new restaurant or selling online jewelry that you design and create yourself.

Entrepreneurial thinking is not easy. More new businesses fail than succeed. Just having a good idea is not enough. Entrepreneurs must be motivated to work hard. They must have the skills to identify challenges and find solutions, including raising money and other resources. They must also be able to determine whether a risk has a better chance of bringing rewards or causing failure.

Career Trend

Many businesses today encourage **intrapreneurship**, which means acting like an entrepreneur while working as an employee. They give employees opportunities to be creative, try out new ideas, and sometimes even to develop a new business within the existing one.

Thinking Like an Entrepreneur

Thinking like an entrepreneur and being aware of how to make a business run more successfully can help you do well at school, home, and in your community, as well as at work for someone else. Successful entrepreneurs know how to come up with an idea and then set up a plan to make it a reality. They set goals, manage their time effectively, and are 100 percent committed to their plan.

When you think like an entrepreneur, you show that you are creative, independent, and responsible. In school, you participate, pay attention, and show respect. As an employee, you treat someone else's business as if it were your own. Here are three ways to think like an entrepreneur even when you are not running your own business.

Why might an employer like to hire and promote an employee who thinks like an entrepreneur?

- *Observe.* Keep on the lookout for chances to learn new skills and accept new responsibilities. At work, staying aware of what goes on around you can help you identify new ideas for business growth, such as new products or services that customers may need or want.
- *Listen.* Pay attention to what others have to say. Your classmates may have a different view of things. Challenges that other employees are facing may give you ideas for making business improvements.
- *Think.* Instead of complaining about a problem, analyze it. Then, suggest possible solutions.

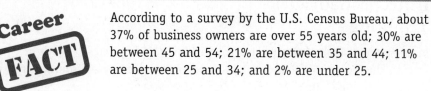

According to a survey by the U.S. Census Bureau, about 37% of business owners are over 55 years old; 30% are between 45 and 54; 21% are between 35 and 44; 11% are between 25 and 34; and 2% are under 25.

Recognizing Entrepreneurial Risks and Rewards

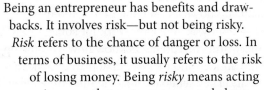

Being an entrepreneur has benefits and drawbacks. It involves risk—but not being risky. *Risk* refers to the chance of danger or loss. In terms of business, it usually refers to the risk of losing money. Being *risky* means acting in a way that puts you or your belongings at risk. In business, being risky usually refers to taking unnecessary chances with your money.

As business owners, entrepreneurs are in control of the money made by their business. They also have the final say in all business decisions. As a result, entrepreneurs are ultimately responsible for the success or failure of their businesses. It is a lot of responsibility, but it can also bring a lot of rewards.

You might think of many reasons to start your own business. Before doing so, however, it's a good idea to consider the pros and cons of being an entrepreneur. The key is evaluating whether the potential rewards are worth more to you than the drawbacks and risks you will take.

NUMBERS GAME

A percentage is a part of a whole. The word *percent* means by the hundred. One-hundred percent equals the whole or all of something. Each percent represents one part of 100.

To find the percentage of a number:

1. Change the percentage to a decimal by replacing the percent sign with a decimal point and moving the decimal point two spaces to the left.
2. Multiply the decimal by the number.

For example, to find 20% of 150:

20% = .20

150 × .20 = 30

Studies show that about 21% of all businesses in the United States are minority-owned. If you live in a town with 200 businesses, how many are minority-owned?

Hint: Find 21% of 200.

International Entrepreneurial Thinking

Entrepreneurs can benefit from international trade. They can sell goods or services that people in other countries need, or buy goods made somewhere else to sell in the United States.

However, there are many risks associated with international trade. For example, governments might place restrictions called **trade barriers** on international trade. Exchange rates—the value of the currency used in one country compared to the value of currency in a different country—change daily and can affect the cost of items sold across borders. Also, cultural differences can impact business decisions.

Using the Internet or your library, research the benefits and drawbacks an entrepreneur might encounter when trying to start an international business. Select one and report on it to your class.

Think Like an Entrepreneur ■ **Chapter 5**

What Are the Rewards?

Of course, entrepreneurs want to make money. A successful business lets you earn enough so that you fulfill your needs and wants without depending on others. If the business grows, entrepreneurs can create even greater wealth by selling it for more money than they used to start it. They may even be able to use the money from the sale to start another business.

In addition to the financial rewards, entrepreneurs can benefit in other ways.

- Entrepreneurs achieve pride and personal satisfaction from starting and growing a business. Not only do you have freedom and independence, you also display your skills and abilities for your family and friends to see. Entrepreneurs also feel excitement and joy from taking a chance and trying something new.

- As an entrepreneur, you get to make your own rules. Depending on your business, you can decide what type of schedule you work, where you work, and how and when you get paid. You also have the final word on which products or services the business provides and how they are provided. For example, when you have a creative idea, you have the power to put that idea into action.

- Being an entrepreneur opens up opportunities that help make your community and the world a better place in which to live. Entrepreneurs often start a business because they see a need in the community. They might start small, but if the business succeeds, they are soon creating jobs, paying taxes, and giving back by donating time and money to worthy causes.

Do you think the benefits of being your own boss would outweigh the drawbacks? Why or why not?

Tech Connect

Entrepreneurs are always looking for ways to use the newest trends and technologies to start or grow a business. You probably think of a social networking Web site as a place to communicate with your friends. Entrepreneurs see it as a business opportunity.

For example, can a fashion designer trying to start a new business draw in customers by tweeting? If a restaurant owner offers discount coupons on a social networking site, will people tell their friends?

Use the Internet to research one way an entrepreneur has tried to use social networking or another type of technology to develop a business. Then, use a word processing program to write a report about what you discover.

Facts About Entrepreneurs

Following are some facts about business owners taken from a survey conducted by the U.S. Census Bureau.

✔ *Of the business owners who were surveyed, 68% had some college education when they started the business.*

✔ *More than 34% of the business owners used money of their own, or from their families, to start or buy the business.*

✔ *Slightly more than 50% of the business owners worked more than 40 hours per week.*

✔ *About 30% of the businesses were home-based.*

How does the saying, "Nothing ventured, nothing gained" relate to entrepreneurs?

Myth Most entrepreneurs take wild and uncalculated risks when they start their businesses.

Truth Entrepreneurs know that taking high risks is a gamble. They are not afraid of risk, but they know how to assess a situation and take on challenges when they feel confident they have the skills for success.

What Are the Risks?

When an entrepreneur starts a new business, risk is involved. **Risk** is the chance of losing something. An entrepreneur invests money, time, and energy in his or her business in the hope of receiving greater rewards, or benefits. Being fully responsible means the success or failure of your business rests on you. Even if you make all the right decisions, the business might fail. In that case, you may walk away with nothing—or worse, **debt**, which is money you owe to others. The failure can affect your family and friends, too. They may lose money they invested in the business, and they may suffer from stress.

Other risks include unexpected problems, unreliable income, and time away from family and friends.

- Unexpected problems are challenges for which you are unprepared. They can be discouraging and frustrating and can cause **stress**, which is a physical reaction to a difficult or demanding situation. Facing these challenges can get scary and lonely, especially if you don't have the emotional support of family and friends.

- Unreliable income is when the amount of money you pay yourself changes month to month. The amount of money you can pay yourself may go up or down, depending on how well your business is performing. Many new businesses don't make much money in the beginning, so you may not always be able to pay yourself. During rough times, you may even have to put more money into the business to pay your business bills, leaving less money to pay your personal expenses..

- It's not unusual for entrepreneurs to work a lot of extra hours to make their businesses successful. This is especially true during the initial start-up process. These long hours can decrease the time you have available for your friends and family. Until you can afford to hire other people to help, you may have to perform many types of tasks. This will require discipline and a willingness to do whatever needs to be done.

Developing the Characteristics of an Entrepreneur

Recall that your *character* is the personal qualities or traits that make you unique. Most entrepreneurs share some similar characteristics that make them suited to starting a business. For example, they tend to be creative, hard working, and willing to take risks.

Entrepreneurs also have certain skills that help them succeed. A *skill* is an ability that you learn through training and practice. For example, they understand how to create and manage a business, and they are able to keep tasks and information organized.

No one is born with all the characteristics and skills needed to be a successful entrepreneur. But if you keep a positive attitude and believe in yourself, you can develop many of them. They can help you succeed as an entrepreneur, and also in other areas of your life.

Qualities of an Entrepreneur

Do you have the qualities of an entrepreneur? Can you develop them?

- *Entrepreneurs have courage.* They are willing to take risks in spite of possible losses. They tend to be creative. They invent new ways of doing things, and they see possibilities where others see only problems.

- *Entrepreneurs are curious.* They are eager to learn, and they know how to ask questions and find answers. They are determined. They refuse to quit even when they are faced with many obstacles.

- *Entrepreneurs are disciplined.* They stay focused, and they are able to follow a schedule to meet deadlines. They are enthusiastic. They are passionate about their goals and see problems as opportunities. Entrepreneurs are flexible. They can adapt to new situations and are not afraid to make changes.

- *Entrepreneurs have empathy.* People who have **empathy** are sensitive to the thoughts and feelings of other people. They have patience and recognize that most goals cannot be achieved in a short time.

- *Entrepreneurs are honest.* They are committed to being truthful and sincere with others. They are responsible. They do their best to live up to others' expectations. They are accountable for their decisions and actions, and they do not blame others for their own mistakes.

Which personality traits do you have that might help you be a successful entrepreneur? Which do you need to develop?

Entrepreneurial Skills

The skills you use to be a successful entrepreneur are the basic skills you need for success in all areas of your life. You can work in school, at home, and in your community to develop these skills through practice, and with the help of adults. You learn about many of these skills in this book. Some of the basic skills entrepreneurs need include the following.

- *Business skills:* Understanding how to create and manage a business
- *Communication skills:* The ability to listen well, write well, and speak well
- *Computer skills:* The ability to use technological tools effectively
- *Decision-making and problem-solving skills:* Knowing how to apply logic, information, and past experiences to new decisions and problems
- *Mathematical skills:* Using math to create budgets, keep accurate records, and analyze financial statements
- *Organizational skills:* The knack of keeping tasks and information in order; the ability to plan well and manage your time
- *Relationship skills:* The ability to persuade and motivate people; knowing how to be a leader and to work as part of a team

Building on Experience

Make a list of activities or tasks you've done in the past, such as hobbies, part-time jobs, volunteerism, clubs, sports, science fairs, school activities, and classes. Then create another list that identifies the kind of personal characteristics and skills you need to perform those activities. Based on the two lists, write an essay that describes how your past experiences could help you as an entrepreneur.

Money Madne$$

Every year the Fourth of July parade passes in front of your house. You and a friend decide this year you will set up a stand to sell bottles of water. You decide to purchase 250 bottles. The bottles come 25 in a case, and each case costs $10.99. How much money will you need to purchase the bottles of water? **Hint:** Calculate the number of cases you need, and then multiply that number by $10.99.

If you sell each bottle for $1.50 and sell all 250 bottles, how much profit will you make? **Hint:** Your profit is the amount of money you earn less the amount of money you spend.

Keys to Success

Don't be discouraged from becoming an entrepreneur just because you don't yet have all the traits and skills you will need. You can increase your entrepreneurial potential by focusing on six specific areas. Even if you never become an entrepreneur, paying attention to these areas will help you be more successful in life.

- *Business knowledge.* Make a habit of reading magazine and newspaper articles on business topics. Use the Internet to research business subjects. Watch films or television programs about successful entrepreneurs. This can help you learn more about business. If you know someone who owns a business, discuss the business with that individual.

- *Financial skills.* Strengthen your math skills by taking a course in accounting, personal finance, or investing. If you find that math is a difficult subject, ask a teacher to spend a little extra time with you before or after school. Team up with a friend who is good at math. Play math games or do math homework together.

- *Career exploration.* It is not too early to begin thinking about your career. You can identify your strengths and weaknesses and work to develop the qualities you think you will need. You can also develop your interests, build a career portfolio, and talk to people who have jobs you think are interesting.

- *Community awareness.* Look for opportunities to become involved in your community.

- *Education.* Learning is important, no matter what career you choose. Take advantage of opportunities to learn new things, ask lots of questions, and work hard to do your best. Obtaining an educational certificate, diploma, or degree not only benefits you personally, it can also help open doors to more career opportunities.

- *Relationships.* Spend time with people who believe in you and inspire you. Being around positive people will help you stay positive and accomplish more. Find opportunities to develop leadership skills and to gain experience working as a member of a team.

How can developing relationships with experienced workers help you develop entrepreneurial skills?

According to the Small Business Administration (SBA), an agency of the U.S. government that provides aid and advice to small businesses, a small business is defined by the number of employees or its average annual receipts, depending on the type of industry. The primary size standard is 500 or fewer employees for most manufacturing and mining industries, and $7 million or less in annual receipts for most non-manufacturing industries.

Chapter 5 ■ Think Like an Entrepreneur

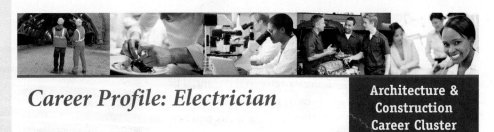

Career Profile: Electrician

Architecture & Construction Career Cluster

Job Summary

An electrician is someone who installs the wires that carry electricity through houses, offices, factories, and other buildings. Some electricians build or repair machines that run on electricity.

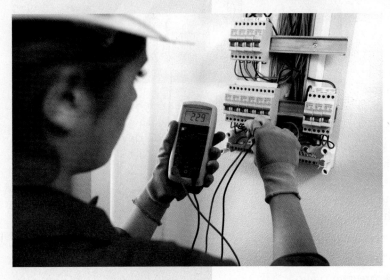

Why is problem solving an important skill for an electrician?

Electricians must have a lot of specialized, technical knowledge. For example, they must understand the dangers of electricity and how to follow rules and regulations controlling wiring. They must read blueprints to see where to put wires, electrical equipment, and outlets for plugs. Some electricians create blueprints to show others how the wiring should be installed.

To become an electrician, people take classes about electricity, often starting in high school. It also helps to take classes in math and science. Once they receive a high-school diploma, they work as assistants or apprentices to get on-the-job training, while taking additional classes toward certification. It takes many years to become skilled enough to be certified or licensed and longer to become a Master Electrician.

Electricians may work for construction companies, or they may work in factories. Many start out working for someone else, and then begin operating their own, independent businesses.

Use the Internet, a library, or career center to research the responsibilities, education and training requirements, and salary range of an electrician. Write a letter of recommendation for someone who might want to become an electrician.

Think Like an Entrepreneur ■ **Chapter 5**

21st Century Learning

CAREER COUNSEL

Camille is quiet and shy. She is a talented artist and particularly enjoys decorating fabrics with original designs. She often makes T-shirts, scarves, and other items to give to her friends and family as gifts.

Camille's friend Anna thinks Camille's art is good enough to sell. Anna has always wanted her own business. She studied marketing in school and has worked as a sales clerk for five years. She has researched ways to start a small business. Anna offers to handle the business side of things, so Camille can concentrate on designing and creating the products. Anna thinks that together they could be very successful. Camille is hesitant. She does not think enough people would buy her products to make it worthwhile.

What are the challenges facing Anna and Camille? What are the potential rewards? Use your 21st Century Skills such as problem solving, decision making, and critical thinking to help Anna and Camille come to an agreement about what to do next. Write an ending to the story and read it to the class, or form small groups and present it as a skit.

Can an artist be an entrepreneur? What type of skills might he or she need to develop to grow a successful business?

Can Your Idea Become an Opportunity?

Once you come up with an idea, it is time to explore its potential to be an opportunity. Which of these factors indicate that an idea might make a good business opportunity?

- 👎 You do not have the resources to put the idea into action.
- 👎 Someone else has already made a successful business from the same idea.
- 👎 You cannot provide the product or service at a price that will earn a profit.
- 👍 The idea fills a need or want that's not currently being met.
- 👍 Your community is a great location to make the idea a business.
- 👍 You can provide the product or service at a price that will attract customers.

Can you think of other ways to tell the difference between an idea and an opportunity?

Identifying Entrepreneurial Opportunities

Usually you start a business because you see an opportunity. A **business opportunity** is a consumer need or want that can be met by a new business. Not every business idea is a good business opportunity. For example, you might have an idea to open a neighborhood restaurant featuring classic French cooking, but if people in the neighborhood don't like classic French cooking, then your business will not succeed.

A successful entrepreneur knows how to tell the difference between an idea and an opportunity. He or she uses available resources to select an opportunity that shows promise, and then turn it into a business.

Finding Ideas

How does an entrepreneur come up with an idea? Maybe someone has a problem that can be solved with a new product, or is using an existing product that could work better. Maybe someone hears a news story or reads a blog that identifies a need. Often, ideas come from everyday life. Here are some resources you might find in your own community.

- Reading newspapers and magazines can help you identify developing trends in your area, or in a different area. Specialized magazines that focus on something that interests you—such as a hobby—can also help you find opportunities.

- Business and government agencies such as the U.S. Census Bureau, U.S. Department of Labor, and the U.S. Small Business Administration provide statistical data that can help you identify trends. Local organizations such as the Better Business Bureau or Chamber of Commerce provide information on businesses in your local area.

- Trade organizations can provide you with information about a specific industry that interests you, and help you meet and network with people in that industry. Trade organizations are groups that provide training and information for people associated with a particular career or industry. They sponsor meetings and shows and publish newsletters and magazines.

- The Internet lets you explore almost any topic from your own computer. Web sites provide information about industries and companies. Blogs help you learn about consumer problems and opinions.

Evaluating an Opportunity

Just because an idea shows promise as an opportunity doesn't mean it is worth pursuing. You can use these three practical methods to determine if an idea is worth the time, effort, and investment that it will take to make it a business.

- *Cost/benefit analysis.* A **cost/benefit analysis** is the process of adding up all the expected benefits of an opportunity and subtracting all the expected costs. If the benefits are greater than the costs, the opportunity may be worthwhile. Costs include money, such as the amount you must spend on supplies and advertising, as well as items such as time and effort. Benefits also include money, such as how much you will earn, and items such as satisfaction and pride. You may have to estimate the costs and benefits at first.

- *Opportunity-cost analysis.* The **opportunity cost** is the value of what you are willing to give up to achieve something else. An opportunity-cost analysis examines the potential benefits that you will give up when you choose one course of action over another. For example, you might choose to accept a position as an unpaid intern in an industry that you find interesting. You give up the wages you might earn in a different job in order to gain experience in the field of your choice.

- *SWOT analysis.* A **SWOT analysis** is a business method used to evaluate four areas: **S**trengths, **W**eaknesses, **O**pportunities, and **T**hreats. Strengths include your skills, knowledge, and resources; weaknesses include the areas where you—or your idea—might be lacking and need improvement. When you assess your opportunities, you look for everything that provides your idea with a business advantage. For example, does your idea fill an unmet need or want, make use of a trend, or surpass the efforts of an existing company? Finally, you analyze the threats, which are obstacles or challenges that might interfere with your success.

Why is it important to decide if an idea will make a good opportunity before you start your business?

Turning an Opportunity into a Business

What action can you take to turn your opportunity into a business? Here are four common ways that entrepreneurs become businessmen and businesswomen.

- *Start a new business.* This is probably the most challenging option. Benefits include the ability to build the business your own way, but there are many drawbacks. You are responsible for making the many decisions and completing the many tasks that are required to set up and run a new business. You will also assume the bulk of the risk. The process may take a long time.

- *Buy a business that already exists.* This option may require a significant amount of money up front, but you will not face the risks and challenges of starting a company from scratch. You will still be responsible for the continued growth and success of the operation.

- *Buy a franchise.* Buying a franchise gives you a legal agreement to sell another company's products or services. Becoming a **franchisee**—someone who buys a franchise—is less risky than starting a new business, or buying an existing business. The **franchisor**—the company that sells the franchise—usually provides training and national advertising. However, you will have less entrepreneurial freedom because you must follow the rules in the legal agreement.

- *Invent something new.* If you invent a new product, design, or process, you can sell or license the invention to someone else, or create a business to use, manufacture, or sell the invention.

Some studies show that most franchise operations have a 90% or better chance of success. Statistics from the Small Business Administration, however, indicate that only about half of all new franchises survive five years or more and roughly a third survive ten years or more, which are the same success rates for independent business startups.

Think Like an Entrepreneur ■ **Chapter 5**

Evan is not old enough for a driver's license, but he loves cars. One summer, he starts a small automotive detailing business. Detailing involves making the car look clean and new by washing, vacuuming, and buffing every part of it. Evan spends anywhere from two to six hours per car, depending on factors such as the type of car and its condition. He charges $25.00 per hour.

One day Evan receives a call from a man running a classic car show. He wants to hire Evan to detail five classic cars in four days. But he does not want to pay. Instead, he offers to put up signs next to each car at the show with Evan's name and contact information.

What are the benefits and drawbacks of the arrangement for Evan? Use 21st Century Skills such as problem solving, decision making, critical thinking, and effective communication to help Evan decide what to do. Write an ending to the story and share it with your class.

21st Century Learning

Case Study

When Jackson was 12 years old, his state passed a recycling law. As part of the law, consumers received 10 cents apiece for returning certain types of glass bottles and aluminum cans. Jackson started collecting returnable bottles and cans and bringing them back to stores to receive the refund. He saved the money he earned in a bank account.

When Jackson turned 17, he used the money that he had saved as a down payment on a pickup truck. He used the truck to start a business picking up unwanted items from neighbors and friends. He was amazed at the items people didn't want—furniture, clothing, books, and even appliances. He sold most of it to stores or recycling facilities, donated some to charity, and disposed of the rest.

Jackson used some of the money he earned to pay for college tuition and saved the rest. When he graduated, the bank gave him a small business loan so he could open his own store selling used goods.

- Do you think Jackson is an entrepreneur? Why or why not?
- What entrepreneurial characteristics does Jackson display?
- What entrepreneurial skills has Jackson developed and demonstrated?

Answer It!

1. What is the difference between an employee and an entrepreneur?
2. List three ways an employee can think like an entrepreneur.
3. What is the difference between risk and being risky?
4. List four benefits of entrepreneurial thinking.
5. What are three things an entrepreneur invests in his or her business?
6. List four risks of entrepreneurial thinking.
7. List five characteristics of an entrepreneur.
8. What is empathy?
9. List seven entrepreneurial skills.
10. List three ways an entrepreneur uses mathematical skills.

Web Extra

You can find a lot of useful information about entrepreneurial thinking and small businesses at sba.gov, the Web site for the U.S. Small Business Administration. Start on the home page, and explore the links to learn more about how to plan, start, and manage your own business.

Write Now

Interview an entrepreneur or small business owner in your community. Ask the person about the challenges, risks, and rewards of being an entrepreneur. Ask which personal characteristics or skills have contributed most to business success. Then ask what the owner would do differently if starting the business today. Use the information from your interview to write a newspaper article about the person.

COLLEGE-READY PRACTICES

College admissions counselors look for students who exhibit many of the same qualities that make entrepreneurs successful, such as creativity, discipline, flexibility, and responsibility. From your research on the 16 career clusters, select a field or pathway that interests you. Conduct research to identify at least three entrepreneurial opportunities in your selected field or pathway.

Select one of the opportunities you identified and write a letter to a college admissions counselor explaining how you are developing entrepreneurial qualities now that you believe will help you achieve your selected opportunity.

Career-Ready Practices

Career-ready individuals recognize the time, effort, and experience needed to pursue their career goals, even if that goal is owning their own business. They recognize the value of building skills and gaining experience that are aligned with their personal and career goals.

As a class, review the 16 career clusters. Then, divide into teams of three or four and select one of the career clusters. Make a list of entrepreneurial opportunities within your selected career cluster. Pick one in a field that interests you, research it, and make a presentation about it to your class.

Be sure to discuss the seven entrepreneurial skills and how they would be of use in your business.

Career Portfolio

Entrepreneurial characteristics and skills can help you succeed in school and as an employee. Make a list of entrepreneurial characteristics and skills. Describe how you exhibit each item or how you can develop it. Include an explanation of how each characteristic or skill can help you succeed. Add the list to your career portfolio.

Part II
Career Preparation

In This Part . . .

Chapter 6 ■ Skills for Success

Chapter 7 ■ Academic Planning

Chapter 8 ■ Communicating with Others

Chapter 9 ■ Building Relationships

Chapter 10 ■ Basic Math Skills

Chapter 11 ■ Technology and Your Career

6 Skills for Success

Skills in This Chapter . . .

- **Making Decisions**
- **Solving Problems**
- **Setting Goals**
- **Thinking Critically**
- **Using Management Skills**

GETTING STARTED

What does success mean to you? Are you successful if you get a B on a math test? Do you feel successful if you beat your personal best at sports, even if your team loses?

Success is the completion of anything intended or finishing what you planned to do. In other words, **success** is when you set a goal for yourself and then achieve that goal. Because we all set different goals, success means different things to different people. But no matter how you measure success, there are basic skills that can help you get there. When you recognize how to set goals, make decisions, and solve problems, you can apply these skills to all areas of your life.

➤ Look through magazines and cut out pictures of at least three people you think are successful. As a class, discuss why you picked the pictures. How can you tell the person is successful? What do your selections say about how you measure success?

Influences

Things that influence your decisions:

✔ *Values*
✔ *Goals*
✔ *Standards*
✔ *Ethics*
✔ *Roles*
✔ *Stage in life*
✔ *Other people*
✔ *Resources*
✔ *Experience*
✔ *Knowledge*

Making Decisions

Any time you make up your mind about something, or choose one option over another, you are making a **decision**. Some decisions are simple—what time will I leave for school? Some are more difficult—what will I do after high school? The results—or **consequences**—of your decisions affect you in big and small ways.

- If the consequences of a decision are positive and contribute to your well-being, it means you made a healthy—or good—choice.
- If the consequences are negative and interfere with your well-being, that means you made an unhealthy—or poor—choice.

Decisions give you power and control over your life. When you make a decision, you are showing yourself and others that you are independent and responsible.

Every Decision Is Unique

Clearly, not everyone will make the same decision in the same situation. That's because we each use our own character qualities, values, and available resources when we make a choice.

You and a friend might both decide to join your community's recycling action committee, but you might have different reasons for doing so. You might be disgusted by the trash you see along the road, but your friend might want to meet new people. A third friend might opt not to join the committee at all.

Six Steps to a Decision

You can take some of the uncertainty and doubt out of decision making by turning it into a process. A **process** is a series of steps that leads to a conclusion.

1. *Identify the decision to be made.* Make sure you recognize and understand the choice. Define the decision as a goal—what do I want to achieve with this choice?

2. *Consider all possible options.* You usually have lots of options for each decision. Try to think of as many as you can, and write them down. Don't just consider the obvious choice; some of the best options might seem pretty bizarre at first. Consider your available resources and what you are trying to achieve.

3. *Identify the consequences of each option.* Each option will have consequences—some positive and some negative; some long-term and some short-term. Recognizing all the consequences will help you predict the outcome of your decision.

4. *Select the best option.* Once you consider the options and identify the consequences, you have the information you need to make your decision.

5. *Make and implement a plan of action.* Making the decision is not the end of the process. You must take steps to make it happen. Until you do, the decision is just an idea or thought in your head.

6. *Evaluate the decision, process, and outcome.* After you have acted on your decision, you can look back and evaluate it, based on your values and standards. Did you achieve the goal you defined in step 1? Did you miss any possible options? Did you correctly identify the consequences? Did you make use of your resources? Was the outcome what you hoped for?

Decision-Making for College and Career Readiness

You will have lots of decisions to make as you prepare for college and a career. The decision-making process will help you identify your options, the consequences of each option, and, to make the best choice you can. Once you are comfortable with the process you can use your decision-making skills to:

- Make decisions in school, such as which elective to take or club to join.
- Make decisions in your community, such as where to volunteer or how to be a good neighbor.
- Make decisions about college and career planning, such as whether to join the military, what to study, and how to obtain an internship.

Can Your Decisions Be Hurtful?

How can you be sure your decisions won't interfere with the well-being of yourself and others? When you are evaluating your options while making a decision, consider these questions.

- Is it hurtful to me?
- Is it hurtful to someone else?

If the answer to the questions above is "Yes," you might want to look for other alternatives. If the answer to the following questions is "Yes," you are on the right track.

- Is it fair?
- Is it legal?
- Is it honest?
- Is it practical?

Some decisions are easy to make, and others are more difficult. What is one of the more difficult decisions that you've had to make? What were the consequences of your decision?

Myth All healthy choices lead to positive consequences.

Truth Sometimes even the best decisions lead to negative consequences. Life is unpredictable. When that happens, we have to try to learn from the experience and make the best possible choices going forward.

How can you reduce the stress you feel when you must make a choice?

Overwhelmed by Options

Decision making can be stressful. Even simple choices can seem overwhelming at times. For example, should you wear a suit coat when you apply for a job or is a clean, pressed shirt appropriate? Once you make a decision, you might *second-guess* yourself, which means doubting your own choice. You wore the clean, pressed shirt, but you continue to wonder whether the suit coat might have been better.

- If you are **indecisive**—unable to make a decision—you might think about what will happen if you choose one option over another and never make a decision at all.

- If you are **impulsive**—inclined to act without thinking—you might make a snap decision without considering the consequences.

Career Tips

When you are faced with a decision that affects your career, discuss it with someone who understands your career plan. A career counselor, a teacher, or your supervisor at work will be able to help you assess your options.

Predicting Consequences

Do you need a crystal ball to predict the long-term consequences of a decision? What if you decide to volunteer to read to children at the local kindergarten?

★ Short-term consequence: You have fun with the young children.

★ Long-term consequence: You consider a career in early childhood education.

★ Consequence that affects others: One of the kindergarteners loves the story you read and is inspired to work hard to learn to read herself.

Can you think of other long-term consequences that might come from a simple decision?

More About Consequences

Consequences of our choices can be short-term or long-term. What you eat for breakfast might affect your energy level for a few hours. Whether you go to college can affect your whole life.

Most decisions actually have both long-term and short-term consequences. Long-term, what you eat for breakfast might affect your health and wellness. Short-term, it might affect how you do on a test.

Sometimes, consequences have both positive and negative results. Take the breakfast decision: On the up side, a doughnut tastes good and gives you a sugar boost of energy. On the down side, a doughnut raises your cholesterol level, which might lead to a heart condition. The good news is that you can learn how to make healthy decisions that turn out positive more often than they turn out negative.

Did I Make the Best Choice?

How can your values, standards, and ethics help you make healthy choices? We all make mistakes. Despite our best intentions, we make poor choices. Most of the time, it doesn't matter too much. If you cut your hair too short, it will grow. You can choose a different style next time or even go to a different shop for the cut.

Sometimes, though, we must live with the consequences of our actions for a long time—maybe even our whole lives. It's part of the deal—if you are independent and responsible enough to make the decision, you're stuck with the results. If you choose to cut class or hang out with friends instead of studying, you might not qualify for the college of your choice. But even when we make a poor choice with long-term consequences, we can learn from our mistakes and try to make better choices going forward.

If you make a decision that has negative consequences, what can you do next time to make a better choice?

NUMBERS GAME

You can use **probability**—the chance that something will happen—to measure how likely it is that a particular outcome will occur. The **formula**—rule or method of doing something—for probability is the number of times the particular outcome occurs divided by the total number of possible outcomes. In fractional form, it looks like this:

$$\text{Probability} = \frac{\text{Number of particular outcomes}}{\text{Total number of possible outcomes}}$$

As an example, consider that you are a clerk at a pet store. You have a box of dog collars. There are five blue collars, three red collars, and one black collar. A customer asks for a red collar. What is the probability that you pick a red collar from the box your first time?

Hint: Divide 3 (the number of red collars) by 9 (the total number of collars).

What is the probability that you pick a blue collar your first time? **Hint:** Divide 5 by 9.

Myth All poor choices lead to negative consequences.

Truth As luck would have it, sometimes things work out OK even when you make really poor decisions. Just breathe a sigh of relief, recognize that you were fortunate that nothing negative happened, and once again, make the best possible choices going forward.

Solving Problems

A **problem** is a difficulty or challenge that you must resolve before you can make progress. In other words, any barrier or obstacle between you and a goal is a problem.

Problems pop up all the time. Mostly, we come up with a solution without thinking too hard. Say you want to meet friends for dinner after work, but you promised your supervisor you would complete a report. You have a problem: The unfinished report is an obstacle between you and dinner with friends. Solving the problem is easy: You finish the report or reschedule the dinner.

Some problems sneak up on us over time, sometimes hidden or obscured by something else. You might want to do well at work, but you fall asleep in staff meetings. Is the problem that the meeting is boring, that the conference room is too warm, or is it that you are staying up late at night playing video games? Recognizing the real problem is the first step in solving it.

Owning the Problem

One difficulty with solving problems is figuring out whose problem it really is. Generally, the person who is blocked from a goal is the one who owns the problem.

If your co-worker leaves his copy of a report at home and wants to borrow yours, is it your problem or his? What if you loan him the report and he loses it? Is it his problem or yours? Who should take responsibility for solving it?

If you own the problem, you are responsible for solving it. If someone else owns the problem, you may be able to help solve it, but ultimately it is not your responsibility. Taking responsibility for your own problems, and working to find solutions, shows that you are independent and capable.

What's Really the Problem?

Sometimes things can distract you from seeing the real problem. If you show up at your career counselor's office to ask her to fill out a reference form and she is busy with another student, what is the problem?

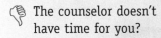 The counselor doesn't have time for you?

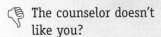 The counselor doesn't like you?

👍 You forgot to make an appointment?

Since you can't go back in time to make an appointment, what steps can you take now to solve the problem?

Why do you think employers value employees who are able to solve problems in the workplace?

Six Steps to a Solution

When problems are harder to identify, or harder to solve, you can use the problem-solving process to figure out the best solution.

1. *Identify the problem.* This is your chance to be honest, acknowledge the problem, and determine what goal it is blocking.
2. *Consider all possible solutions.* There may be one obvious solution or there may be many possible solutions. Write down as many as you can think of. You will need to consider your values, standards, and resources, too. Some solutions might be harder to make happen or take longer than others. Some might cost money, and some might be free. Some might solve only part of the problem.
3. *Identify the consequences of each solution.* As for decisions, each solution will have consequences, and it is important to recognize how the consequences will affect you and others. Again, write them down.
4. *Select the best solution.* The best solution offers the best possible opportunity for you to continue your progress toward your goal.
5. *Make and implement a plan of action.* Recognizing and selecting a solution are only part of the process. You must take the necessary steps to make the solution real.
6. *Evaluate the solution, process, and outcome.* Did your solution work? Did you achieve your goal? Would you do anything differently if you had the same problem again?

Asking for and Giving Help

You do not have to make every decision or solve every problem on your own. You have many people around to help, including family, friends, and teachers. You might think asking for help is a sign of weakness, but it's really a sign of strength. Recognizing that you need help shows that you are responsible. It proves that you understand your limitations. Face it—you don't know all the answers. Asking for help shows that you want to learn from the experience of others.

However, when someone comes to *you* for help, remember it is his or her problem to solve, not yours. With that in mind, you should listen carefully and critically, and then share your thoughts and opinions. Try not to make the decision or solve the problem yourself—it's not your responsibility. If you don't think you are qualified to help, say so, and then suggest someone who might be. You can probably help a friend decide what to wear on a date, but a parent, teacher, or counselor might be better equipped to offer advice on how to prepare for a job interview.

Career Profile: Computer Support Specialist

Information Technology Career Cluster

Job Summary

Why is problem solving an important skill for a computer support technician?

When someone has computer trouble, he or she asks a computer support specialist for help. A computer support specialist works to identify the problem and then tries to fix it.

In addition to knowledge about computer systems, computer support specialists must be able to communicate effectively with the people who come to them for help. They often work over the phone or by using e-mail to help people solve the problem themselves. Some do the actual repair work. Many computer support specialists also set up computer systems, and then train people on how to use them.

To become a computer support specialist, people take classes about building, installing, repairing, and using computer systems. They also study math and develop problem-solving skills. Most employers look for computer support specialists who have a college degree and experience working with computers.

Use the Internet, a library, or career center to research the responsibilities, education and training requirements, and salary range of computer support technicians. Write a want advertisement for a computer support specialist job opening.

Money Madne$$

You are responsible for purchasing supplies for a meeting. There will be 50 people at the meeting. You need pens, pads of paper, and folders for each attendee. Pens come 25 in a box. Each box costs $3.25. The paper costs $1.25 per pad. Folders cost $2.10 per box of 50. There is a 6% sales tax in your state. How much money do you need to purchase all of the supplies?

Setting Goals

Recall that a **goal** is something you are trying to achieve. Goals help direct your actions and guide your decision-making because they give you something to work toward. They help give your life meaning, because you know that there is a purpose in what you do. When you achieve a goal, you can be proud and express satisfaction.

If all you do is think about a goal, it's just a dream. You make goals real by deciding what you want to achieve and then planning how to get there. A successful entrepreneur, for example, has a vision of what he or she expects to accomplish. To achieve that vision, it is important to set realistic and attainable goals. Goals help an entrepreneur plan, develop, and grow a business. They let the entrepreneur identify the challenges ahead, so he or she is prepared when problems arise. While you should set goals that are within reach, there is nothing wrong with challenging yourself to push harder.

Goals are not written in stone. As you progress through different stages of your life, you will learn more about yourself, your values, and your standards. Your resources will change. You can change your goals at any time and develop new goals. Most people do.

Five Steps to a Goal

You can use the following process to help identify, assess, and set goals.

1. *Identify the goal.* Write down the goal using as much detail as you can. This helps you understand and recognize the goal. Be positive, not negative: "I will attend the FCCLA meeting" rather than "I won't skip the FCCLA meeting."
2. *Assess the goal.* Determine if it is something you really want and if it is realistic and attainable. It might be a fad or something that sounds good or even something someone else wants for you.
3. *Make an action plan for achieving the goal.* What resources will you need? If you cannot come up with a plan that works, it may mean the goal is not realistic or attainable. If so, you may need to go back to step 1.
4. *Write down your action plan.* Be as specific as possible.
5. *Reevaluate your goals.* Every once in a while, make sure they are still important to you and, if so, that you are on track to achieve them.

Goals in All Areas of Your Life

You can set goals for all your roles in life that help you prepare for and achieve a satisfying career. What if you are interested in a career in health sciences?

★ At home: Investigate the items necessary for a home first aid kit and enlist your family's help in preparing one.

★ At school: Enroll in classes and extracurricular activities that will help you develop the skills you need to pursue an education in health science.

★ With friends: Help your friends recognize the importance of safety precautions such as wearing bike helmets and buckling seat belts.

★ In your community: Volunteer at a hospital or medical center.

★ At work: Consider part-time or volunteer opportunities that will contribute to your knowledge about the health sciences industry.

Can you think of other ways to integrate your career goals into many areas of your life? Make a list and discuss it in class.

Make a Plan

Just because a goal is short-term doesn't mean you can achieve it without a plan. What if you have a short-term goal to meet friends at the mall?

★ *Where will you meet?*
★ *What time will you meet?*
★ *How will you get there?*
★ *What time should you leave?*

Considering all of the factors helps you make decisions and form a realistic plan. The more specific the plan, the easier it will be to achieve your goal. How can you make a plan for long-term goals?

Short-Term Goals

When you want to achieve something quickly, you set **short-term goals**. You can accomplish short-term goals in the near future—maybe even today. For example, finishing your homework on time is a short-term goal. It is usually easy to define short-term goals because they are specific and not very complicated. If you keep a to-do list, it is full of short-term goals—meet friends at the mall, call your grandmother, make your bed.

Long-Term Goals

A **long-term goal** is something you want to achieve in the more distant future—maybe a year from now or maybe even more distant than that. Graduating from college is a long-term goal. So is buying a car.

Defining long-term goals may be more difficult than defining short-term goals. You might know you want to get married some day, but you don't know when or to whom. You might know you want to travel, but you don't know where or how.

What Are Milestones?

Sometimes it's harder to stay focused on a long-term goal—it seems far away. Breaking the long-term goal down into a series of short-term goals—or **milestones**—makes it easier to stay on track.

Becoming a nurse might be a long-term goal. To achieve that goal, you can set a series of realistic and attainable milestones, such as:

- Volunteering at a hospital or senior care center
- Working part-time in a clinic
- Graduating from high school
- Attending college

What can you do now to make long-term goals a reality?

Thinking Critically

Critical thinking is the ability to be honest, rational, and open-minded. It can help you evaluate your options in many situations. You can use it when you are making decisions, setting goals, and solving problems. When you think critically, you try not to let emotions get in the way of choosing the best course of action.

- Being honest means acknowledging selfish feelings and preexisting opinions.
- Being rational means relying on reason and thought instead of on emotion or impulse.
- Being open-minded means being willing to evaluate all possible options—even those that are unpopular.

Thinking critically doesn't mean you should ignore emotions or any of the other influence factors. It just means you should consider all possibilities before rushing to judgment.

Critical Thinking in Action

When you think critically, you consider all possible options and other points of view. You look objectively at information. **Objective** means fairly, without emotion or prejudice. Then, you use your values, standards, and ethics to interpret the information subjectively. **Subjective** means affected by existing opinions, feelings, and beliefs.

Looking at things both objectively and subjectively can help you make choices that are right for you. For example, you can look at a candidate for class president objectively and see that she is smart, hard-working, and honest. Subjectively, you can disagree with everything she stands for and vote for someone else.

You can think critically about a lot of things, not just decisions and problems. You don't have to believe everything you hear or read. You can question a news report, look more deeply into the meaning of a magazine article, or investigate the truth behind a rumor.

Myth You can't trust your gut and think critically at the same time.

Truth Trusting your gut is not the same as being impulsive. Sometimes your gut—or that little voice inside your head—is really your values, standards, and ethics pointing you in the right direction. If you use the decision-making process and your gut is pointing out the best choice, go with it.

Are Emotions Always a Problem?

Emotions can affect critical thinking. That's not necessarily bad, but it's not necessarily good, either. Skipping class because you are angry about a low grade might seem like a good choice at the time, but the consequence might be more low grades.

If you think critically, you acknowledge that you are angry and are honest about why. Then, you can let the anger go and assess your options with a clear head. Maybe it would be better to go to class and ask the teacher what you can do to improve. The consequences might be better grades, an improved relationship with the teacher, and—best of all—no more anger.

How can thinking critically help employees make decisions or solve problems in the workplace?

21st Century Learning

Alyssa is friendly and nice and likes being around people. In school, she enjoys classes in business math and technology. She is thinking about studying finance in college. She applies for a summer job as a bank teller. The branch manager hires someone else, but tells Alyssa she is a strong candidate, and he will call her if another position becomes available.

Alyssa accepts a job as a data records assistant at a hospital. She is assigned to work on a project to convert paper files to electronic files. She spends her days alone at a computer.

After two weeks, the bank branch manager calls. A teller position has become available, and he offers the job to Alyssa.

What should Alyssa do? Use your 21st Century Skills such as decision making, problem solving, goal setting, and critical thinking to help Alyssa examine her situation and move forward. Write an ending to the story. Read it to the class, or form a small group and present it as a skit.

Using Management Skills

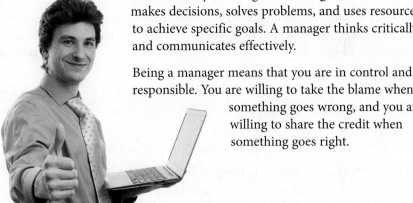

You are already a manager. A **manager** is someone who makes decisions, solves problems, and uses resources to achieve specific goals. A manager thinks critically and communicates effectively.

Being a manager means that you are in control and responsible. You are willing to take the blame when something goes wrong, and you are willing to share the credit when something goes right.

What a Manager Does

A good manager uses a three-step process to make thoughts and ideas become reality. He or she knows how to use goal setting, decision making, and critical thinking to find a way to get things done. You can use these steps to solve problems the way a manager does.

- First, you figure out what needs to be done. You set a goal. You consider your resources. You assess problems that might be in the way. And then you develop an action plan.
- Next, you put the plan into action. You monitor the plan to make sure it is going the way you expect. You might even need to make changes along the way.
- Finally, you look back to assess and evaluate. Did the plan work the way you expected? Did you achieve your goal?

What Makes a Manager Effective?

A manager is effective when he or she achieves a goal. An effective manager knows how to recognize and use available resources, including:

- Time
- Energy
- Technology
- People

An effective manager knows how to make healthy choices in all areas of life. For example, an effective manager knows how to:

- Communicate with family members
- Think critically about friends and peers
- Set goals for school and career
- Solve problems in the community

> ### Cross-Cultural Management
>
> Management culture—a companywide attitude toward workplace values and standards—is often influenced by the culture of the country where the business is located. For example, in some countries, it is not important to be on time for meetings. In other countries, businesses close for an hour or more in the middle of the day.
>
> In today's global economy, it is important for managers working internationally to understand and respect differences in cultural management. Managers who do not may have trouble earning the respect of their international employees.
>
> Use the Internet or the library to research management culture around the globe, then make a chart comparing at least three countries. What cultural differences do you think would pose the biggest challenge to international managers?

What Does the Title "Manager" Mean?

At work, some people have the job title "Manager." Usually, a manager supervises other employees to make sure all responsibilities are met. For example, at a bank, the branch manager supervises the bank tellers and others who work at the branch. At a supermarket, the bakery manager supervises the bakers and others who work in the bakery department.

Managers generally earn more money because they have more experience and more responsibility in the workplace. You become a manager at work by showing you are responsible, independent, and capable of making decisions and solving problems. You increase your chances of becoming a manager by taking classes in school to learn how to manage effectively. You can also attend graduate school after college to earn a Masters in Business Administration—MBA—degree.

Becoming a Leader

A **leader** is a type of manager. A leader is someone who unites people to work toward common goals. If you are a leader, it means others trust you and trust your judgment. In the world of work, leaders supervise other employees and manage projects.

Although you might have heard that someone is a "born leader," that's not usually the case. Becoming a leader takes time and patience. Leaders have to prove that they can make healthy decisions, set goals, and solve problems. They have to know how to use their resources to help others achieve their goals. Leaders usually start out as good employees. They are promoted into management positions as they develop and display leadership qualities. For example, a teacher may continue his or her education, assume leadership roles, and become a principal or superintendant.

Leaders are successful in their different roles and meet their responsibilities. For example, in school, you might participate in clubs and organizations. At work, you might accept responsibility for new projects. In the community, you might attend meetings and help organize events. At home, you might show your siblings how to cooperate with each other.

What Makes a Leader?

Leaders exhibit positive qualities that other people respect, such as self-confidence. They use skills such as goal setting and critical thinking to make healthy decisions for the benefit of the group. If you are an effective leader you:

- Respect others
- Know how to compromise
- Know how to communicate

As a leader, you understand your own strengths and weaknesses. You also know how to accept responsibility when something goes wrong and how to give credit to others when something goes right. Finally, a strong leader is willing to take a stand, even if it's unpopular.

What skills and resources can you use to be an effective leader?

Tech Connect

In most businesses, managers are responsible for keeping projects on time, on track, and on budget. Even small projects may become overwhelming at times. To make the job easier, many use project management software tools for collecting data, tracking tasks, and enabling everyone involved to communicate effectively.

Project management software programs vary in terms of complexity. Most include such features as scheduling, assigning tasks, file storage, file sharing, to-do lists, time tracking, and security.

Use the Internet or magazines or visit a local computer/electronics store to research project management software. Imagine you are a department manager for a small business. Select a product and prepare a list that outlines its features and capabilities that you will share with your employees. Present the list to your class.

Case Study

Jamal has been taking karate classes after school since the age of 11. Now he is a sophomore in high school. He has earned a first-degree black belt and works at the dojo (karate school) to earn money to pay for his lessons.

Jamal's friend has asked him to join the wrestling team at school. The practices are fun, and Jamal is good at wrestling. The coach thinks he can start on the varsity team if he works hard. The problem is, practices are at the same time that Jamal usually works at the dojo.

- What decision does Jamal have to make?
- Are there any problems he has to solve?
- What resources can Jamal use to make his decision?
- What do you think Jamal should do? Why?

Answer It!

1. Explain the difference between a healthy choice and an unhealthy choice.
2. List the six steps in the decision-making process.
3. List at least three things that influence your decisions.
4. How can you tell who owns a problem?
5. List the five steps in the goal-setting process.
6. What is the difference between short-term goals and long-term goals?
7. List three qualities of critical thinking.
8. What is the difference between being objective and being subjective?
9. What is a manager?
10. List at least three qualities of a leader.

Web Extra

acrn.ovae.org/links.htm is the Web site for America's Career Resource Network, a group of state and federal organizations that provide information, resources, and training about career and education planning. Use the link for Students to explore the topics and resources and to access the site's Career Decision-Making Tool.

Skills for Success ■ Chapter 6

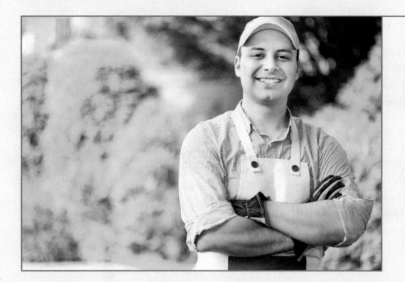

Write Now

What do think you are going to do after high school? Do you have a goal? Are there obstacles between you and that goal? Imagine yourself ten years from now. Write an autobiography—essay about your life—explaining the goals you set, decisions you made, and problems or challenges you overcame during those ten years to become your future self.

COLLEGE-READY PRACTICES

College-ready individuals understand that using critical thinking to evaluate the consequences of positive and negative personal choices can help you learn, grow, and mature.

Many college admissions counselors look at the social networking pages of applicants as part of the admissions process. The way you use social media now might affect your options in the future. With a small group, write a skit that shows the impact of positive and negative choices, including the use of electronic communications such as social networking sites. Present your skit to the class.

Career-Ready Practices

Career-ready individuals recognize that problems arise in the workplace. They understand the first step in solving a problem is to identify its cause. Then, they consider options for solving the problem. Once they determine the best solution, they follow through to ensure it is solved.

In teams of four or five, brainstorm a list of different types of businesses. Pick one business, and, on a separate piece of paper, write down one problem or challenge that employees of that business might run into. Do not identify the type of business. Exchange problems with another team. Work as a team to identify the type of business and brainstorm solutions to the problem. Discuss your results with your class.

Career Portfolio

Setting goals is an important part of the education and career planning process. Make a list of long-term and short-term goals for yourself. Select one, and use the five-step goal-setting process to develop the goal and make an action plan for achieving the goal. Include a timeframe and assess the plan to make sure it is reasonable and attainable. Add both the list of goals and the action plan to your career portfolio.

7 Academic Planning

Skills in This Chapter . . .

- **Recognizing the Value of School**
- **Analyzing Your Learning Style**
- **Transitioning to High School**
- **Developing an Academic Plan**
- **Developing Career-Related Skills in School**
- **Planning for Postsecondary Education**
- **Identifying Career-Related Opportunities for Students**

GETTING STARTED

Can you name one thing you can do to increase the amount of money you will earn in your lifetime? How about graduating from high school? Even better, graduating from college!

High school graduation might seem very far away. The thought of studying for so many years, of taking—and passing—so many classes, might feel like a heavy burden. But the benefits of success are worth the effort. Graduates are more likely to be hired than nongraduates. They earn more money, are healthier, and live better lifestyles. Completing your education is one of the best things you can do for yourself and your career. Academic planning can help you reach that goal.

➤ Fold a piece of paper in half. On one side, make a list of jobs you think you could get without a high school diploma. On the other side, list jobs you think you could not get without a high school diploma. As a class, discuss which jobs offer better benefits and rewards.

Recognizing the Value of School

Finishing school is an investment in your future. Studies show that people who stay in school have higher self-esteem, better physical health, and a more positive attitude. They are more likely to have a job, and they earn more money.

School is an opportunity. When you stay in school, you can learn new things, meet new people, and try new activities. You can practice skills you will need when you graduate, such as getting along with others, speaking up for yourself, and being on time for appointments.

Why School Matters

School also provides an opportunity to prepare for a career. Core subjects such as reading, writing, and math are vital for the career-search process. Science, social studies, music, art, technology, family and consumer sciences, and sports all help you gain knowledge and build skills you will need to succeed at work, such as teamwork, leadership, and problem solving. School clubs and organizations also help you build skills for future success.

Why Dropping Out Is Not Cool

You might think dropping out of school to get a job is a good idea. Maybe you need money. Maybe you are tired of going to class. Maybe you think you are ready to live like a grownup. However, when you are not in school, you will be expected to pay your own rent, pay for your own transportation, and pay for your own food. If a company does hire you, it will likely pay you about 35 percent less than if you had graduated.

A **dropout**—someone who leaves school without receiving a diploma—is more likely to be poor, receive assistance from the government, and be a single parent. Dropouts are more likely to commit crimes and go to jail, to suffer from depression, and to have substance abuse problems.

According to the U.S. Census Bureau, the average dropout can expect to earn an annual income of $20,241. A high school graduate, on the other hand, can expect to make about $30,500 a year, and those with a bachelor's degree can expect to make more than $50,000 a year.

Academic Planning ■ Chapter 7

How Education Affects Employment

Education teaches you skills and information you need to get and keep a job. Think about this: Most companies will not hire an employee who has not graduated from high school. Some will not hire an employee who has not graduated from college. If a company does hire dropouts, it usually pays them less than it pays graduates.

Employers assume that graduates have certain characteristics that dropouts don't have. If a graduate and a dropout apply for the same job, the employer thinks that the graduate:

- Knows how to read and write
- Understands basic math
- Can communicate with others
- Knows how to solve problems
- Has a good work ethic
- Has a positive attitude
- Is self-disciplined
- Is motivated to succeed

If you were an employer, what characteristics would you look for in someone you were going to hire? Would you rather hire a graduate or a dropout?

Meeting Career Goals

Developing strong academic skills can help you in any career you choose. What skills can you acquire and strengthen in school in order to achieve your career goals?

- 👍 Prioritize assignments and manage your time to complete them.
- 👍 Listen carefully and follow instructions.
- 👍 Enroll in courses related to the career you want to pursue.
- 👎 Cram for a test the night before it's scheduled.
- 👎 Socialize and talk a lot during class.
- 👎 Enroll in easy classes so you can get a good grade and improve your GPA.

What other academic skills can help you meet your career goals?

Money Madne$$

Research shows that high school dropouts in the United States earn a lot less than high school graduates. In fact, they earn 35% less. What exactly does that mean? It means that if a high school graduate working as a chef earns $27,000.00 a year, a high school dropout would only earn $17,550.00.

How much more does the graduate earn?

To find the **difference**, or how much more the graduate earns than the dropout, subtract the dropout's salary from the graduate's salary.

What if the graduate earns $32,000.00 a year? To find the dropout's salary, you multiply the grad's salary by 35%, and then subtract the answer from 32,000.00. If you have a calculator, press 32000 * 35%. Then, subtract the result from 32000 to get the amount of the dropout's salary.

If you don't have a calculator, convert the percentage to a decimal so you can multiply it by the graduate's salary. To convert to a decimal, divide the percent by 100: 35 ÷ 100 = 0.35.

Analyzing Your Learning Style

A **learning style** is a way of taking in information so that you remember it and can use it in a practical way. Researchers have identified three major learning styles—visual, auditory, and tactile—although every person learns in his or her own unique way. Some do better when they can see information and examples; some do better when they hear it. Some people learn by doing—which means they need practical experience with tasks.

No single learning style is better than any of the others. Each has its own strengths and weaknesses. Recognizing your own learning style can help you succeed in school, as well as in other areas of life. Your teachers, parents, and guidance counselors can help you identify your learning style and find methods of learning that work best for you.

Visual Learners

Visual learners take in information best when they see it. They tend to have excellent imaginations. They think in pictures and remember things best if they see them. They are usually good spellers and may have art skills.

- Visual learners may need to hear spoken instructions more than once before they understand them.
- Visual aids such as charts, illustrations, maps, notes, and videos help visual learners absorb information.
- Visual learners often benefit from marking or highlighting written information.
- Visual learners might find it helpful to sit in the front of a class so they can see the teacher's facial expressions clearly.

Auditory Learners

Auditory learners take in information best when they hear it. They listen well and like to talk about what they learn. They follow spoken instructions better than written instructions. They are usually good at giving speeches and presentations.

- Auditory learners remember information when they hear it in a lecture or on a recording.
- **Reciting**—speaking—information out loud helps auditory learners remember it.
- Auditory learners benefit from a spoken explanation of visual material, such as charts, diagrams, and maps.
- Auditory learners might find it helpful to use rhymes, raps, or musical jingles to remember information.
- **Dictating**—speaking ideas aloud while another person writes them down—is a good way for auditory learners to organize their thoughts.

Studying Works!

Choosing the correct study methods for your own learning style is important, but some study methods work for everyone. How can you use studying to improve your chances of academic success?

- 👎 Study while watching television
- 👎 Cram in all your studying right before a test
- 👎 Stay up all night to study
- 👍 Study a little bit every day
- 👍 Organize a study group
- 👍 Ask your teacher for study tips

What other study methods do you think might improve your chances of academic success?

Tactile Learners

Tactile learners learn best by touching things. They tend to express themselves physically. For example, they might move their hands or gesture when they speak, and they might like to walk around when they are listening. They usually are not effective listeners and lose interest in lectures or speeches. Tactile learners often learn best by practicing what they learn and relating information to real-life situations.

- Tactile learners may find it helpful to take frequent breaks and move around while studying.
- They may benefit from studying while standing up or eating while studying.
- Acting out scenarios can help tactile learners remember information.
- Tactile learners remember information better if they take notes while listening.
- Tracing written words with their fingers can help them remember the text.

Study Strategies for Success in School

People who do well in school usually work very hard. They know how to use study **strategies**—careful plans and methods—that improve their ability to remember new things. Once you identify your learning style, you can select study strategies that work for you.

- *General strategies.* Sit near the front of the class so you pay attention; take notes; set a daily study goal, such as completing your homework.
- *Strategies for memory.* Study your notes soon after class; make flash cards with a key idea on one side and a definition or explanation on the other side.
- *Strategies for listening and note-taking.* Listen for clues that the speaker is giving a key point; underline or star the main points; use abbreviations for commonly used words.
- *Strategies for planning and organization.* Keep a calendar, schedule, or to-do list; write assignments and due dates in an assignment notebook and on a calendar; prioritize your responsibilities and tasks; break a large project into smaller steps, and set a deadline for each step.
- *Strategies for tests.* Ask your teacher what the test will cover and what type of questions it will include; don't wait until the night before to study—study a little bit each evening in the days leading up to the test; read test directions carefully; check your answers; skip questions you are unsure of and go back to them after you have completed the others.

What study strategies do you find most useful?

> **Tech Connect**
>
> Schools take advantage of technology in many ways, including to communicate with parents and the community and to cut costs. However, the main reason for using technology in school is to teach.
>
> Some technologies such as interactive white boards, digital text books, online videos, and collaborative Web sites can make learning more engaging. Some make teaching more effective. For example, specialized technologies such as voice recognition software help students with disabilities succeed.
>
> Do you think you need technology to learn? Do you think technology makes school more effective? Interview someone in your school community about technology in the classroom. Then, write an article about it to print in your school newspaper or post on your school Web site.

Transitioning to High School

Are you ready for high school? You might be excited to leave middle school behind. After all, high school offers more opportunities, including more courses, more independence, and more extra-curricular activities. There are also more people, a bigger building, more homework, and more responsibility.

Knowing what to expect when you get to high school can help you know how to prepare for the change. There are things you can do now that will make the transition easier.

How to Prepare for High School

The most important thing you can do in middle school to prepare for high school is to develop strong study habits. Successful study skills will lead to good grades in middle school, and help you keep up when you get to the more competitive classes in high school.

You can also explore co-curricular activities to discover your interests and abilities. These activities help you make friends with similar interests, build teamwork and leadership skills, and gain experience that will come in handy on your college applications. If your middle school does not offer a wide range of elective courses, sports, or clubs, you can find things to do in your community that will help you figure out what you like to do best. Then, when you get to high school you will be prepared to become involved and even take on a leadership role.

Determining Academic Requirements for High School

Before you can transition from middle school to high school, you must meet specific academic requirements. Your middle school guidance counselor can help you identify the specific requirements in your district. In general, in order to start high school, you must achieve an overall average of 70 on a scale of 100 based on course-level, grade-level standards (essential knowledge and skills) for all subject areas and a grade of 70 or above in language arts, mathematics, science, and social studies. You must also meet the passing standard on the Texas Assessment of Knowledge and Skills (TAKS) tests in reading and mathematics. If you do not meet these requirements, you have an opportunity to attend summer school and try again. If you have an Individual Education Program (IEP), it should identify the requirements for transitioning from middle school to high school.

Choosing a High School

Most school districts offer different options for high school. There may be the local public high school, charter schools, career and technical schools, schools that specialize in a particular area of study, and even private schools that you can attend. Sometimes you must apply for a high school when you are in 7th or 8th grade. The application might include an interview at the high school, and even taking a test.

Choosing a high school involves:

- Talking to a parent, guardian, or counselor about the options that are available.
- Meeting with a guidance counselor to learn about the high school requirements.
- Identifying your own interests and academic needs and matching them to a high school.
- Visiting high schools to learn more and see what they are like.
- Applying to two or three where you would be happy and able to succeed.

Once you know where you will be going to high school, you can learn as much as possible about the school so you will be prepared. For example, you can attend orientation sessions to meet teachers and counselors, go to events such as games or concerts, and explore the school's Web site.

What to Expect in High School

Starting high school can be scary. There are so many unknowns:

- Will you get lost?
- Will you make new friends?
- What courses should you take?
- Should you join a club or team?

It can also be exciting. There are so many new opportunities:

- Classes might be harder, but you learn interesting things.
- More students means more and diverse people to meet.
- More independence gives you the chance to prove you are responsible.

Developing an Academic Plan

Recall that part of the goal-setting process is to make a plan for achieving the goal. A **personal academic plan** is a document that you use to set goals for the things you want to accomplish while you are in school. Some schools call it a *personal education plan*, a *personal career plan*, or even a *personal life plan*. It serves as a map that you can follow and helps you achieve your educational, career, and lifestyle goals.

Goal-setting skills can help you manage the academic planning process from start to finish. It can help you stay focused on the educational objectives that you need to achieve your career and lifetime goals.

What to Put in an Academic Plan

Some things that you might put in your personal academic plan include:

- ✔ *Goals beyond high school*
- ✔ *Assessment of your skills, knowledge, and experience*
- ✔ *Assessment of factors that will contribute to your success*
- ✔ *Assessment of factors that might interfere with your success*
- ✔ *Graduation requirements*
- ✔ *Plan for achieving graduation*
- ✔ *Plan for achieving goals beyond high school*

The Purpose of the Plan

A personal academic plan can help you make sure you graduate from high school, take advantage of opportunities to develop your skills and abilities while you are in school, and help prepare you for life after graduation.

At a very basic level, a personal academic plan will help you keep track of your progress toward achieving your school's graduation requirements. An academic plan can help you assess your achievement at each grade level. You can use the information to prepare for and select appropriate courses for the next year.

At a higher level, a personal academic plan can help you identify areas of opportunity in school and out. It can:

- Help you identify your academic strengths and weaknesses
- Help you find ways to overcome challenges, obstacles, and problems
- Assist you with developing strategies to improve your success in all areas
- Encourage you to participate in **co-curricular** activities such as athletic teams, clubs, and career student organizations
- Inspire you to become involved in your community
- Help you identify the skills you need to meet the requirements for certification in the career of your choice

Academic Planning ■ Chapter 7

How Do I Create an Academic Plan?

Your teachers and school counselor can help you develop an academic plan. Some schools have forms you can use; if not, you can organize your plan in a way that maps your academic achievement and future plans. You may also be able to find an academic plan form online.

Most personal academic plans include goals in four basic areas.

- Career-related planning, skills, and experiences
- Academic preparation and planning
- School and community involvement
- Plans and goals for after high school

When you create your academic plan, make it realistic and attainable. You want to feel confident that you can put the plan into action and achieve your goals. Keep in mind that your plan may change over time. Your goals might change, or your available resources might change.

Setting Long-Term and Short-Term Academic Goals

Start your academic plan by writing a statement that describes your long-term ultimate goal. The goal should be specific and should include career, education, and lifestyle goals. For example, you might write: *I will be an occupational physical therapist in California by the time I am 30 years old.* Or, *I will be a commanding officer in the Navy by the time I am 30 years old.* Or, *I will own and operate an Italian restaurant in Chicago by the time I am 30 years old.*

Why is it important to include co-curricular activities as part of your academic plan?

Meeting High School Graduation Requirements

Before you create an academic plan, make sure you understand your high school's graduation requirements.

Each state defines the minimum course requirements a student must complete in order to graduate. Individual school districts may adopt additional requirements beyond those defined by the state.

Explore your school's graduation requirements. Make a list of the requirements or print an existing one and create a schedule or plan for how you will meet them.

Add short-term goals that define how you will gain the skills, knowledge, and experience that you need to achieve your ultimate goal. Include a timeline for achieving each one. For example, for the first long-term goal listed on the previous page, you might set the following short-term goals:

- Get an A in health class this semester
- Enroll in advanced science classes next semester
- Volunteer at a physical rehabilitation center this summer
- By the end of the school year, research colleges and universities in California that offer physical therapy degrees

Developing Career-Related Skills in School

How will you find a job if you cannot read the job description? How will you prepare a cover letter if you cannot write a paragraph? How will you compare salaries if you cannot do basic math?

Luckily, you have the opportunity in school to develop the basic skills you need to succeed in a career, by taking core courses such as math, language arts, social studies, and science. You also have the opportunity to explore courses that relate to your interests and abilities, such as music, art, carpentry, textiles, nutrition, and other electives.

Building Math Skills

Basic math skills such as addition, subtraction, multiplication, and division are vital for success in careers ranging from nursing to inventory control. Most schools require students to pass a basic math test in order to graduate.

- If you are a nurse's aide, you count a pulse for ten seconds, and then multiply the number by six to find a patient's heart rate.
- If you are installing drywall, you measure the square footage of a room to determine how much building material to buy.

You might benefit from a business math class if you are interested in a career in business. Business math covers topics such as accounting, tax management, reading and creating charts and graphs, and personal finance.

You will need advanced math skills for a career in technology, business, engineering, or science. Advanced math includes subjects such as advanced algebra, calculus, and trigonometry.

NUMBERS GAME

An **average** is the sum of two or more quantities divided by the number of quantities. A **grade point average** (GPA) is the average of all grades that you receive. Colleges and universities use GPA as one measure of an applicant's qualifications for admission. You can use your grade point average to assess your overall academic achievement.

To calculate your GPA, you must convert letter grades to their numeric equivalent, and then find the average.

A+ = 4.0   A = 4.0   A– = 3.7
B+ = 3.3   B = 3.0   B– = 2.7
C+ = 2.3   C = 2.0   C– = 1.7
D+ = 1.3   D = 1.0   D– = 0.7
F = 0.0

For example, to find your GPA if you received an A–, a B+, a B, and a C+:

1. Convert the grades:
   A– = 3.7
   B+ = 3.3
   B   = 3.0
   C+ = 2.3

2. Add the values of the grades:
   3.7 + 3.3 + 3.0 + 2.3 = 12.3

3. Divide the sum of all grades by the number of grades:
   12.3 ÷ 4 = 3.075

What would your GPA be if you received a B+, B–, C+, and a C– for the first semester, and then a B+, B, B–, and a C for the second semester?

Developing Communication Skills

Almost every employer wants workers who can read, write, and speak effectively. Communications skills are the tools you need to share information with others.

- You might need to write an order for supplies.
- You might need to explain the features of a product to a customer.

In school, language arts classes help you develop communications skills. You learn to organize written information, understand the things you read, and speak out loud in front of an audience.

Clubs and organizations also help you build communications skills. You can join the debate team to learn how to argue politely and effectively. You can write for the school newspaper or Web site. You can perform with the drama club.

Choosing Electives

Most of the classes you take in school are required. But, as you move into higher grades, you will have more opportunity to take **electives**—classes you choose because you are interested in the subject. When you take an elective, you can discover how your interests might lead to a career. Electives often align with the career clusters. For example, you might take a Foods and Nutrition elective and discover that you would be interested in a career as a dietitian. Or, you might take a drafting elective and learn that you have a talent for architecture.

Electives allow you to explore new subjects outside the standard core courses. The number and type of electives vary from school to school. Your school counselor can help you select courses that match your career and academic goals.

Choosing Advanced Courses

If you have ability in a subject, you might choose to take advanced-level courses. Advanced-level courses challenge you to work harder and learn more about a subject. Teachers or counselors may recommend you for advanced courses, or you could ask to enroll.

Some advanced courses have the added benefit of letting you earn college credit while you are still in high school. Advanced Placement (AP) and International Baccalaureate (IB) courses are offered at most high schools. At the end of the course, you have the opportunity to take an exam; if you do well, most colleges give you course credit for it.

In some states, dual credit courses let you earn high school and college credit. Some colleges have *articulation agreements* with local high schools that enable students to earn credit for certain courses. For example, you might receive college credit for taking a Culinary Arts course or an Introduction to Agriculture course.

Access to Electives

Not all schools have the resources to offer a wide range of electives. What if your school doesn't offer electives in areas that interest you?

★ Take a class outside of school.
★ Explore virtual or online learning options.
★ Read books on the subject.
★ Find a work-based opportunity to learn more about the subject.

What electives does your school district offer? Which do you find most interesting?

How can electives help you learn more about your skills and interests?

Learning Beyond High School

Can you think of ways to continue achieving your career goals even if you do not go directly from high school to college?

After you receive your high school diploma—the document that certifies that you successfully completed your high school course of study, you may enroll in college to continue your education or earn a professional degree. A degree is awarded by an institute such as a college or university when you complete the requirements of an academic or vocational program. Many careers require a minimum of an **associate's degree**—two years after high school—and many require a **bachelor's degree**—four years post high school (see the table on page 129). Professions such as doctor, lawyer, and teacher require additional education. But even if you don't go directly to college, you have opportunities to continue your education and develop career-related skills.

- You may enlist in the military.
- You may enroll in a technical or vocational program to earn a professional certificate for a career such as master electrician or certified nurse's aide (CNA).
- You may become an **apprentice**—someone who works with a professional to learn a skill or trade.
- You may find an on-the-job training program.

Military Service

When American students turn 18, they have the option of enlisting in the military. Other countries *require* citizens to serve in the military. Austria, Brazil, Denmark, Finland, Israel, Mexico, Norway, Russia, South Korea, Sweden, Switzerland, and Turkey are all countries that have mandatory military service requirements for citizens when they turn 18 years old. The U.S. Armed Forces are all-volunteer. That means citizens serve voluntarily, not because they have to.

Do you know anyone in the military? Do you know why they volunteered? Use the Internet or the library to learn more about the differences between mandatory—required—and voluntary military service. Which do you think is better? Write an essay explaining your point of view.

Planning for Postsecondary Education

Is a high school diploma enough education to land you the career of your choice? If not, you will want to start thinking about **postsecondary education** or school after high school. For most people, postsecondary education means college, but it can also include military training and apprenticeships.

Planning for postsecondary education involves:

- Selecting a school or program
- Making sure you have the necessary qualifications
- Applying for admission
- Obtaining financial aid

Myth I need to decide on my career before I can choose a college.

Truth Most students start college without knowing what they will do after graduation. Having career goals might help you select a college, but it is not necessary.

Researching Colleges

You can find information about different postsecondary opportunities online, in your school guidance center, and at the library. As you research the programs, ask yourself the following questions:

- Does it offer a degree in my field of interest?
- Where is it? Will I live at home or on campus? Is it in a city or a small town?
- How many students are there? Will I be more comfortable in a small school or a large school?
- How much does it cost? What is the annual **tuition**—cost of education? How much is **room and board**—the cost of meals and a dorm room? What is the cost of books? Are there lab fees or other additional fees?
- Am I qualified for admission? Does my grade point average meet the school standards? Did I score high enough on the standardized college entrance exams? Do I have the right co-curricular activities?

Career Tips

You can prepare for a future career by researching the skills you will need to receive a license or certification. Then, you can select a program that best meets your needs. Look for the information on industry association Web sites, or contact the government agency that awards certification.

What type of college interests you?

Which College Is Right for Me?

There are more than 3,500 colleges in the United States. They range from technical colleges that award certification for a specific job or career, to universities that grant bachelors' and graduate degrees. Use the chart at the bottom of this page to compare postsecondary programs.

Applying to College

A college application form asks for information about yourself and your family. It has space for listing your co-curricular activities and for a personal essay that tells the college about you, your goals, and your abilities. In addition to the application form, you will also need:

- An official **transcript**, which is a record of the courses you took in high school and the grades you earned
- An official score report, which gives the results of the standard college entrance exams
- Recommendations, which are forms that you ask one teacher and one counselor to fill out, describing your qualities as a student and a person
- Financial aid forms, which provide information about your ability to pay

	Technical College/ Vocational School	Community College/ Junior College	Four-Year College	University
Focus of program	Specialized training for particular occupation	Two-year degree in career area or academic course credit that transfers to a four-year college	Four-year degree; general academic courses plus focus on major	Four-year degree plus graduate programs
Length of program	Nine months to two years	Two years	Four years or more	Four years or more

Academic Planning ■ Chapter 7

College-Ready Assessments

Preparing for postsecondary opportunities includes taking one or more entrance or placement exams. Your performance on these exams will impact your options after high school graduation. They can help you set attainable personal academic and career goals that match your interests and abilities. Entrance exams, such as the Scholastic Aptitude Test (SAT) and the American College Test (ACT) are used by college admissions counselors to compare you to other applicants who also took the same exam. Placement exams, such as Texas Success Initiative (TSI®) are used by the college or university where you will enroll to help identify your strengths and weaknesses to determine your readiness for college-level work. The results help place you in the appropriate course of study. The Armed Services Vocational Aptitude Battery (ASVAB), which is used by the military, is both an entrance and placement exam that helps determine if you are fit for military duty. You can find information about how to prepare for these exams, what to expect, and even practice questions online.

- The SAT consists of four timed parts: evidenced-based reading, writing and language, and mathematics. Scores for each part range from 400 to 800, so the best possible score is 800 on each part, or 1600. But, very few students earn a perfect score. There is also an optional essay, which is scored separately. This format was introduced in 2016, so there are no average scores available, yet.

  Find out more online: https://sat.collegeboard.org/home

- The ACT consists of four timed parts: English language, reading, mathematics, and science. Each part is scored from 1 to 36, and then the scores are averaged to get a composite score. The average composite score is 21. There is also an optional writing test, which is scored from 1 to 12. The average score for writing is between 7 and 8.

  Find out more online: http://www.act.org

- The ASVAB is a series of timed tests used by the U.S. Military to determine whether you are qualified to enlist, and which Military Occupational Specialties (MOS) you qualify for. Subjects include general science, arithmetic reasoning, word knowledge, paragraph comprehension, mathematics knowledge, electronics information, auto and shop information, mechanical comprehension, and assembling objects.

  Find out more online: http://official-asvab.com/index.htm

- TSI, for incoming college students in Texas, consists of untimed computer-adaptive assessments in mathematics, reading, and writing. Some students are asked to take additional tests in specific subject areas. The assessments include multiple-choice questions that align with the Texas College and Career Readiness standards. Because the assessments are adaptive, the program adjusts the difficulty of the questions based on your skill level. That means your response to one question affects the difficulty of the next question. Some students are exempt from the TSI based on their performance on other tests or coursework.

  Find out more online: https://accuplacer.collegeboard.org/sites/default/files/accuplacer-tsi-assessment-student-brochure.pdf

Myth I will only be accepted to college if I earn all A grades.

Truth Colleges look at the complete student, not just the grades. Admission is also influenced by factors such as letters of recommendation and participation in co-curricular activities. As for grades, showing improvement and taking advanced courses that challenge you to work hard help make a positive impression.

Paying for College

Postsecondary education is expensive. Tuition alone may range from a few thousand dollars for a year at a community college to more than $50,000.00 at a private university. Add room and board, books, and other fees, and the amount seems overwhelming. While a few people might have enough money saved in a bank account, most must try to find other resources to pay.

- First, it is never too late to start saving, and no amount is too small. You can use a standard savings account or a special college savings account. College savings accounts usually have restrictions on how the funds can be used and may also have tax benefits.
- Most schools offer financial aid to qualified students. **Financial aid** is money to help you pay for education. It may be in the form of **grants**, which you do not have to repay, or **loans**, which you do have to repay along with interest. It may also include a **work-study program**, which is the opportunity to work while you are in school so you can earn money to help pay your costs.
- Schools and other organizations offer scholarships to qualified students. **Scholarships** are grants, so you do not have to repay them. Often, they are merit-based, which means you must meet specific academic or other criteria.
- You can take out a loan from a bank. Most education loans are supported by the government; others are funded by financial institutions such as banks.

These resources may also be available to support career advancement opportunities outside of college. For example, there are programs that provide scholarships, loans, or financial aid for internships that help you obtain work experience.

Identifying Scholarships to Support College and Career Advancement

As mentioned above, scholarships are grants that you do not have to repay. Colleges and universities award scholarships to incoming students who meet specific criteria, such as academic excellence, musical achievement, athletic success, and more. You can learn about available scholarships at the schools you are interested in by contacting the admissions office or looking at the school Web site online. In addition, there are many other resources for scholarships that support college and career advancement. Most private, public, and government organizations offer scholarships to students who meet their criteria. You usually have to fill out an application and write an essay explaining why you deserve consideration. Awards can range from $50.00 to thousands of dollars so it is worth seeking out these opportunities. Your school college counselor can help you find the ones that match your interests and abilities. The U.S. Department of Education is also a good resource. Its Office of Federal Student Aid has information online at https://studentaid.ed.gov/sa/types/grants-scholarships.

Things to Do on a College Visit

You can take virtual college tours online, but the best way to get a sense of whether you would be happy at a college is to visit. On a college visit, you can:

- ✔ Tour the campus
- ✔ Talk to students
- ✔ Stay overnight in a dorm
- ✔ Eat meals in a dining hall
- ✔ Sit in on a class
- ✔ Talk to a professor
- ✔ Attend an information session

Career Tips

Talk to your parents as soon as possible about the cost of college and how you and your family will pay for it. The conversation will make it easier for you to develop your academic plan.

Academic Planning ■ Chapter 7

Career Profile: College Admissions Officer

Education & Training Career Cluster

Job Summary

A college admissions officer is an education administrator responsible for making decisions about whether or not to admit an applicant. He or she spends a lot of time reading applications and discussing them with other admissions officers.

A college admissions officer represents the college on and off campus by leading information sessions or presentations for potential students and their families. Other duties may involve traveling to high schools to promote the college and conducting one-on-one interviews with applicants.

Admissions officers usually have a bachelor's degree and may obtain advanced degrees in college student affairs, counseling, or higher education administration. They benefit from courses in computers and statistics. Experience with computer systems is helpful because much of the application process is completed electronically.

Use the Internet, a library, or career center to research the responsibilities, education and training requirements, and salary range of a college admissions officer, or another education administrator position that interests you. Write a sample job description for the position.

What personal characteristics do you think would benefit a college admissions officer?

Identifying Career-Related Opportunities for Students

Employers like to hire people who have experience. But most teens don't have much work experience. Some states have laws restricting the number of hours a teen can work, and many companies won't even hire people under 16.

Gaining job experience is an important part of developing an academic plan. Experience serves two main purposes.

- It gives you the opportunity to learn about a career first-hand. While you are working, you can find out if you enjoy the environment, feel comfortable with the responsibilities, and see a future for yourself in the industry.

- It gives you the opportunity to build a work history that you can include on job and school applications. A **work history** is a list of jobs you have held from the past through the present, showing your experience as an employee.

You can explore career-related opportunities with the help of your parents, teachers, or school counselor. They can show you how to identify opportunities that relate to your skills, interests, and abilities, and how to apply for positions.

What Is an Internship?

An **internship** is a temporary job, usually for students. Internships may or may not pay a salary, but they provide other useful benefits. You have the opportunity to work in a field that interests you and to build skills. You meet people who might act as **references**—people who will provide a recommendation for you when you apply for jobs or school in the future.

- An internship might be part-time after school. For example, you might intern at a law office two afternoons a week.
- Many internships are in the summer. You might intern full-time at a television station during summer vacation.
- Some internships are for a specific project or time period. You might intern at a company to develop a Web site or to digitize records. When the project is complete, the internship ends.

Usually you apply for an internship using a process similar to applying for a job. Your school counselor can help you find internship opportunities.

Volunteer!

Have you ever considered volunteering? Many volunteers are looking for ways to help others and contribute to their communities. But volunteering is also an opportunity to gain career-related experience, meet people who work in a field that interests you, and do something you can put on job and college applications.

Volunteers are unpaid workers. Many volunteer positions are in service organizations, such as homeless shelters, health clinics, animal shelters, and food pantries. These organizations are usually short on money to pay workers, so they depend on volunteers to perform a full range of responsibilities. They need people to do everything from answering telephones to maintaining mailing lists to communicating directly with clients.

- If you are interested in the arts, you might volunteer as a **docent**—guide—at a museum.
- For health career experience, you might volunteer as a receptionist in a hospital, clinic, or nursing home.
- If you are considering a career in communications, you might volunteer as a production assistant at a public access cable station, public television station, or public radio station.
- If you are thinking about a career in recreation, many nature centers and state parks hire volunteers to run gift shops, clean, maintain trails, lead tours, and answer visitors' questions.

The opportunities for volunteering are nearly endless. Select an organization that interests you and ask what positions are available. The organization may even offer training.

Have you ever volunteered? What job-related experience did you gain?

Other Ways to Gain Experience

In addition to part-time jobs, internships, and volunteering, there are other ways you can gain valuable career-related experience.

- Mentoring programs pair experienced workers with students or less-experienced employees. The mentor acts as an advisor, providing advice, job contacts, and tutoring.
- Apprenticeships allow a worker to learn a trade by working with someone who has already mastered that trade. An apprenticeship combines on-the-job training and classroom education. Some trades require a worker to complete an apprenticeship in order to qualify for certification.
- Job shadowing allows you to follow an experienced worker through his or her work day to see the specific responsibilities and tasks required on the job.
- School clubs and organizations offer opportunities to gain management and leadership skills.

- Entrepreneurial thinking can help you create your own job and gain valuable experience. Are you patient, responsible, and good with kids? You can babysit. Do you love animals and exercise? You can walk dogs. Do you know how to play the guitar? You can teach others. If you have a skill or ability that other people will pay for, you can gain work experience by being an entrepreneur.

Joining a Student Organization

Career and Technical Student Organizations (CTSOs) are groups of students who are enrolled in a career and technical education program that engages in career and technical activities as part of the curriculum. There are different CTSOs for different career and technical programs; students join the organization related to the CTE program in which they are enrolled.

CTSOs provide members with a range of individual and group programs and activities. They promote career education and training and provide opportunities for leadership, teamwork, competition, and citizenship. CTSOs offer a unique program of career and leadership development, motivation, and recognition for students in grades 6 through 12. Members benefit from:

- Belonging to a positive and supportive group of their peers
- Examining firsthand the relationship between academics and the world of work
- Identifying career opportunities
- Building confidence and knowledge through competition
- Exercising leadership and teamwork skills
- Practicing personal skills for success
- Developing employability skills
- Recognizing and developing interests and abilities that align with career goals

Common Career and Technical Student Organizations

There are CTSOs to support every career cluster and pathway. The national CTSOs supported by educators in most states include the following:

- Family, Career and Community Leaders of America (FCCLA) for students in Family and Consumer Sciences programs (fcclainc.org)
- Future Business Leaders of America-Phi Beta Lambda (FBLA-PBL) for students in business, government, and public service programs (fbla-pbl.org)
- Business Professionals of America (BPA) for students in business and marketing programs (bpa.org)
- DECA for students in marketing programs (deca.org)
- HOSA (Health Occupations Students of America) for students in health services programs (hosa.org)

Academic Planning ■ Chapter 7

- National FFA Organization for students in agriculture programs (ffa.org)
- National Young Farmer Educational Association (NYFAE) for students in agriculture programs (nyfae.org)
- SkillsUSA for students in technical, skills, and service programs (skillsusa.org)
- Technology Student Association (TSA) for students in technology, innovation, design and engineering programs (tsaweb.org)

For more information on CTSOs at the national level, contact:

U.S. Department of Education
Office of Vocational and Adult Education
Washington, DC 20202
(202) 205-5440
National Website Address: ed.gov/about/offices/list/OVAE/index.html

21st Century Learning

CAREER COUNSEL

Julio's favorite subject in school is biology. He loves watching medical shows on television. He thinks he is quick to react and solve problems. He thinks he might be interested in a career in health care—specifically as an emergency medical technician. He knows he will need money for an education, so he starts looking for his first job.

His mother works as a dispatcher for the local emergency service center. She says they need temporary office help to catch up on filing. If Julio is interested, she can set up an interview with the office manager.

Julio thinks office work is boring. He would like to find a part-time job that relates directly to emergency medical care. He visits the hospital to see what part-time openings they have, but they tell him they don't hire anyone under 17. He contacts a medical clinic, but they say he needs training and experience.

What problems are facing Julio? Use your 21st Century Skills such as decision making, problem solving, goal setting, and critical thinking to help Julio decide what to do next. Write an ending to the story. Read it to the class, or form a small group and present it as a skit.

Case Study

Sophia moved to a new city three months ago and started at a new school. She thinks the work is much harder than at her old school, and she is having trouble keeping up. Sophia's teachers think she is not trying very hard. One told her that if she didn't improve, she would have to repeat the grade.

She misses her old school and her old friends. She has made only one good friend in her new neighborhood—Selena—but Selena goes to a different school. Sophia will be 16 in three months. She is thinking about getting a job and quitting school.

- Do you think Sophia is right about getting a job and quitting school?
- What problems does she have to deal with?
- What can she do to solve the problems?

Answer It!

1. List at least three benefits of staying in school.
2. List the three major learning styles.
3. List three types of visual aids that might help visual learners.
4. What is one study method that might help auditory learners remember information?
5. What type of learner tends to gesture when speaking?
6. What is a personal academic plan and why is it important to have one?
7. What are four ways to earn college credit while in high school?
8. How can you build communications skills in school?
9. List four items you need to apply to college.
10. What are the two main purposes of gaining job experience?

Web Extra

There are many Web sites that offer information about how to make a successful transition from middle school to high school. One useful site is www.ownyourownfuture.com. With your teacher's permission, visit the Own Your Own Future Web site and explore the resources for your grade level. Look for information about the academic requirements for high school, as well as how to prepare socially and emotionally. You may also want to look for academic requirements specific to your school district on the district's Web site. Then, use the information you have found to make a calendar or timeline listing of what you can do to make a smooth and successful transition from middle school to high school.

Academic Planning ■ Chapter 7

Write Now

Many organizations and programs that offer scholarships and financial aid require that you fill out an application and write a personal essay. Take the time to research financial aid, scholarships, and other sources of income for college and career advancement. Then, write a personal essay explaining how you would use the funds to support your academic and career goals.

COLLEGE-READY PRACTICES

How well you perform on standardized college entrance and placement exams can impact where you go to school and what you study. College-ready individuals take the time to compare the available exams so they know which ones they must take, which ones they should take, and when to take them.

Conduct research to learn about the different entrance and placement exams and which ones are required by the college, university, or branch of the military that you are considering. Create a chart to compare the exams. Then, write a paragraph explaining which exams you will take, and why, including how the results of the exams might impact your personal academic and career goals. Read your paragraph to a partner or to the class.

Career-Ready Practices

Maya is interested in a career in recreation management. She was offered a volunteer position at a professional golf tournament at a golf course in her area. She would work as an assistant to the Volunteer Coordinator. The position would be part-time for three months, and then full-time for the week of the tournament. It does not pay any wages. In fact, Maya would have to pay $50.00 to buy a uniform! She has asked you to help her decide whether she should accept the position or not.

Working in teams of three or four, use the decision-making process to help Maya decide what to do. Analyze the situation, look for all possible options, and consider the consequences of each option. Make sure you consider the benefits of acquiring career-related skills. Career-ready individuals use volunteer opportunities to gain knowledge and experience. Likewise, they can strengthen their academic skills by applying them to real-world, workplace situations. Write a report or prepare a presentation that explains your advice to Maya. Be specific about the reasons for your decision.

Career Portfolio

Your career portfolio should include a personal academic plan. Working with your teacher, a counselor, or a family member, develop a personal academic plan that starts now and takes you through to high school graduation. If your school has a form for the plan, use it. Otherwise, use the Internet to locate a form that you might be able to use or adapt for your needs. Be sure to include your academic and career goals and an assessment of your skills, interests, and abilities. You can include your school report card or transcript, as well as a list of your co-curricular and community activities. Add the plan to your career portfolio.

8 Communicating with Others

Skills in This Chapter...

- **Using Written Communication**
- **Using Verbal Communication**
- **Developing Listening Skills**
- **Identifying Nonverbal Communication**
- **Recognizing Obstacles to Communication**

GETTING STARTED

Human beings are born with a desire to communicate. Over millions of years, we have developed many ways to communicate, from gestures and body language to pictures to verbal and written language. Communication is not only important in your daily life. It is also the key to career success. It prevents misunderstandings. It gives you a way to share ideas. It even makes it easier for you to appreciate and respect other's opinions.

At its most basic, communication is an exchange between a sender and a receiver. The sender transmits the message with a specific intent. The receiver interprets the message and responds. Being able to identify types of communication will help you master the art of communicating.

➤ Using a piece of paper, create a mind-map of the different forms of communication. At the center of your paper write the word "Communication" in a bubble. Branching from that word, make bubbles for as many different types of communication you can think of. Share your mind-map with the class and discuss how many types of communication the class identified.

NUMBERS GAME

Often, authors who write books are paid a royalty, which is an amount based on how many books are sold. That means that a percentage of the amount you pay for a book goes to the author, while the rest goes to the publisher.

The word *percent* means by the hundred. Each percent represents one part of 100; 100 percent equals the whole or all of something—no matter how much is in the whole.

To find the percentage of a number:

1. Change the percentage to a decimal by replacing the percent sign with a decimal point and moving the decimal point two spaces to the left.
2. Multiply the decimal by the number.

For example, to find 10% of 150:
10% = .10
150 × .10 = 15

If you are an author earning a 7% royalty, and a book costs $15.50, how much of that money will you earn, and how much will go to the publisher?

Using Written Communication

Many historians believe that the printing press, invented and built by Johannes Gutenberg in 1440, was the single most important invention of all time. Most agree that its invention marks the transition from the Middle Ages to the **Renaissance**, a period of artistic and intellectual revival in Europe. Why is that? The printing press made it easier and cheaper to create printed communications such as books and newspapers, and this in turn allowed for information to spread more easily. When people began to read books for themselves, they began to think for themselves.

Before the printing press, creating written communications was a time-consuming and costly business. Books were usually written by hand. Paper was scarce and expensive. Written communications were therefore rare. These days, you are surrounded by written communications, from the textbook in front of you to the notes in your notebook to the information written on your class whiteboard. Even text messages, tweets, and e-mail are types of written communications. Everyone now has the ability to use written communications to transfer ideas. This ability will be **integral**—necessary or essential—to almost any career you might enter.

Reading

Why do you think it is important to learn how to read? If you did not know how to read, you would have to learn all of your ideas from people speaking to you. Reading allows you to learn on your own. You can learn ideas that were developed and written down in the past, even centuries ago.

Reading is vital for success in high school and college. Teachers and professors expect you to keep up with the assigned reading so you are prepared for class. Reading will be essential for your career. Whether you want to be a doctor, car mechanic, astrophysicist, or firefighter, you will need to read every day you are on the job.

There are different ways to read, depending on what you are reading. You can read *passively* or *critically*.

- **Passive reading** is the kind of reading you do for entertainment, such as when you read a magazine or a comic book. In this kind of reading, you are just taking in information or following the plot.
- When you read critically, you take time to really think about what is written and why. **Critical reading** happens when you are actively engaging the subject as you read.

Communicating with Others — Chapter 8

The difference between passive and critical reading is like the difference between listening to your friend talk and having a debate with her. When you read critically, you are having a debate with the text. Here are the steps to critical reading:

- Determine the best reading strategy for the text. This means that you need to figure out whether to skim the text, read closely for detail, or read for meaning or critical analysis.
- As you read, stop and think about what is being said and try to put it in your own words or rephrase it from your own point of view.
- Take notes while you're reading to make sure you are understanding and remembering the main points.

Money Madne$$

You find two jobs that you think you would be qualified for. One pays an annual salary of $57,000.00, with no bonus. The other pays $45,000.00, but if you meet all goals set by your employer, you can earn a bonus equal to 15% of your annual salary. Which job would you rather have? What factors other than pay might influence your decision?

Writing

Like peanut butter and jelly, reading and writing go together. In fact, you cannot do critical reading without also writing. But not all writing is the kind you do in English or Creative Writing class. Jotting down notes and texting a message to a friend also count as writing.

Writing is something you get better at with practice. You can think of words as being like tools. Writers who are just learning to write use simple words to communicate their ideas, just as beginning builders start with basic tools to put together small projects. The more words you learn and the more you learn how to put them together, the bigger the projects you can build.

Just like learning any skill, learning to write is an ongoing process. You should always expect to get better, and accept **constructive criticism**—suggestions from others on how to improve your writing—and consider how it can make you a better writer.

Like reading, writing well will be essential in high school, college, and almost any career you choose. Most colleges require you to write a personal essay as part of your application for admission. And, even before you begin work at a job, you will usually have to write three important documents:

What If You Couldn't Read?

In this country, most—but not all—people learn to read as children. What if you were an adult and you couldn't read?

★ *You could not get a driving license, because you would not be able to read the written part of the license test.*

★ *You could pick out items at the grocery store, but you would not be able to read the nutritional information and you might have trouble comparing pricing.*

★ *You would have to depend on other people to tell you what was written in instructional manuals and brochures.*

★ *The Internet, the largest source of information on the planet, would be relatively useless to you, since it is mostly text-based.*

Can you think of any more reasons why knowing how to read is important in modern life?

What are the benefits and drawbacks of having a career as a writer?

How Do You React to Constructive Criticism?

Even the best writers in the world have editors. And this means that writers accept and use constructive criticism in developing their writing. How do you respond to constructive criticism?

👎 You ignore it, because you think your way is best.

👎 You get angry and assume that the constructive criticism is meant to hurt you personally.

👎 You accept it all without thinking hard about whether it is right or not.

👍 You take each suggestion as a part of the writing process, not a personal attack.

👍 You consider each criticism one by one and put yourself in the shoes of your editor.

👍 You accept the criticisms that make the most sense to you, revise your writing, and resubmit to your editor, with explanations of which criticisms you agreed with, which you did not, and why.

What do you think are other ways that constructive criticism can be taken? List three reasons why it is important for any piece of writing.

- Applying to most jobs requires submitting a resume. You write a resume to explain your skills, tell where you went to school, share information about your other talents, and give the names of references—people who can verify your experience.
- When you submit a resume, you usually also submit a cover letter. A **cover letter** is a short one-page summary of who you are, why you are applying for the job, and why you are the best person for it.
- After you have had an interview, you should submit a follow-up letter to emphasize your continued interest in the job and review why you would be the best candidate for the job.

While you are employed, you may have regular writing tasks such as preparing letters to clients, creating written reports for customers, or writing memos to your co-workers. You may write content for a Web site or for a marketing e-mail. You might also have to write a letter of resignation, if you have decided to change from one job or career to another. This is a letter that expresses your reasons for leaving and your appreciation for your former employer.

Career Trend

In an increasingly digital age, distributing written information has never been easier. This means that there are more opportunities for careers in writing. Blogging and self-publishing are two great examples of this.

Electronic Communication

Electronic communication includes all the ways you use electronic devices to communicate, including e-mail, text messages, videos, Web sites, and videoconferencing. Increasingly, colleges and employers expect you to use electronic communication. You may be asked to fill out an application form online, interview on the phone or by videoconference, and correspond using e-mail. You will make a positive impression if you can demonstrate your ability to use communication technology effectively and professionally.

Some ways you can use electronic communication effectively include:

- Be direct and to the point.
- If there is a subject line, make it clear and relevant to the message.
- Use proper spelling and grammar. Only use abbreviations and shortcuts if they are well known and you are certain the recipient will understand.
- Be polite. Remember that unless the recipient can see your face or hear your voice, it is easy to misinterpret the tone of the message. The recipient may not realize you are trying to be funny or sarcastic.
- Consider the time of day before you send a message. No one wants to be awakened by the ding of an incoming text that could have waited until morning.
- Reply promptly, but take the time to consider your response without being emotional.

Using Verbal Communication

You were born with the ability to learn how to speak and understand others. This ability to use and understand language is something that all human beings have, and it is something that makes us unique among all other animals on the planet.

Of course, just because you know how to talk does not mean you know how to speak *well*. Even if you were the smartest person in the world, it would be very hard for you to get the career you want if others did not understand you.

Mastering the art of verbal communication is a very important part of getting the career you want, because you will have to use this skill to convince an employer to hire you. You will also have to use verbal communication every day on your job, no matter what it is.

Using the Communication Process

You can communicate effectively by using a six-step process:

1. *Be clear.* The receiver is more likely to get your message if you deliver it in a way he or she can understand. Speak slowly. Consider who you are talking to. You probably use different language when you talk to a teacher than when you talk to a friend. You might even use a completely different type of communication. For example, you probably talk to your teacher face to face, but you might text your friend.

2. *Be personal.* Use the other person's name so there's no doubt who you are talking to. Use an "I" statement—a statement that starts with the word "I"—to frame the statement in terms of you and your goals. An "I" statement indicates that you are taking responsibility for your thoughts and feelings. It helps the receiver understand your point of view and respond to you.

3. *Be positive.* Phrase your message in positive terms. Say what you want, not what you don't want. For example, say, "I want to leave at 3:30," instead of saying, "I don't want to leave at 4:00."

4. *Get to the point.* Follow the "I" statement with an explanation of the message you are sending. Explain how or why you feel a certain way, or how or why you think a certain thing. For example, say, "I want to leave at 3:30. I think that will give us enough time to get there without having to rush."

5. *Listen to the response.* Pay attention and use active listening techniques to make sure you hear the response.
6. *Think before you respond.* Make sure you understand the message. Repeat it, if necessary, and ask questions for clarification. Use critical thinking to make sure you are not letting emotions and preconceived ideas get in the way.

Different Types of Verbal Communication

As you already know, the ways you speak to your friends, your parents, your teachers, or your siblings differ greatly. For instance, you might get in trouble with your teachers if you spoke with them as if they were your friends. On the other hand, speaking to your friends as if they were your teachers wouldn't be any fun.

Just as you use different types of verbal communication when speaking to your friends and your parents, you use different types of verbal communication when speaking to people with whom you relate on the job, such as your boss, your co-workers, your customers, and your **subordinates**—the people you supervise.

The way you speak can determine the outcome of a conversation. Which of these examples of verbal communication is more likely to have a positive conclusion? Why?

- Your boss has a higher rank in the company than you do and has control over your job. When communicating with a boss, you usually use a formal style of verbal communication. You show **deference**—courteous regard or respect—by using her title and refraining from jokes or casual remarks you might use with a friend. You must also recognize that your boss has the final say on most matters, and so you must know when to agree with her decisions, even if you disagree with them personally.

- Your co-workers, or **colleagues**, have the same rank in the company that you do. This does not mean that you should joke around with them as you do with your friends. It also does not mean you should use the same style of verbal communication you would use with your boss. You can speak to them more casually.

- Your customers are the source of your business. When communicating with customers, remember the old saying, "the customer is always right." This means that in order to be successful, you must respect your customers and speak with them positively.

- If you supervise others in the company, you communicate with them using the respect that is due to any worker. You can also use an assertive or positive style of verbal communication, so that they know exactly what is expected of them.

Not only is what you say important, but so is how you say it. In addition to the different ways you can verbally communicate with others in the workplace, you can use different tones to make your meaning clear. These include:

- Assertive speech, which includes action verbs such as "should," "does," and "will."

- Aggressive speech, which you should avoid in the workplace, since it can lead to arguments or hurt feelings. An example of aggressive speech would be, "You better help me now!"

- Passive speech, which is usually best to use when you also want to show respect or deference. Examples of passive speech include words such as "can," "might," and "could."

What do you think it takes to be an effective communicator?

21st Century Learning

CAREER COUNSEL

Laura has an after-school job at a local coffee shop. The job allows her to earn some extra money and interact with different people. Her co-workers are all very nice and there are even a few regular customers whom she sees every day. She really enjoys having the extra spending money and not having to ask her parents for an allowance.

But Laura is not happy. She feels her boss does not respect her. His tone of voice is usually harsh and aggressive. He makes sarcastic remarks about her work. And he has asked her to work more hours, even though Laura is in school and has many other responsibilities besides work. Even though Laura likes everything about her job except her boss, she is thinking about quitting.

What problems are facing Laura? What can she do to fix them? Use your 21st Century Skills such as identifying different types of communication, problem solving, and critical thinking to help Laura. Write an ending to the story. Read it to the class, or form a small group and present it as a skit.

How Verbal Communication Helps You Think

Often, the best way to think something through is to talk about it with someone else. Why is that? The answer is that one mind is only so powerful, but two minds can think of twice as much. In other words, when you take a difficult problem and talk about it with someone else, you are doubling the brainpower devoted to solving it. And sometimes, having to explain something to someone else helps you understand it yourself.

Talking or debating about something is a great way to learn. But this strategy can be an even better way to problem solve in the workplace.

There is a difference between speaking out loud and thinking to ourselves. How might speaking without thinking cause problems or conflict?

- Talking critically about a difficult problem is not so different from thinking critically about a difficult problem. One benefit of talking over a problem with someone is that you get a different perspective from the other person.
- Using verbal communication to problem-solve allows you to take sides and debate the problem. You may even take the position of being **the devil's advocate**. This is a label used to describe taking a position you do not necessarily agree with, but pretending you do for the sake of argument. Playing, or having someone play, the devil's advocate can be a powerful way to work through a difficult problem.
- Using verbal communication to help you think and problem-solve only works if you strike a balance between caring too much or too little about the topic under debate. You must be actively engaged in order to move the debate along. But you must not get too passionate about your side. After all, you are both striving towards the same goal, which is to solve your problem!

Myth All debates must be personal and must involve two passionate advocates who oppose each other.

Truth Debating can be a productive way to solve a difficult problem, as long as both sides keep in mind that they are actually on the same team!

Delivering an Oral Presentation

The ability to stand in front of an audience and present information is essential for succeeding in college and in most careers. In college, you may have to present a paper or defend a thesis. At work, you may have to market a product or idea, or train co-workers.

Most people are nervous when they speak in public. The best way to deliver an effective presentation is to take the time to prepare and practice. In addition to knowing what you are going to say, it helps to be prepared for the unexpected, such as questions from the audience. If you are using tools such as a slide projector or monitor, you must also be able to start, control, and stop the device.

Steps to Becoming a Great Speaker

Great speakers are rare and easy to recognize. Many of our U.S. presidents were great speakers. Why is that? Because being able to communicate is one of the most important skills needed in a career in politics. The best among the political communicators often become president.

Communicating with Others ■ Chapter 8

You don't have to be the president to be an expert at using verbal communication. Here are a few steps on the way to becoming a great speaker.

- It is said that great speakers "know their audience." But do you know what this actually means? It means you must know what your audience wants to hear. It also means that you must be considerate of the person (or persons) you are speaking to. What the president says to the American people—and how he says it—is usually different in tone and content from what he might say to a foreign leader.

- You must know what you are talking about. This seems obvious, but it actually takes a lot of work. Being a great speaker means that before you speak, you take the time to really consider what you say, what it means, and how it will be received. This means thinking before you speak.

- Never stop learning! The greatest speakers would all admit that they are always learning. You might sometimes think that you have learned all you need to know on a certain topic. But if you keep an open mind, you might find that you learn something new every day. The ability to keep an open mind and continually learn new things can make you not only a better speaker, but a better person as well.

Why is being a great speaker an important skill for a career in politics?

Communication in and out of the Workplace

Being an effective verbal communicator is essential to being good at your job. Here are some characteristics of a good communicator.

✔ You avoid passive communication when speaking.

✔ You do not automatically treat your co-workers as your friends. You speak with them with the degree of respect and formality that is common in a workplace.

✔ You speak to your superiors with deference.

✔ Your tone of voice is appropriate for your audience. This means that you do not use aggressive speech when assertive speech is best.

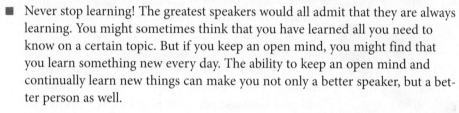

Career Profile: Politician

Government & Public Administration Career Cluster

Job Summary

Whether they are local, regional, statewide, or national, politicians are representatives of some group of people. What can you expect of a career in politics? Since democracy involves citizens voting for their representatives, you can expect to talk with a lot of people. This means you must communicate well, not only in speaking but in writing.

A career in politics can be very rewarding. Politicians speak and act on behalf of all those who they represent. Politicians are usually well respected. Keep in mind, however, that the job is always hard work!

Use the Internet, a library, or career center to research the responsibilities, education and training requirements, and salary range for a career in politics. Then, imagine you are running for a political position. Use a word processor to write a campaign speech announcing that you are running and why.

Developing Listening Skills

Consider the question, "If a tree falls in the woods and no one is there to hear it, does it make a sound?" Do you think the answer is yes or no? In fact, the answer depends on how you define "sound." But the question raises an important point. In verbal communication, half of the process is speaking. The other half is hearing, or listening. It may surprise you to learn that there is a difference between hearing and listening.

The Difference Between Hearing and Listening

How can listening be different from hearing? Sound is all around us, all the time, so we often cannot help hearing sounds. It may even seem unavoidable to hear certain people speak. But there is an important difference between merely hearing and actively listening. That difference is up to you. It depends on how much you are thinking about what you are hearing. If you are really thinking, considering, and engaged with the words you hear, then you are actively listening. If, on the other hand, the words you hear go into your head and you soon forget them, then you are merely hearing.

A good way to consider the difference is to think of hearing as *passive* and listening as *active*. In other words, "hearing" is something that happens to you. You cannot avoid hearing, unless you plug or cover your ears. On the other hand, listening is something you actually do. It involves thinking about what you are hearing.

Active listening is an important part of effective communication. When you are an active listener, you pay attention to the speaker and make sure you hear and understand the message. Active listening is also a sign of respect. It shows you are willing to communicate and that you care about the speaker and the message. When you listen actively, the other person is more likely to listen when you speak, too.

Which group of students is listening, and which is hearing?

Communicating with Others ■ Chapter 8

Listening on the Job

How is listening relevant to your career? No matter if you are a boss, a subordinate, or a co-worker, you will need to listen at some point. The better you are at it, the more successful you will be. Here are a few reasons why:

- Good listeners often pick up **subtle**—unmentioned or implied—cues in conversation. This means that a person's tone of voice or his word choice tell you more about what he thinks than what he says. In your career, you might have a great advantage if you can pick up these clues.

- Good listeners are often more trusted than bad listeners. If you listen carefully to what people tell you, you deliver the message that they are important to you. They are more likely to trust you and like you.

- Being a good listener will make you a better worker. When you listen carefully to instructions, for example, you are more likely to be able to carry out your tasks without having to be reminded what to do. You are able to act to resolve problems that people tell you about so that your workplace is a more positive environment.

Can you think of a career in which your primary responsibility is listening?

Active Listening Skills

Practice your active listening skills by working in groups of three.

- One person will be the active listener
- One person will discuss a topic of interest (such as a planned vacation, a new job, or an upcoming college visit)
- The third person will observe to see if the active listener is using active listening skills

Take turns playing each role. Develop a list of the active skills your group members applied and which ones they didn't.

Share your list with the rest of the class. As a group, discuss how active listening can make you a more powerful communicator.

Skills for Active Listening

Use these skills to be an active listener:

✔ Show interest using eye contact and positive nonverbal messages.

✔ Let the other person finish speaking before you respond.

✔ Ignore distractions such as cell phones and other people.

✔ Set your preconceived opinions and emotions aside.

✔ Repeat the message that you hear out loud to make sure you received it correctly.

Identifying Nonverbal Communication

Besides reading, writing, and speaking, what other types of communication are there? Two other important ways of communicating are sign language and body language.

Did you know that sign language is just as powerful as spoken language? You can communicate everything you would with speech using sign language, without saying a word. Using sign language requires that all parties in the conversation know how to understand the various signs. This problem is often solved using sign language interpreters who sign as someone is speaking to deliver spoken messages to those who cannot hear them.

Body language is a different type of nonverbal communication that everyone uses every day, sometimes without knowing it. Did you ever notice that if you didn't do your homework somehow your teacher knows it and calls on you anyway? Your body language may be giving you away. Body language can be either positive or negative. It can show you are confident and prepared, or that you are unsure or afraid.

Finally, one other way to communicate without speaking or writing is to use **visual aids**—pictures, graphs, charts, or anything without words that conveys your ideas. Visual aids can often deliver a clearer or more easily understood message than written or verbal communication.

Cross-Cultural Communications

Nonverbal communications can be tricky when you are communicating with people from different cultures. Personal space, touching, and gestures frequently mean different things depending on where in the world you live.

For instance, in the United States we beckon one another by crooking our index finger. In Asian cultures, that gesture is used to call dogs and other animals and is considered insulting if used to a person.

Do you know of any other types of communication that mean different things to different people? Can you find examples on the Internet? Make a poster that shows an example and display it in your classroom.

Positive Body Language

Have you ever come away from a conversation feeling good about how you communicated with the other person? Chances are that most of your good feeling was caused by what she said. But equally important is how she said it. Along with verbal cues such as a pleasant tone of voice, you may have picked up other important cues from that person's body language. Did he or she smile during the conversation or look you in the eye without staring intently? These are both aspects of positive body language. Here are some other ways you can communicate using positive body language.

- *Watch your posture.* Stand up and sit straight without bending backward, and keep your shoulders square with the person to whom you are speaking. If sitting, you may lean slightly forward to show that you are interested in what the other person has to say.

- *Open arms communicate an open mind.* Standing or sitting with your arms at your sides or in front of you indicates that you are open and receptive to your conversation partner's ideas.

- *Active hand gestures indicate active debate.* This means that you move your hands to show and demonstrate ideas. Why is this important? Because it shows that you are actively engaged in the conversation, not just passively hearing and responding.

- *Maintain adequate eye contact.* This does not mean you must lock eyes with the person you are speaking with at all times. It does mean that you must at least meet his eyes when he looks at you.

In your career, you will probably encounter situations where positive body language will serve you well. Calming an upset customer and reassuring a nervous co-worker are two situations when positive body language can help you succeed.

How can you use body language to make a positive impression when you are communicating?

Negative Body Language

Just as positive body language can affect your interactions with others, so too can negative body language. Negative body language can portray you as hostile, combative, and unwilling. Here are some examples of negative body language.

- *Avoid unnecessary movements.* Don't repeatedly tap your fingers or a foot, scratch, put your head down on your desk, or look out the window. These actions give the impression you are impatient or uninterested.

- *Avoid closed or folded arms.* When you close your arms across your chest, you signal that your mind is closed to new ideas.

- *Don't stare intently.* While you should make eye contact with the person to whom you are speaking, you should not stare a long time. A prolonged stare may seem threatening to another person.
- *Don't signal that you've stopped paying attention.* Failing to maintain some eye contact and letting your eyes wander are telltale signs that you are no longer paying attention.

Being aware of and avoiding negative body language will make you a far more effective communicator.

Career Tips

Practice being a positive body language communicator. Stand in front of a mirror and act as if you are speaking with another person. Look at how you stand, where your arms fall, and what your body language says. Mastering of your body language can be your secret weapon in getting the career you want!

Using Visual Aids

Have you ever heard the expression, "A picture is worth a thousand words"? Do you know what that means? Not only pictures, but any **graphical representation**—or way of conveying information without words—can be a great way of communicating ideas without saying them directly. And there are times when this is the best way to communicate. For instance, visual aids can sometimes make us see ideas that would be difficult to explain otherwise. Graphs and charts are excellent examples of this.

Graphical representations can also make us feel emotions in ways words sometimes cannot. For instance, reading a description of the first moon landing might make you think the achievement was interesting and worthwhile, but actually seeing a picture of it can make you well up with national pride.

Can you think of a time when pictures are more effective than words?

In your career, you might sometimes find that using visual aids will be the most effective way to communicate. This includes putting a picture or graph in your presentation, making a chart instead of describing a trend in your data set, or drawing a picture to explain what you are talking about. Visual aids are often helpful in communicating ideas that you have so far been unsuccessful at communicating with words.

The possibilities for using visual aids are endless. Can you think of other examples of when using a visual aid is the best way to communicate?

Recognizing Obstacles to Communication

Because human beings are unique individuals, not everyone communicates in the same way. Differences in culture, values, and personal characteristics can affect communication, making it difficult for even the best communicators to get their message across.

The best approach for solving this problem is to learn to recognize obstacles to communication. You can then try to avoid or overcome them by changing the way you communicate. Being sensitive to communication obstacles, and being able to avoid problems caused by miscommunication, will make you a great candidate in any career field.

Tech Connect

Modern technology is helping people overcome obstacles to communication more than ever before. People with disabilities often have special challenges to overcome. Using technology such as voice-recognition software, which converts spoken words to digital text, many disabled individuals can communicate with ease.

Use the Internet to find out more about the ways modern technology helps us overcome obstacles to communication. Then, use your computer to prepare a list of three examples, including links to the research you have done.

Communication Skills for the Entrepreneur

While large companies have human resources departments to manage the process of hiring new employees, entrepreneurs often must handle advertising, interviewing, and hiring on their own. What communication skills do you think an entrepreneur might need to successfully identify and hire a suitable employee? What communication skills do you think an entrepreneur might look for in an employee? Create a poster with a T-chart that shows communication skills that are important for the entrepreneur and the employee. Discuss how these are different and similar.

Recognizing Communication Barriers

In the workplace as in school, people are all different. A good word for this is **diversity**, which means an environment that includes variety, such as people of different ages, cultural backgrounds, and race. Diversity is generally a good thing. Diversity allows for different ideas and opinions. You can learn from people who are different from you, because they can bring different perspectives to your attention.

However, diversity can also bring special challenges to communication. When people have ideas and opinions different from yours, it can sometimes be difficult for you to communicate with them. The trick is to figure this out before your miscommunication becomes a problem.

Have you ever felt there was a barrier between you and the person you are speaking with? How could you overcome this problem?

Here are some common differences in communication that can become obstacles to effective communication. We call these differences **communication barriers**, because they can sometimes be like a wall between you and the person you are trying to communicate with.

- *Language barriers.* When two people who have different native tongues try to communicate in a single language, problems may arise if one person does not understand the language well enough.

- *Cultural barriers.* People of different cultures may have differing expectations or express themselves in different ways. Many cultures avoid direct eye contact as a sign of respect, for example, whereas direct eye contact is considered a sign of active listening in the United States.

- *Gender barriers.* Women and men communicate in different ways. For example, men may communicate by means of short statements and facts, while women often prefer more detail and the "big picture." This means that there can sometimes be problems when men and women do not recognize their different communication styles.

- *Value/belief barriers.* You do not have to be from a different culture to have a different set of values or beliefs. Even though the United States is a single culture, it is also a diverse country. This means that barriers can arise when people do not have the same ideas about morality, religion, and other life issues. Differences of opinion are perfectly fine, but sometimes these differences cause problems in communication.

Remember that in all cases of communication barriers, it is not about who is right and who is wrong. Rather, it is about recognizing and accepting diversity for what it is. You will confront some or all of these barriers. If you can recognize them, then you can take steps to resolve or avoid them.

Overcoming Miscommunication

Once you recognize that there is a communication barrier, you can try to overcome it. One method is to change the tone of voice you are using. For instance, a gender barrier might arise from too aggressive a tone of voice in a conversation between you and someone of the opposite gender. Changing your tone of voice to be more neutral or passive can overcome the barrier.

Another way to overcome a communication barrier is to change your body language. As you learned already, negative body language can turn a conversation sour. But it can also create a communication barrier. For instance, language barriers can be made even worse by negative body language. On the other hand, positive body language can often help you get your point across, even when there is a language barrier.

Finally, value/belief barriers present problems for communication, because they often arise from issues that people usually feel very passionate about. In cases where you recognize a value/belief barrier, the best way to overcome it is to stop the conversation, take a step back, and re-evaluate. Even when two people have opposing beliefs, there may be some common ground on which they can agree. A good way to overcome a value/belief barrier is to figure out what this common ground is and restart the conversation from there.

Do you go to school with people whose families are from different cultural backgrounds? Are there certain aspects of communicating that might seem inappropriate to one culture, but not another?

Case Study

Peter works for a large international clothing chain. He is in charge of offices all over the world. This means that he must go to meetings with people from different countries who speak different languages, have different cultural expectations, and have different values and beliefs.

One of the more difficult annual meetings Peter must attend takes place in a country that is very different from the United States. At this meeting, Peter must give a presentation that explains the sales for the year and discusses where the company is heading in the future.

Peter does not believe he is a good speaker. Although he always prepares a draft of his presentation, he never asks anyone to review it because he doesn't want any more criticism. In years past, he has simply stood at the front of the meeting and spoken, with his hands in his pockets. These meetings have not gone well. This year, Peter is dreading having to present at the meeting.

- What are the reasons that Peter's past presentations have not gone well? Prepare a list with at least five reasons that you could give to Peter.
- What advice can you give Peter on getting help with his speech in advance?
- How can Peter make his presentation more interesting?
- What are some of the ways Peter can avoid communication barriers that arise from the meeting that takes place in a culture so different from Peter's?
- How can Peter best get the difficult sales information across?

Answer It!

1. Why is good communication key to your success in finding a career?
2. Why is Gutenberg's printing press considered by many to be the most important invention of all time?
3. What are two different ways to read?
4. Why do we say that reading and writing go hand-in-hand?
5. List three types of documents you will write while you search for a job.
6. Why is it a good idea to speak with deference to your boss?
7. Why can talking about a difficult problem with another person sometimes help solve the problem better than just thinking about it alone?
8. What is the difference between hearing and listening?
9. When is it a good idea to use a visual aid to communicate?
10. How can different cultures create a problem for communicating?

Communicating with Others ■ Chapter 8

Write Now

You are applying for a job as a teacher in a school like yours. You consider yourself to be well qualified for the job. Write a four-paragraph cover letter that expresses your interest in becoming a teacher. In the first paragraph, introduce yourself and express your desire to apply for a job. Be specific about the job you are applying for, including what subject you want to teach. In the second paragraph, list your qualifications and explain why they make you the best candidate for the job. In the third paragraph, include some specifics about the school, the subject, and your teaching style. In the fourth paragraph, summarize your interest, your qualifications, and the reasons why you should be hired. Try to summarize your work without repeating anything.

Submit your cover letter to your teacher, and she will act as your potential employer and let you know if you got the job.

Web Extra

The Internet is the world's biggest forum for communication. Social network sites such as Facebook.com, Twitter.com, and Pinterest.com are places where people from all over the world communicate with one another. And the best part is, this communication comes in all forms: written, verbal, visual, etc. With a parent, teacher, or librarian, go to one of these social network sites. Find one example of positive and one example of negative communication. These examples can be in any of the communication forms. Then, make a short report on your findings and why they were positive and negative.

COLLEGE-READY PRACTICES

Most colleges and universities, scholarships, and internships require applicants to submit a personal essay of up to 650 words. College-ready individuals know that a personal essay is a short, autobiographical piece of writing that provides insight into the author's abilities, interests, and personality.

Conduct research to learn what makes a personal essay effective. Then, write one that is no more than 650 words long. Read your essay out loud to a partner, small group, or to the class.

Career-Ready Practices

Imagine that your school is planning on changing the school hours to start much later in the morning and go until later in the afternoon/evening. Divide into groups of eight to ten people to debate this issue.

There will likely be some in your group who are for this and some who are against it. Split into two sides along these lines. Make the groups even. Then, the group that was for your school changing the hours will argue that the hours should not change. The group that was against the change will argue in favor of it. Make sure you pay attention to your tone and presentation skills as you share your opinions. Career-ready individuals are skilled at interacting with others and articulating ideas. They speak clearly and with purpose.

Use this debate to make a list of pros and cons. Which side won? As a class, compare the outcomes for each group.

Career Portfolio

Developing strong communication skills will help you find, get, and keep a job. Do you have strong communication skills? Are there some you could work to improve? Make a table listing different communication skills, and rank your strength in each on a scale of 1 to 10—with 1 being weak and 10 being strong. Make a written plan for improving the communication skills you consider to be the weakest, and for practicing those you consider strong. Add the table and the plan to your career portfolio.

It is also a good idea to add documents that illustrate your successful communication skills. For example, you might include a list of skill competencies you have achieved; licenses or certificates you have obtained; any awards or scholarships you have received; projects you have completed as a volunteer or member of a career and technical student organization; and an evaluation from a teacher. Prepare a multimedia presentation of your portfolio to deliver to the class, a teacher, a career counselor, or an employer to demonstrate your effective communication skills.

9 Building Relationships

Skills in This Chapter . . .

- **Relating to Others**
- **Building Team Relationships**
- **Managing Conflict**

GETTING STARTED

People are naturally social. We live in groups, socialize in groups, and work in groups. Like success in school, career success has a lot to do with your ability to get along with the people around you. When we have healthy relationships with other people, we are happier than when we have unhealthy relationships. We can use skills such as decision making, problem solving, and—most of all—communicating to build healthy relationships with all the people in our lives.

➤ Write down one person in each area of your life—family, school, peers, community, and work if you have a job. Next to the name, write whether you think the relationship you have with that person is healthy or unhealthy, and list at least one reason why. As a class, discuss the reasons you think a relationship is healthy or unhealthy.

Relating to Others

Whenever you interact with someone else, you are **relating**. Relating includes the way you communicate and behave when you are together. You might be friendly, you might be mean, or you might be competitive.

A **relationship** is a connection or association between two or more people. It describes the way you relate. You have personal relationships with people such as family and friends, but you also have relationships with others—a stranger who bumps into you and apologizes or the mail carrier who delivers your mail.

Developing a healthy relationship with someone contributes to your well-being. It is satisfying and makes you feel good. When a relationship is unsatisfying, it can interfere with your well-being. Recognizing factors that contribute to healthy relationships can help you succeed in all areas of your life.

Relating to People at Work

Relationships you build at work—like relationships you build at school or at home—can have a major impact on your overall well-being.

- You have relationships with your co-workers and your supervisor. As a team, you work together to achieve common goals. For example, you might work as a flight attendant. The other members of the flight crew are on your team. You work together to make sure that you depart on time and arrive safely.
- You may also have relationships with customers or clients. You work to provide them with goods or services that satisfy their needs. The passengers are your customers. You work to provide them with information about the flight, drinks or snacks, and other things to make them comfortable.

You may have relationships with other people in your work environment, such as the baggage handlers and cleaning crew. Knowing how to get along and avoid or solve conflicts is an important part of your career success. You can start building these skills today by how you build relationships with other students and teachers in your school or how you relate to family members.

Building Relationships in Your Own Business

Entrepreneurs are more involved in the day-to-day operations of their business than owners and managers are at a large company. Entrepreneurs interact with their employees on a regular basis and must be able to build relationships with everyone. Often, the success of the start-up business depends on the entrepreneur's ability to communicate, resolve conflict, and work together with his or her employees. How do you think building relationships is different at a small start-up company than at a large corporation?

Building Relationships ■ Chapter 9 161

Career Tips

Being nice is a good way to start developing positive relationships with people in all areas of your life. For example, if you offer to help a co-worker or classmate, others will recognize you as supportive. If you compliment or praise someone's work or effort, they will recognize you as appreciative.

Qualities of a Healthy Relationship

A relationship is healthy and functional if both parties give and receive what they need. A healthy personal relationship has these important characteristics:

- Respect
- Trust
- Honesty
- Responsibility
- Open communication

Relationships that are full of conflict or are not satisfying are unhealthy, or **dysfunctional**—they are not working correctly. One way to tell if a relationship is unhealthy is that you are unhappy. You might not feel respect or trust for the other person, or you might fight a lot.

What can you do if you have an unhealthy relationship with your employer?

A Two-Way Street

All relationships are a two-way street. That means both people involved share the responsibility of keeping the relationship healthy.

When you have a relationship with someone, you exchange resources. That means you give and take. For example, you might exchange respect and support with co-workers. You might give advice to customers in exchange for money. You might give labor—your work—to your employer in exchange for salary and benefits.

When the exchange is equal—you give as much as you receive—both people are satisfied. When the exchange is unequal—you are either giving or receiving more than the other person—it may lead to problems.

- If you don't have time to help your classmate with a project, but you expect her to help you, you are receiving more from the relationship than you are giving.
- If your employer asks you to work late but does not pay you for the additional hours, you are giving more to the relationship than you are receiving.

Expecting Too Much

Expecting too much means you are taking more than you are giving. You are too needy—you need more resources than you give in order to feel satisfied with the relationship. What are the consequences of expecting too much from a relationship, compared with expecting equality?

- 👎 You expect others to listen to you complain all the time.
- 👎 You expect others to provide constant reassurance about your appearance, personality, and abilities.
- 👎 You expect others to put up with you even when you are dishonest, irresponsible, or disrespectful.
- 👎 You expect others to share your values and standards.
- 👍 You expect to exchange resources equally and to give as much as you receive.

How can a positive attitude and high self-esteem help you build healthy relationships at work?

Communicate, Communicate, Communicate!

Effective communication is the foundation for a healthy relationship. When you communicate openly and honestly, you understand each other. You avoid misunderstandings.

- When you listen and respond to someone, you show respect.
- When you share your thoughts and opinions, you show trust.
- When you honor a confidence—listen to someone's private thoughts without telling anyone else—you show that you are trustworthy.
- When you are willing to discuss differences, you show openness.

Using communication to build a healthy relationship helps you enjoy being with other people, as well as be more productive at work. You are more likely to be comfortable and happy when you are with people who you respect and trust, and who respect and trust you.

NUMBERS GAME

How are decisions made in teams? Sometimes, the leader makes decisions, taking the suggestions and ideas of the team members into consideration. Another method of team decision making is majority rule, which means the decision is made based on what most people want. A **majority** is a greater number or part. For example, if there are eight people on a team, and five prefer the color pink while three prefer the color yellow, the group preferring pink is the majority. The opposite of majority is **minority**—the lesser number or part. In this example, the group preferring the color yellow is the minority.

- If a team consists of ten people, how many need to agree to make a majority?
- What if the team consists of eleven?

Of course, when there is an even number of people in a group, there might be a tie. How might you make a decision if there is a tie?

Building Team Relationships

A **team** is a group of two or more people who work together to achieve a common goal. When you are part of a team, you have access to all the knowledge, experience, and abilities of your teammates. Together you can have more ideas, achieve more goals, and solve more problems. A healthy relationship as a team member contributes to your well-being. Your teammates respect you and value your contribution to their success.

A successful team relationship depends on all team members working together. They rely on each other. They trust one another. If one team member does not do his or her share, the entire team suffers.

Career Trend

More and more companies today organize their employees into teams made up of people from different departments. For example, a team might include representatives from research, development, design, marketing, and finance. The goal is to bring different thoughts and ideas together from the start to produce higher quality products and services.

Challenges of Teamwork

The challenges of a team relationship come from having different people working together. Even if everyone agrees on a common goal, they may not agree on how to achieve that goal. Your friends might agree to celebrate the end of the school year together, but:

- Some of you might want to go to the lake.
- Some might want to play Frisbee in the park.
- Some might want to have a picnic.

To be successful, a team needs all members to agree about how to achieve your goal. To achieve agreement, teams must be able to communicate and negotiate, which means to give and take until you find a solution that satisfies everyone. They must also be ready to resolve conflicts in an open and honest way. You might also have to make personal sacrifices for the success of the team.

What If You Hate Your Team?

You might not want to be part of a team. Maybe you were placed on a team that you did not choose. Maybe you think you are the only one on the team who is making an effort. How might your behavior influence the team's success?

- 👎 Refuse to cooperate.
- 👎 Do not participate in team discussions.
- 👎 Do what you think is best, even if the rest of the team has a different plan.
- 👎 Yell at your teammates until they do what you want.
- 👍 Listen respectfully, keep a positive attitude, and present your opinions in a clear and concise way.

Sometimes being an effective member of a team does not depend on the actual outcome of the project. Sometimes it depends on how well you get along with your teammates and whether you give your teammates a chance to succeed.

What Makes a Team Successful?

What can you do to encourage team members to cooperate?

Teams are influenced by different things, including the personal qualities of the team members, the resources available, and the purpose or goals of the team.

When you first join a team, you and your fellow team members might feel nervous and uncomfortable. You might not know each other very well. You might wonder what to expect.

At first, you might misunderstand each other or misinterpret communications. One teammate might say she can't come to a meeting, and you might think she means she doesn't want to be part of the team. Another teammate might have lots of plans and ideas, and the rest of you might think he is trying to take over and be the boss.

As you get to know each other, and learn how to communicate, you might feel a sense of belonging. You might develop common bonds. You will be able to work together to achieve your goal.

21st Century Learning

CAREER COUNSEL

Jesse is 13. He signed up as a volunteer for his town's community clean-up day. When he arrives, he is assigned to work with three other people picking up trash in a neighborhood park. The group is diverse. There are two high school girls who Jesse thinks are Asian and an African American man about the same age as Jesse's dad.

At the park, Jesse puts on his gloves, grabs a trash bag, and looks around for a good place to get started. The girls take out their cell phones and text their friends. The man yells at the girls to get to work. Then, he turns his attention to Jesse. He points out trash and orders Jesse to pick it up. Jesse feels like he is the only one doing any real work. He begins to regret his decision to volunteer.

How can Jesse make the best of the situation? Use 21st Century Skills such as problem solving, critical thinking, communications, and decision making to help Jesse identify the challenges he is facing and to find solutions. Write an ending to the story and read it to the class, or form small groups and present it as a skit.

Building Relationships — Chapter 9

Developing As a Leader

Even when all members of a team have an equal role in decision making and problem solving, it is important to have a leader. Recall that a *leader* is a type of manager who knows how to use available resources to help others achieve their goals.

Leaders exhibit positive qualities that other people respect, such as self-confidence. They use skills such as goal setting and critical thinking to make healthy decisions for the benefit of the team. If you are the leader, you take on the responsibility for:

- Organizing the team's activities
- Encouraging everyone to share ideas and give opinions
- Motivating all team members to work toward established goals

Being the leader does not mean you are always right. The leader's opinion does not count more than the opinions of the other team members. An effective leader keeps the team on track and focused on achieving its goals.

What personal qualities does a leader display?

Career Profile: Video Director

Arts, A/V Technology & Communications Career Cluster

Job Summary

A video director is responsible for directing the work of the cast and crew during the production of a video. Directors interpret scripts, select the cast, and manage rehearsals. If necessary, they approve the sets, costumes, choreography, and music.

A few directors succeed in the motion picture industry, but many work for businesses creating training and marketing videos. Many are self-employed, which means they must work hard to promote themselves to get new jobs.

Video directors must prove they have skill and talent. They must also manage a diverse team of people, including actors, producers, and writers, and keep track of the budget. Most attend college to study theater arts, while taking advantage of opportunities to direct videos while in school. They may start their careers as an assistant director.

Use the Internet, a library, or career center to research the responsibilities, education and training requirements, and salary range of a video director. Write a video script for a biography of a successful video director. If possible, turn the script into an actual video, or perform it with a team of classmates.

What leadership qualities might help a video director succeed?

Qualities of an Effective Team

An effective team cooperates to achieve a specific goal. Members of an effective team:

✔ Listen to each other
✔ Respect each others' opinions
✔ Recognize each others' skills and abilities
✔ Share the work load
✔ Share responsibility

What About Team Members?

While a strong leader is important to the success of a team, team members must also be committed to the group's success. An effective team member helps teammates if they need help, does not blame teammates for problems or mistakes, and offers ideas and suggestions instead of criticism.

You are a good team member if you are:

- Open minded
- Cooperative
- Trustworthy
- Willing to compromise
- Friendly

Recognizing Peer Pressure

Peer pressure is when your peers such as friends, classmates, or co-workers influence you to do something. They might actively try to convince you to do something you haven't done before or to stop doing something you are doing. For example, a co-worker who is a vegetarian might try to convince you to stop eating meat.

Or, you might decide to do something or stop doing something because of what you think your peers might think. For example, you might start drinking soda instead of juice because everyone else drinks soda. No one actually told you to drink soda, but you think you will fit in better and they will like you more if you do.

- Peer pressure can be positive if it contributes to your well-being and does not interfere with your values, standards, and beliefs. For example, a friend might love mountain biking, so you try mountain biking and find out you love it, too.

- Peer pressure can be negative. It might interfere with your well-being or go against your values. For example, you might know taking more than one hour for lunch is wrong, but your co-workers convince you to do it anyway.

Recognizing peer pressure can help you make healthy choices and achieve your career goals. You can decide for yourself whether it is positive or negative and whether you want it to influence you.

Money Madne$$

You like a turkey sandwich on whole-wheat bread with lettuce for lunch. At the supermarket, the sliced turkey costs $5.75 per pound. A head of lettuce costs $1.25. A loaf of whole-wheat bread costs $3.15. You can use the items to make more than enough sandwiches for a whole week.

Your co-workers eat in the cafeteria. They buy freshly made sandwiches for $4.25.

How much does it cost you to bring a sandwich for lunch five days a week? How much does it cost to buy a sandwich for five days? Which option do you think you should choose? Is cost the only thing to consider? Write a paragraph explaining your decision to your co-workers.

Building Ethical Relationships

Recall that *ethics* are a set of beliefs about what is right and what is wrong. When you build ethical relationships, you treat others with respect and honesty and they trust you in return. Your ethics help you set standards for relationships in all areas of your life.

You don't leave your ethics behind when you enter a workplace. As a co-worker, employee, or supervisor, you have a responsibility to behave ethically. For example, you should not lie, steal, or make fun of other people. Employers are always looking for ethical employees.

Work ethics are beliefs and behaviors about what is right and wrong in a work environment. The business where you work may have a set of corporate or business ethics. Business ethics set standards for how the company relates to others, including customers, employees, competitors, and even the government. For example, the company might believe that it has a responsibility to protect the environment, so it takes steps to conserve natural resources by turning off lights, keeping the air conditioning low, and buying recycled paper products.

How can your ethics help you develop strong relationships in a work environment?

Not everyone has the same set of ethics. For example, you might think it is wrong to gossip about others, but your co-workers do it all the time. You cannot control someone else's ethics. All you can do is make sure you relate to others in a way that is true to your own ethics.

However, if someone behaves illegally, or against your company's code of ethics, you have a responsibility to your employer to report it to your supervisor or to the human resources department.

Tech Connect

In general, technology improves worker productivity and makes it easier for companies to succeed. It also brings new ethical problems and decisions.

For example, technology has made it easier for people to steal intellectual property, such as written words or recorded music. It has made it easier for thieves to steal credit card and bank account information. It has even made it easier to edit photographs so people cannot be certain if the picture they are looking at is real.

Use the Internet or your library to research some of the ways developments in technology have impacted ethical behavior in the workplace. Select one and write a newspaper article about it. Use a desktop publishing or word processing program to make a newsletter that includes articles written by your classmates.

Is the Conflict Your Fault?

Your co-workers are angry at you. They have been "forgetting" to tell you about meetings. They do not listen to your suggestions or ideas. What do you do to affect their attitude towards you?

- 👎 Make excuses to cover up your mistakes.
- 👎 Blame your co-workers for your errors.
- 👎 Take credit for others' work.
- 👍 Find ways to cooperate and compromise.
- 👍 Show appreciation for their efforts.
- 👍 Take responsibility for your work.

How can unhealthy relationships with your co-workers interfere with your career success?

Managing Conflict

Conflict is a disagreement between two or more people who have different ideas. Conflict occurs in all relationships at one time or another. It can cause stress and interfere with your well-being if it is not resolved.

- When a relationship is healthy, people talk about what is causing the conflict and work together to find a solution.
- When a relationship is not healthy, conflict can go unresolved. Unresolved conflict interferes with well-being and can cause stress, depression, anger, and resentment.

Managing conflict does not always mean eliminating the conflict completely. It means you are able to recognize what is causing the conflict and that you can cope with it in an honest and respectful way.

Types of Conflict

There are different types of conflict. The way you respond to conflict depends on what it is and how it affects your well-being.

How can you use respect and honesty to manage conflict at work?

Some conflict is minor and short-lived—such as disagreeing with a co-worker about who should clean up the lunchroom. You can usually resolve minor conflicts easily by talking about them and coming to a compromise or agreement.

Some conflict is serious or ongoing—such as refusing to promote a product your company sells or disagreeing with the way your supervisor treats co-workers. Serious conflicts often come from conflicting values and could be difficult—or even impossible—to resolve. You may not want to compromise your ethics or values. Sometimes, you can agree to disagree. Other times, you may have to end the relationship.

Causes of Conflict

Any disagreement can cause conflict. Sometimes understanding what causes the disagreement can help you resolve the conflict more quickly. When you recognize the cause of the conflict, you can take steps to find a solution.

- Differences in values and standards cause conflict. Your supervisor might schedule you to work on Sunday, but it is important to you to spend Sundays with your family.
- Ineffective communication and misunderstandings can cause conflict. If your supervisor doesn't tell you she will be late for work, you might schedule a meeting for her. Then you are angry when you must change the schedule.
- Conflict may be caused by personal qualities such as stubbornness or conceit. Your co-worker might refuse to turn down the music volume when you are trying to talk to a customer on the phone.

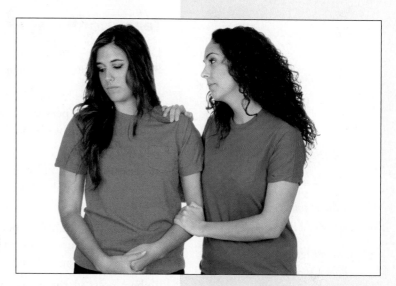

How can an apology show you understand that the other person is hurt or angry?

Resolving Conflict

At the root of all conflict there is a problem that blocks you from achieving a goal. You can use the problem-solving process to find a solution. You will be more successful in resolving or minimizing the conflict if you work with the other person or people who are involved in the conflict. Together you can identify the disagreement or problem. You can then **negotiate**, or discuss options that will lead to an agreement that satisfies everyone. Finally, you can compromise and select and implement—put in action—the best solution.

When you work together to resolve conflict, keep these tips in mind:

- Show respect and understanding for the other person.
- Remain calm and objective.
- Take responsibility for your own actions and feelings, without blaming anyone else.
- Remember to use the six key factors of effective communication, including sending a clear, concise message; listening carefully; and using "I" statements to focus on your point of view.
- Insults hurt and cause new conflicts. You can attack the conflict without attacking the other person.

The hardest thing of all might be to apologize. Even if you don't think you caused the conflict or did anything wrong, you can apologize for making the other person feel bad.

Sometimes you might not be able to get the other person to cooperate. He or she might not want to resolve the conflict or might refuse to admit there is a conflict. Then, you will have to decide whether you need to take action on your own or just walk away.

Anger Management

Conflict can make you mad. When you are angry, you might not be reasonable or realistic about the conflict. You might say or do something you will regret later. What if you are too angry to talk to the other person?

★ *Take a deep breath.*
★ *Count to ten.*
★ *Exercise.*
★ *Identify your role in the conflict.*
★ *Brainstorm solutions.*
★ *Consider what you could say to start resolving the conflict.*
★ *Ask someone you trust for advice.*

Once you calm down, you will be better able to think critically about what is causing the conflict and how to resolve it. Can you think of other ways to control your anger and resolve the conflict?

Team Conflicts

Conflicts in a team can interfere with the team's ability to achieve its goals. Because there are many people involved, the type of conflict can vary.

- Sometimes, the conflict might have nothing to do with the team. One member might have a personal conflict that is getting in his way. For example, your classmate might not be able to concentrate on a project because he has a family problem.

- There may be conflict between two team members. You and your co-worker may not agree about whether you should contact a customer using e-mail or the telephone.

- There may be conflict between one team member and everyone else in the team. Your supervisor might want the team to meet in the morning, but the rest of you want to meet in the afternoon.

To be successful, the whole team should work together to resolve the conflict. Each team member can contribute to identifying the problem and brainstorming solutions.

Resolving Team Conflicts

Follow these important steps for resolving conflicts as a team.

- ✔ *Give all team members a chance to express their views.*
- ✔ *Be respectful of all team members.*
- ✔ *Deal with one problem at a time.*
- ✔ *Focus on your common goals.*
- ✔ *Find ways for all team members to be involved in the solution.*
- ✔ *Encourage all team members to take responsibility for their actions—positive and negative.*
- ✔ *Avoid blame.*

Asserting Yourself

One positive way to cope with conflict is to be assertive. Being **assertive** means that you stand up for yourself. You express your feelings and thoughts with confidence, in a strong, honest, and direct way. When you are assertive, you show others that you have the right to be treated with respect.

One negative way to cope with conflict is to be aggressive. Being **aggressive** means that you force your opinions on others. You express your feelings and thoughts in a hostile, or angry, way. When you are aggressive, you show others that you have no respect for their thoughts or feelings.

Finding Common Bonds

In order to succeed in your career, it is important to build relationships with all of your co-workers. In most work environments, you will have the opportunity to work with people from different cultures and backgrounds. One way to develop a relationship with people who seem different is to find common bonds. When you look for common bonds, you focus on the things that make you like someone else. You do not spend time worrying about what makes you different.

For example, you might work with someone who was born in Brazil. You might have different traditions or eat different food, but you might both enjoy playing soccer. You might both have brothers or sisters. You might both play the guitar. Discovering common bonds helps you understand the other person so you can work together successfully as a team.

Select a country and use the Internet to research its cultural traditions. Make a three-column chart comparing your cultural traditions with the cultural traditions of your selected country and identifying areas with common bonds.

Unresolved Conflict

When you let conflict go unresolved, it can slowly eat away at a relationship. You might not want to speak up about what is bothering you. You might think it is silly. You might worry the other person will be hurt or angry. When you don't speak up, you can start to **resent** the other person, or blame him for your unhappiness.

Do you expect the other person to know what you are thinking or feeling? You might be annoyed because your classmate did not respond to your text message. You might be hurt because you think she is ignoring you. It is possible she did not receive the message, or that she accidentally hit delete instead of send. Unless you speak up, you will not be able to resolve the conflict and move forward toward your common goals.

How can unresolved conflict interfere with your success at work?

Odors in the Office

The co-worker who sits at the next desk wears very strong cologne. It gives you a headache and makes it difficult for you to concentrate on your work. What steps can you take to successfully solve the problem?

- 👎 Take lots of ibuprofen to make your headache feel better.

- 👎 Sit at your desk and complain about it in a loud voice, hoping he will overhear you.

- 👎 Leave an anonymous note about it on his desk.

- 👎 Buy him a brand of cologne that you like.

- 👍 Ask to speak with him privately, explain the situation, and see if you can work together to find a solution.

- 👍 Ask your supervisor to speak with him about changing his behavior.

- 👍 If he will not change his behavior, request a different desk.

Does wearing strong cologne show respect for your co-workers? What other personal behaviors might interfere with a co-worker's ability to do her job?

Myth Only bad employees cause conflict.

Truth While negative attitudes and inflexible opinions can cause conflict, there are lots of other causes, too. Very good employees can cause or contribute to conflict if they don't understand how to use effective communications, teamwork, leadership, and problem solving to manage and solve conflict when it occurs.

Case Study

Kali babysits for her neighbor's two children every Friday night. She arrives at 6:00 p.m., gives the children dinner, cleans up, plays with them until bedtime, and then puts them to bed. She spends the rest of the time watching television and chatting with her friends. The parents come home at 10:00, pay Kali $25.00, and walk her home. Kali likes her neighbors and the kids, and enjoys babysitting. She is happy earning $25.00 each week.

One Friday when she arrives, her neighbors tell her that their niece and nephew are sleeping over. Kali prepares dinner for all four kids and cleans up after them all. They have a hard time deciding what to do before bed and argue a lot. It takes a long time to get them all settled and asleep. Kali barely has time to do anything before the parents arrive home. They pay her $25.00, thank her, and walk her home. Kali feels used and resentful. She tells her neighbors that she cannot babysit anymore.

- What do you think caused Kali to feel hurt?
- Do you think Kali is right to quit the babysitting job?
- How could Kali have acted differently?

Answer It!

1. List at least three groups of people you might have relationships with at work.
2. List at least five characteristics of a healthy relationship.
3. What is the foundation for a healthy relationship?
4. Explain how you can model characteristics of an effective team leader.
5. List at least five qualities of a good team member.
6. Explain the difference between positive and negative peer pressure.
7. Define conflict.
8. List at least three reasons for conflict in the workplace.
9. Explain the difference between assertive and aggressive behavior.
10. Explain how you can model characteristics of effective conflict management.

Web Extra

There are many Web sites that publish team-building advice and activities for teens, as well as for professionals. Use the Internet to identify some reputable sites you think might be useful for yourself and your classmates. Make a directory of the sites to leave in your classroom or school guidance office.

Building Relationships — Chapter 9

Write Now

Are the relationships you have with your family different from the relationships you have with friends? Are the relationships you have with teachers different from the relationships you have with neighbors? Select two relationships that are important in your life, and then write an essay comparing and contrasting them.

COLLEGE-READY PRACTICES

In college, as in a career, many projects are accomplished by teams. College-ready individuals understand that it is necessary to know how to be an effective leader and team member, and how to communicate and resolve conlict when necessary.

Working in teams of four or five, develop a presentation about the options available after high school. Each team member should be responsible for one aspect of the presentation. Take turns being the team leader. Use appropriate technology to plan, develop, create, and deliver your presentation. Model effective team characteristics and, when it is your turn, effective team leader characteristics. When conflicts arise, model effective conflict resolution characteristics. Deliver your presentation to the class and respectfully watch the presentations developed by the other teams.

Career-Ready Practices

Form groups of five to eight. Take five minutes to write a list of things that everyone in the group has in common. For example, everyone is in the same grade or everyone lives in the same town. Next, using a different piece of paper, take five minutes to write at least two things that are unique about each person in the group. For example, one person might be the only one who plays the violin or one person might be the only one who has an older brother. Keep in mind that recognizing unique abilities or characteristics of each group member can help build strong team relationships. Career-ready individuals do not allow differences among team members to create barriers. Instead, they focus on positive interaction with others and encourage all team members to contribute in a productive manner. Share your lists with the other groups in the class.

Career Portfolio

Employers want to hire ethical employees who they can trust to represent the company in a fair and responsible way. Make a list of the personal and work ethics that are important to you, and explain why these ethics would make you a valuable employee. Add the list to your career portfolio.

10 Basic Math Skills

Skills in This Chapter . . .

- **Adding, Subtracting, Multiplying, and Dividing**
- **Working with Percentages, Fractions, and Decimals**
- **Working with Ratios and Proportions**
- **Analyzing Data and Probability**
- **Analyzing Charts**

GETTING STARTED

Every year since you started school, you have taken math. You count. You add. You subtract. You memorize your multiplication tables. Why?

Math is essential. You use math every day in many situations. You calculate how much time you have to get to class. You count your change at lunch. You dole out portions of a snack to friends. Math not only helps you in your daily life, it also helps you succeed in other classes in school. You use math in core courses such as social studies and science. You use it in electives for tasks such as measuring ingredients for a recipe or constructing sets for drama.

And, of course, developing basic math skills will help you succeed in whatever career you choose. For example, in business you use math to create budgets and analyze financial statements. In health care occupations, you use it to measure medicine. In construction trades, you use it to calculate how much material you need and what it will cost.

➤ Make a list of everyday tasks for which you use math. Then, write down how you would accomplish each task if you couldn't add, subtract, multiply, or divide.

Adding, Subtracting, Multiplying, and Dividing

The basic mathematical operations are addition, subtraction, multiplication, and division. You use them to manipulate numbers to find answers to common questions. For example, if you want to know how much money you earned selling clocks today, you would add together all of the sales receipts for clocks. If you want to know how much each share of company stock is worth, you would divide the total value of the company by the number of shares of stock.

Knowing how to add, subtract, multiply, and divide is a necessary skill for everyday life, as well as for any career you choose. You can develop these skills in math class at school and by practicing them at home, at work, and in your community.

What Good Is Addition?

Addition is the totaling of two or more numbers. Each number you are adding is called an **addend**. The result is called the **sum**, or **total**. For example, two computers in the reception area + three computers in accounting = a total of five computers in the office.

At work, you use addition for tasks such as calculating income, preparing expense reports, totaling sales, and ordering **inventory**—goods you have available to sell. If you work in health care, you might use addition to total the amount of fluids administered to a patient during the day or to count medication in a pharmacy. If you work in construction, you might add the total linear feet of lumber you need. If you work in transportation, you might add the number of miles you drove.

When Do I Subtract?

Subtraction is the opposite of addition. Subtracting numbers means taking one number away from another number. The result is called the **difference**. For example, if you have six sweaters in stock and sell four of them, you have two sweaters left in the store.

At work, you need subtraction for tasks such as tracking inventory, checking payroll deductions and withholdings, monitoring debit and checking account balances, and calculating net profits. If you are a police officer or a nurse, you might use subtraction to calculate how many hours you have left on your shift. If you are a hotel manager, you might use subtraction to determine how many rooms are vacant.

Basic Math Skills ■ Chapter 10

Memorize Those Multiplication Tables!

Multiplication is a quick method of adding numbers. It lets you repeat an addition problem in one step. For example, if a customer wants to buy four sweaters for $12.50 each, you could add $12.50 + $12.50 + $12.50 + $12.50 = $50.00, or you could multiply $12.50 × 4 to find the answer of $50.00. The numbers you multiply—in this case $12.50 and 4—are called **factors**. The answer is the **product**.

Some tasks for which you use multiplication at work include calculating hourly pay, finding the total cost of multiple items, and managing inventory. If you are a lawyer, you might use multiplication to calculate your billable hours. If you are a scientist, you might use multiplication to forecast the growth of cells.

How would you use math if you worked as an air traffic controller?

Do I Need Division?

Division is the opposite of multiplication. It is the process of separating a whole into parts. For example, if you have four employees and want to schedule them each for an equal number of hours, you divide the total number of hours in the week by four: 40 hours ÷ 4 = 10 hours each. The whole—in this case 40 hours—is called the **dividend**. The number you divide the dividend by—in this case, the four employees—is called the **divisor**. The result is the **quotient**.

You use division at work for tasks such as sharing profits, deciding how to spend your budget, deciding how to use floor space in a store or warehouse, and calculating per unit costs. A teacher might use division to organize a class into groups.

Career Trend

Many companies are screening job applicants for basic math skills before hiring. They administer tests to verify that candidates can add, subtract, multiply, and divide. The trend is occurring in many industries, but it is particularly common for jobs that require employees to handle money, such as cashiers, bank tellers, and waitstaff at a restuarant.

International Measurements

In the United States, we use a measurement system called the U.S. customary system, which is sometimes called the conventional or traditional measurement system. Most other countries use the International System of measurement, or the metric system.

The metric system is a decimal system, which means it is based on multiples of ten. Meters, grams, and liters are the basic units of measurement. Prefixes—letters added to the beginning of a word—indicate the size of the units. For example, a kilometer is 1,000 meters. A kilogram is 1,000 grams, and a kiloliter is 1,000 liters.

The customary system is not based on any one unit. Distance is measured in inches, feet, yards, and miles. Weight is measured in ounces, pounds, and tons. Volume is measured in cups, pints, quarts, and gallons.

Why do you think it is important for employees in the United States to understand the metric system? Use measurement tools in your classroom to measure distance, weight, and volume of common items with both systems. Make a chart showing the comparison.

What's Geometry?

Geometry is math that deals with the size, position, and shape of 2- and 3-dimensional objects. Some occupations that require the use of geometry include:

- ✔ Carpenter
- ✔ Astronomer
- ✔ Interior designer
- ✔ Surveyor
- ✔ Structural engineer
- ✔ House painter
- ✔ Robotic engineer

Working with Percentages, Fractions, and Decimals

Percentages, fractions, and decimals are different ways to represent the same thing: a number less than a whole. The main difference is the context in which you use each one. For example, ¼, .25, and 25% all represent the same value, but you use them in different situations:

- 25% to represent a sales discount
- ¼ teaspoon to represent an amount of an ingredient in a recipe
- .25 to represent part of a dollar

You will benefit in any career by knowing how to work with percentages, fractions, and decimals in different situations. The first step is to understand the difference between them and when to use one or another.

Basic Math Skills ■ **Chapter 10**

What's a Fraction?

A **fraction** is a *ratio*—or comparison—of two whole numbers, or one whole number divided by another whole number. You write a fraction in a way similar to writing a division problem. A dividend, called the **numerator**, is written above a line, called the **fraction bar**. A divisor, called the **denominator**, is written below the fraction bar.

For example, if you have three-quarters of an hour to travel to a meeting, you can write the fraction like this: ¾. The number 3 is the numerator, or dividend. It represents the number of quarter-hours you have. The number 4 is the denominator, or divisor. It represents the total number of quarter-hours in a whole hour.

A mix of whole numbers and fractions is called a mixed fraction. For example, 1¼ is a mixed fraction.

At work, you most often use fractions to measure quantities. For example, a chef uses fractions to measure ingredients and a tailor uses fractions to measure fabric or thread. You might also use fractions to explain statistics. For instance, a market research analyst might notice that half—½—of a group prefers one product over another. A third use of fractions is to refer to time. For example, you might schedule an appointment for a quarter—¼—of an hour.

What's a Decimal?

A **decimal** is a different way of writing a fraction. You write a decimal using a decimal point, or dot. Numbers to the left of the decimal point are whole numbers. Numbers to the right of the decimal point are less than one. Each place in a decimal represents a multiple of ten. The first place to the right of a decimal is tenths. The second place is hundredths. The third place is thousandths. For example, .25 means there are two tenths and five hundredths. The number 2.25 means there are two wholes, two tenths, and five hundredths. One of the most common uses for decimals is to write dollar values. The numbers to the left of the decimal point represent whole dollars, and the numbers to the right represent the cents.

Many occupations use decimals. In health care, decimals indicate body temperature—98.6 is normal. Mileage is measured using decimals. Often, people convert fractions to decimals so they can work with the numbers more easily. For example, it is easier to add .25 + .33 than it is to add ¼ + ⅓.

Could you make change if you did not know how to work with decimals?

Money Madne$$

Counting change is an important skill for many workers as well as for consumers. Bank tellers count change for customers cashing checks or withdrawing money from an account. Cashiers count change when a customer pays more than an item costs and receives the difference back. Consumers count change to make sure they receive the correct amount.

You and a friend are going to the movies. The tickets cost $10.75 each. You buy both tickets online. At the theater, your friend gives you a $20.00 bill to pay for his ticket. How much change do you give him? What if the tickets cost $11.50 each?

What's a Percent?

A **percentage** is also part of whole. Specifically, 1 percent is one part out of 100. That means that 100 percent equals the whole, or all, of something—no matter how much is in the whole.

You can quickly calculate a percentage by dividing the amount you have by the total amount, and then multiplying the quotient by 100. For example, if there are 500 sweaters in the stockroom, and 45 are blue, you can calculate the percent of blue sweaters by dividing 45—the number of blue sweaters—by 500—the total number of sweaters. Then you multiply the quotient—.09—by 100. The answer is 9%. If 55 are red, the percent of red sweaters is 55 ÷ 500 = .11 × 100 = 11%.

To calculate the amount of a percent, multiply the total number by the percent. For example, if you know 6% of the 500 sweaters are brown, multiply 500 × 6% to learn that you have 30 brown sweaters. You can convert the percent to a decimal first, to make the multiplication easier: Replace the percent sign with a decimal point and move the decimal point two spaces to the left. So, 500 × .06 = 30.

How do you use percents at work? To determine market share; to track inventory; to calculate interest; or to compare sales of different products, sales to different customers, sales by region, sales by store, or sales by salesperson. A salesperson uses percentages to calculate commission. A farmer might use percentages to decide how many acres to use for a specific crop.

> **Myth** There is only one way to solve a math problem.
>
> **Truth** As for any problem, there are always multiple options for coming up with a solution to a math problem. The method that works best for you might depend on your own learning style. For example, if you are a visual learner, you might illustrate the problem in order to see it clearly. If you are a tactile learner, you might use manipulatives—or objects that you can handle or count.

Career Tips

Studying math is an excellent way to prepare for many careers. It provides a foundation for all areas of science and engineering as well as social sciences such as anthropology, sociology, and psychology. Lawyers, business professionals, and health care workers all require a solid background in math and statistics.

Basic Math Skills ■ Chapter 10

181

Tech Connect

A spreadsheet is a grid of columns and rows used for storing numerical or financial data. Once, spreadsheets were written on paper; now, almost all spreadsheets are created using computer programs such as Microsoft Office Excel. Spreadsheet programs are used in any job that works with numbers—which is pretty much every job. For example, you can use one to set up a budget, create charts, track data, forecast sales, analyze statistics, and record the results of scientific experiments.

Spreadsheet programs have made workers more productive by automating tasks they used to do by hand. For example, they have features that let you set up equations or formulas so your computer can automatically perform calculations. The formulas might be simple, such as totaling a column of numbers, or complex, such as generating a chart based on data in multiple columns or rows. Workers can use the same spreadsheet in different situations by changing the values but keeping the formulas in place.

How much time do you think a spreadsheet saves? Write a column of ten numbers, with at least three digits in each number. Exchange your paper with a partner and see how long it takes to add the column without using a calculator or computer. Change three of the numbers. How long does it take to recalculate the sum? Using a computer, enter the numbers in a spreadsheet program such as Excel. Use the SUM formula to find the total. Change three numbers and recalculate the sum. Which method is faster? Which is more likely to result in an error?

Can I Use a Calculator?

A calculator can make coming up with the right answer to a math problem easier—as long as you know what operations to use! What are some of the benefits and drawbacks of relying on a calculator?

👎 You might not have the calculator with you.

👎 You might not know how to set up the equation to get the right answer.

👎 You might be able to perform the calculation faster in your head.

👍 You will get the correct answer to basic math problems.

👍 You can perform more complex problems faster.

Can you think of other benefits and drawbacks of using a calculator?

Working with Ratios and Proportions

Have you heard the term *scale* used for changing the size of something? For example, you might scale a digital photograph to make it bigger or smaller. You might scale a recipe to make more or less. When you scale, you use ratios and proportions. A ratio lets you compare two values—even when the units are different. For example, you can compare the number of apples in a display case to the number of oranges. Then, you can use proportions to adjust the values an equal amount.

Ratios are useful in industries where you combine different components to create a product or when you want to compare two different units. For example, a chef who combines ingredients uses ratios to make sure the dish tastes good. An entrepreneur uses ratios to track the amount of profit compared to the amount of loss.

Proportions come in handy for people who adjust the sizes of things: That same chef might want to feed four people using a recipe written for two—so he increases the amount of ingredients while keeping the proportions the same. An architect builds a scale model of a house by keeping the measurements in proportion.

What's a Ratio?

A ratio compares two numbers. For example, if there are three employees in the sales department and one manager, the ratio of employees to managers is 3 to 1. You often write a ratio using a colon to separate the two numbers, like this: 3:1. You can also write the ratio as a fraction, like this: 3/1.

Ratios are often used to measure rate. The basic difference between rate and ratio is that the two things either have the same unit—a ratio—or a different unit—a rate. For instance, when a person runs 7 miles per hour, it indicates a rate of 7:1, which is commonly written as 7 miles/hour. The different units are miles and hours.

Ratios help workers perform a variety of tasks. An artist might combine a ratio of red to blue paint to create a desired shade of violet. If the ratio of red to blue changes, the shade of violet changes, too. A baker might use 2 parts sugar to 1 part flour, which is a ratio of 2:1. If he changes the ratio—say he uses 1 part sugar to 2 parts flour—1:2—the dough will not come out correctly.

Ratios are also useful in other careers. A grocer might notice that she sells four dozen McIntosh apples for every one dozen of Delicious apples, a ratio of 4:1. She can use that information to decide how many apples of each type to buy in the future.

What might happen if the concrete is not mixed using the proper ratios of ingredients?

Basic Math Skills ■ Chapter 10

What's a Proportion?

A **proportion** is when you have two ratios that equal each other. For example, one class might have 1 teacher for every 15 students (a ratio of 1:15), and another class might have 2 teachers for every 30 students (2:30, which is the same as 1:15). The ratios are equal.

You use proportions to maintain equality between two things, even when the number of things changes. For example, a window washer might use 1 cup of ammonia for every 2 gallons of water—a 1:2 ratio. If he wants to use 4 gallons of water, how much ammonia should he use? Since he doubled the amount of water, to maintain the ratio he must double the amount of ammonia—2 cups of ammonia for 4 gallons of water—still a 1:2 ratio.

Anyone who uses ratios in a career is also likely to use proportions. For example, if a baker wants to increase the number of rolls she gets from a recipe, she can increase the amount of dough. To make sure the dough comes out right, she must keep the ingredients in proportion to each other. If she uses 1:2 parts sugar to flour in the smaller amount, she must use 1:2 parts sugar to flour in the greater amount.

Proportions are also used to measure equality in other areas. A school might want to keep the ratio of teachers to students in proportion—or the same. A city planner might want to keep the ratio of residential homes to businesses in proportion.

NUMBERS GAME

Measuring with a ruler or tape measure is an important skill for many careers, such as tailor, carpenter, or architect. An inch ruler or tape measure is divided into inches, and each inch can be divided into 16 parts. Marks on the ruler indicate the parts: the shortest lines mark 1/16, the next shortest mark 1/8, then 1/4 and 1/2. Each inch line is marked with a number (refer to the illustration).

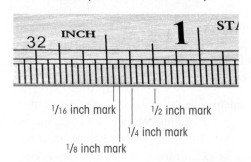

To measure, you line up an inch mark at the edge or point where you want to start measuring, use the lines to count off the length, and mark the spot where you want to stop measuring. For example, to measure 3/8 from the edge of fabric, you line up the edge with the inch mark, count three 1/8 inch marks, and draw a line to mark the spot. To measure 5/8 from the edge, you line up the edge, count five 1/8 inch marks, and mark the spot.

Using a blank piece of paper and a ruler, tape measure, or the illustration, measure 1/4 inch from the left edge of the page, and draw a line to mark the spot. Then, measure the distance from the 1/4 mark you made to the right edge of the page. Write the measurement next to the line marking the 1/4 spot.

Measure 5/8 from the right edge, mark the spot, then measure the distance from the mark to the left edge of the page. Write the measurement next to the 5/8 mark.

Measure 1 1/8 from the top edge, mark the spot, then measure the distance from the mark to the bottom edge. Write the measurement next to the 1 1/8 mark.

Developing Math Skills in School

Math class is your opportunity to develop basic math skills. What if you stopped going to math?

★ You would not do well on college entrance exams such as the SAT or ACT.

★ You would not be prepared for common after school or summer jobs, such as cashier or wait person.

★ You may not meet your high school graduation requirements.

★ You might have trouble performing everyday math tasks, such as counting change, measuring time, and preparing recipes.

What steps could you take if you have difficulty keeping up in math?

Analyzing Data and Probability

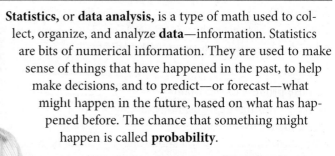

Statistics, or **data analysis,** is a type of math used to collect, organize, and analyze **data**—information. Statistics are bits of numerical information. They are used to make sense of things that have happened in the past, to help make decisions, and to predict—or forecast—what might happen in the future, based on what has happened before. The chance that something might happen is called **probability**.

Data and statistics help to prove facts, illustrate goals, and forecast possibilities. Almost all careers use statistics and probability. Meteorologists collect statistics about weather and use the information for forecasting. Geologists collect statistics about mineral deposits and use the information to predict the probability of locating oil. Market researchers collect statistics about people's buying habits and use the information to predict the probability of a product's success. Insurance actuaries collect statistics about risk factors and use the information to predict the probability of illness, accidents, or death.

Common Statistics

Some common statistics include:

✔ *Median: The middle value in a set*

✔ *Average or mean: A typical value calculated by adding all values and dividing the total by the number of values*

✔ *Mode: The number that occurs most often in a set*

✔ *Range: The difference between the lowest and highest values in the set*

✔ *Frequency: The number of times an event occurs*

Data Analysis

To collect data, you ask questions or conduct a survey. For example, you can conduct a survey of your classmates to find out whether they prefer cats or dogs or how many siblings they have or if they like the taste of mushrooms. You can also collect data from history. For example, a retail chain might collect sales data from the past three years for each of its stores.

Once you collect the data, you analyze it and turn it into statistics. For example, a dietitian might ask, "How many servings of vegetables did you eat each day this week?" You might answer two on Monday, three on Tuesday, none on Wednesday, one on Thursday, two on Friday, three on Saturday, and three on Sunday.

From the information, the dietitian can calculate many things: the average number of servings you ate each day, the total number of servings you ate this week, the least number of servings you ate, the most number of servings you ate, and so on. If he has asked other people the same question, he can combine all of your answers and start to look for patterns. For example, he might be able to identify whether people eat more servings on Fridays than on Wednesdays. He can use the information to make decisions such as how to get people to eat more vegetables on Wednesdays.

You might use data analysis at work in marketing, pharmaceutical research, or retail sales. A politician might use it to decide whether to run for office, a banker might use it to decide where to invest funds, and an entrepreneur might use it to convince investors to back his start-up. Schools use statistics to decide how many teachers to hire, and colleges use statistics to help determine which applicants to admit.

Basic Math Skills ■ Chapter 10

> ### International Currency
>
> **Currency** is the unit of money used in a country. For example, the dollar is the currency in the United States. The euro is the currency in the European Union. When you travel to a different country, you exchange the currency you have—dollars—for the currency you need. The **exchange rate** determines how much of the foreign money you can buy for one dollar.
>
> So, if the current dollar exchange rate for euros is 0.731608, it means you can exchange $1.00 for 0.731608 euros.
>
> To find out how much you will get for more than $1.00, you multiply the amount you have in dollars by the exchange rate. To exchange $25.00:
>
> $25.00 × 0.731608 = 18.29 euros.
>
> If the current exchange rate for the Mexican peso is 13.1, how many pesos will you get if you exchange $25.00?
>
> The exchange rate changes constantly and it is different for each currency around the world. You can find the current exchange rates online at sites such as x-rates.com.

What's the Chance of That?

You can use probability—the chance that something will happen—to measure how likely it is that a particular outcome will occur. The formula for probability is the number of times the particular outcome occurs divided by the total number of possible outcomes.

For example, a piano teacher can assess the probability of a particular student being late for class by dividing the number of times the student has been late by the total number of classes the student has had. If she has been late three times out of nine, the probability is 3 ÷ 9, or .33, which is the same as 33%. So, there is a 33% chance the student will be late.

How often do you consider probability in everyday life?

An event that is certain to happen has a probability of 1—or 100%. If your birthday is March 10, and today is March 9, the probability that tomorrow is your birthday is 100%. An event that cannot happen has a probability of 0—or 0%. If your birthday is March 10, and today is April 6, the probability that tomorrow is your birthday is zero.

How is probability used at work? Almost every business uses probability to assess risk. Insurance agencies assess the risk of loss. Manufacturers assess the risk of product failure; farmers use probability to determine whether it will rain; military leaders use probability to determine whether a campaign will succeed or fail; school counselors use probability to determine if a student is at risk of dropping out; and parole boards use probability to determine if an inmate will reoffend if released.

Career Profile: Actuary

Finance Career Cluster

What are the benefits and drawbacks of working as an actuary?

Job Summary

An actuary is a business professional who analyzes the financial consequences of risk. For example, an actuary might analyze data to estimate the probability of an event such as death or illness. Actuaries also determine the potential cost of such an event and help companies set policies to minimize the risk.

Most actuaries work for insurance companies, but some work in other businesses, including the government, hospitals, banks, and investment firms. They must have a strong foundation in mathematics, statistics, and general business and should know how to use computer programs such as spreadsheets and databases. Actuaries must have a bachelor's degree and are required to pass a series of exams in order to become certified.

Use the Internet, a library, or career center to research the responsibilities, education and training requirements, and salary range of an actuary. Write a **cover letter** to a potential employer applying for a position as an actuary. (A cover letter is a letter explaining your interest and qualifications for a job that you send along with a resume.)

21st Century Learning

Olivia wants to be a pediatric nurse. She knows she needs strong math skills to be accepted into a training program. She works hard in math class. She sits in the front row and takes notes. She does all her homework, every day. She studies for tests and quizzes. But she has never received a grade higher than a C.

Olivia's friends think she should consider a career that does not require as much math. Olivia loves children and thinks she would be an excellent pediatric nurse. She volunteers at the hospital and sees how important pediatric nursing is. She cannot think of another career that would be as rewarding. Her mother tells her to keep trying, but Olivia is very discouraged. She doesn't even want to go to math class anymore.

What should Olivia do? Use 21st Century Skills such as problem solving, critical thinking, communications, and decision making to help Olivia identify the challenges she is facing and to find solutions. Write an ending to the story and read it to the class, or form small groups and present it as a skit.

Analyzing Charts

A **chart**—also called a **graph**—is a picture that displays data. You use charts to compare data, identify patterns, or show trends over time. For example, an entrepreneur might use a pie chart to compare amounts provided by different investors.

A doctor might use a line chart to show changes in a patient's temperature over time.

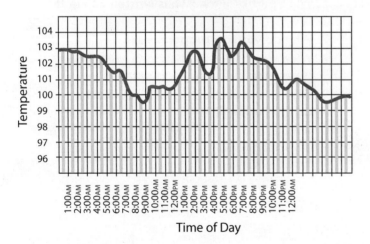

Most people respond better to visual information than to numbers. Charts help people understand data quickly because they can see what the numbers mean. For that reason, charts are useful in any career where you use data to make a point or to communicate.

Jobs That Use Charts

If you can create and interpret charts, you will be an asset in any career. Health care professionals use charts to track patient progress. Business managers use charts to display financial information. Human resources managers use charts to track employee data such as salary and staffing levels. Zookeepers use charts to schedule feedings, warehouse managers use charts to monitor inventory, and wholesalers use charts to track sales.

Types of Charts

To create an effective chart, you must select the appropriate chart type. For example, when you want to compare parts of a whole, such as sales by salesperson relative to total sales, you use a pie chart. When you want to show trends over time, such as how total sales change from month to month, you use a line chart. Here are some common chart types.

- ✔ A column chart shows data changes over a period of time or illustrates comparisons among items.
- ✔ A bar chart illustrates comparisons among individual items.
- ✔ A line chart shows trends over time.
- ✔ A circle—or pie—chart shows the relationship of parts to a whole.
- ✔ A scatter chart compares pairs of values.
- ✔ An area chart displays the magnitude of change over time.

Case Study

Viet is very creative. He has artistic talent for painting, drawing, and design. He plans to study art and design in college, and to pursue a career as a graphic artist.

Although Viet has always done well in math, he does not enjoy math class. He does not think he needs math skills to achieve his career goals. To meet his school's graduation requirements, he enrolls in a low-level math class. He puts in very little effort, preferring to spend his time working on his art. At the end of the year, he has all A's on his report card except for a B– in math.

- Do you think Viet made the right choice by taking a low-level math class?
- What problems might Viet encounter because of his decision?
- Do you agree that Viet does not need math skills to achieve his career goals?

Answer It!

1. List the four basic mathematical operations.
2. What is addition?
3. What is the opposite of addition?
4. What do you call the answer to a multiplication problem?
5. What do you call the answer to a division problem?
6. What part of a fraction is written above the fraction bar?
7. Why do you use proportions?
8. What type of math deals with collecting, organizing, and analyzing data?
9. What is the probability of an event that is certain to happen?
10. What is the probability of an event that can never happen?

Web Extra

There are many Web sites that provide enrichment math activities for students or let you ask math-related questions. For example, mathforum.org provides information about many math topics, has a question and answer section, and lists activities illustrating many math concepts. Use the Internet to find a math activity you could do at home or in school.

Basic Math Skills ■ Chapter 10

Write Now

What would you say to a friend who was thinking of skipping math class? Write a letter explaining why skipping math might have a long-term effect on his or her career and life goals.

COLLEGE-READY PRACTICES

Most people recognize that math is an important component of careers in fields such as finance, engineering, and technology. College-ready individuals know having basic math skills is important for all careers.

Request a meeting with your math teacher or school counselor to discuss how you can improve or enhance your math education. Discuss the classes and electives that are available and plot a "math career" that outlines the path you will take to graduate with a solid foundation of math skills.

Career-Ready Practices

As you have learned, math skills are important for just about any career you choose. Career-ready individuals use the knowledge and skills they have acquired in school to be more productive on the job. They apply what they have learned to solving problems and making decisions.

Form teams of three or four. Select a job in one of the career clusters. Think of a situation in which a person in that job would need to use math. Write a math problem for that situation. Exchange your problem with a problem developed by another team. Work together to come up with the solution.

Career Portfolio

Create a math academic plan for yourself, mapping out the courses you will take from now until graduation and the skills you will develop. Explain how each course will help you achieve your career goals. Add the plan to your career portfolio.

11 Technology and Your Career

Skills in This Chapter . . .

- **Recognizing the Impact of Technology**
- **High Technology in Industry**
- **Using Technology in Your Career**

GETTING STARTED

We all use technology, but can you tell which kinds of technology might be useful in your career? What makes a certain technology useful? Does a cell phone make communication easier? Does a computer make you more productive? How do you know which technology to use?

Technology is the use of resources such as scientific knowledge and methods to make our lives easier or better. This includes devices such as smartphones that allow you to send e-mail messages, documents, and pictures via wireless networks and video games that allow players to enjoy the experience of virtual reality, a computer-generated simulation of the world. But there is a key difference between technologies that make our lives and jobs easier and those that make our lives and jobs more fun. Those that make our jobs easier, and therefore allow us to produce more products, are likely to also be the most important technologies of the 21st Century. They might have an impact on you one day by expanding the range of possible careers available.

➤ Take out a piece of paper and fold it in half. In the top half, write five careers. In the bottom half, list one type of technology that would make you better at each career. As a class, discuss your choices, and consider the various ways technology makes it easier to accomplish everyday tasks in your career.

191

Recognizing the Impact of Technology

What would you do without your cell phone or the Internet? Technology makes our lives easier, and as soon as a new technology is introduced, it becomes hard to imagine life without it. For instance, before the Wright brothers invented an airplane, most people never even thought about traveling to far off countries. Now, companies have offices around the globe, people travel for business, education, and vacation, and the world is a much smaller place.

Remember: Although technology can be very useful, it cannot do your work for you. The most it can do is make it easier for *you* to do your work. Sometimes this means you can get your work done more quickly. Sometimes this means you can do more work or do your work more efficiently.

How Technology Impacts Career Options

Before the high-tech revolution that began in the 1980s, deciding on a career and finding a job required a lot of manual labor. In the past, finding a career usually meant going to the library to research various industries or spending a lot of time with a career counselor. Job seekers had few options for finding specific jobs other than reading the want ads in the newspaper. After identifying a potential job, the job seeker would usually have had to type a resume on a typewriter—retyping the whole page if he or she misspelled a word. Communications between the applicant and the employer usually required letters sent by the U.S. Postal Service, which took days or longer.

But thanks in large part to technology, finding a career and working at a job today are much different.

- You can use computer programs to help you determine which careers you might like or might be good at.
- You can use the Internet to research careers with a specific employer quickly and efficiently, because nearly every company has its own Web site.
- You can find jobs in your area—or anywhere in the world—using job search Web sites.

Videoconferencing uses technology to make long-distance meetings less expensive and more convenient for businesses. Can you think of other ways technology has changed business in recent years?

- You can use business social networking sites to get the word out that you are looking for a job, and use online contact lists to network with people who might be able to steer you to a new job.
- You can fill out application forms online or send your resume by e-mail, saving time and money.
- If you have a phone, a computer, and Internet service, you can become a **telecommuter** who works from home. This means that some geographically distant career choices could be possible options for you without having to relocate or travel.

High-tech devices such as smart phones and computers and the Internet have expanded the range of possible careers for you, and thanks to them, the U.S. Bureau of Labor Statistics estimates that you will likely have as many as three different careers in your working life.

New Careers Driven by Technology

Not only has technology changed the way you find a career and do your work, it has also launched many new careers. Some that have emerged in the last decade include:

- **App developer**, who designs computer programs that can run on mobile devices.
- **Data scientist**, who uses statistics and computer programming to analyze huge amounts of data and identify useful information.
- **Social media manager**, who uses social networks to identify customers, generate sales, and build awareness.
- **Nursing informatics specialist**, who oversees the integration of data, information, and medical knowledge to help patients and their health care providers make decisions.

Technology Trends in the Workplace

You will encounter new trends in the workplace when you start your career. Typically, new technologies are used in the workplace when they save time and money compared with traditional ways of working. Just as word processing programs have replaced typewriters, so new technology trends have begun to replace more time-consuming or inefficient ways of working in many areas of business. Consider some of these cutting-edge technologies and how they speed up our work.

- **Videoconferencing** enables people to meet "face-to-face" without having to travel any distance at all. By setting up two video cameras and establishing an Internet link, employees at distant offices are able to see and speak to each other without spending time and money on traveling.
- **Collaborative software** allows more than one person to work on a document at one time. For example, more than one person at a time, working at different computers in different locations, can work on the same word processing document, spreadsheet, or presentation. Collaboration software increases productivity. It also saves time, money, and trees (paper).

Does the Internet Help or Hurt My Career Search?

Technology can be an incredibly useful tool, but it can also slow you down if you are unfocused. It is your responsibility to use technology wisely, to recognize when it becomes a distraction, and to avoid letting it waste your time.

- 👎 You e-mail a company your resume when the want ad states that the company will not accept e-mailed applications.
- 👎 You instant message with your friends when you are supposed to be doing career research.
- 👎 You play Internet games and visit social networking sites instead of industry and company Web sites.
- 👍 You find salary and employment data for careers you are considering.
- 👍 You e-mail your resume to potential employers that accept online applications.
- 👍 You use a job placement Web site to find and research career options.

What are other ways that the Internet can both help and hurt your career search? List two new examples of each.

NUMBERS GAME

Graphs (or charts) let you display numerical information in a visual way that is easy to understand. A graph shows the relationship between different types of data, such as number of items sold each month or trends in sales over a year's time. A graph has a vertical line, the Y-axis, which shows values such as prices, and a horizontal line, the X-axis, which shows categories such as years, months, or departments.

One of the easiest types of graphs to create is a **bar graph**, which is useful for showing comparisons among individual items. In a bar graph, categories appear along the vertical axis and values display on the horizontal axis. This is a good graph to use if your category labels are long or you want to show how items compare over time.

The following bar graph makes it easy to compare the prices of three notebook computers.

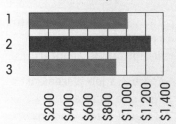

Make a bar graph comparing the costs of the following three printers:

Printer 1: $129.99

Printer 2: $85.49

Printer 3: $149.99

Which printer would you buy? Is cost the only consideration?

- **Wi-Fi** is a technology that allows you to connect to the Internet without having to be connected to any wires or cords. This means that you can be online in a coffee shop, airport terminal, or even in a public park, so long as there is a Wi-Fi "hotspot," or source. Like a radio, Wi-Fi broadcasts a signal through the air to your computer or other device. Wi-Fi allows your office to be virtually anywhere you are with your computer.

- Not all repetitive tasks must be done by people anymore, thanks to **automation software**. Skilled programmers can create scripts that use the power of a computer to complete tasks that are repetitive, time-consuming, and boring. For instance, most computers can now perform data backups, software updates, and virus scans automatically. Another example is software that lets a computer calculate large data-sets using only a few bits of user input.

Money Madne$$

Your parents pay the monthly charge for your no-contract prepaid cell phone, but you pay for your data usage. The plan charges you $19.99 per month for unlimited talk and text. Data costs $3 a month extra for 50 megabytes (MB). If you download 22 songs at 3.5MB per song and 18 photos at 5MB per image, how many megabytes will you use? How much will you pay for the data services?

High Technology in Industry

Have you ever seen a printing press or a steam engine? Do you think of them as old or modern? High-tech or low-tech? In their day, these inventions were certainly fine examples of high technology. Yet now, high technology, what you might call "high-tech," is a term for only the most advanced or specialized technological developments. Examples of high-tech products range from tiny microchips inside computers to unmanned robots that roam the surface of mars. High technology is all about "pushing the envelope," or developing technologies that make things that might seem impossible, possible.

Many industries have seen significant growth because of the development of new technologies. Retail has grown thanks to online shopping. Health care has improved thanks to diagnostic systems. Agriculture has become more efficient due to irrigation systems.

Technology and Your Career — Chapter 11

Technology has reduced the drudgery and unsafe working conditions of many manufacturing jobs. Instead of a person tightening one bolt on each piece of machinery that goes by or breathing paint fumes for eight hours a day, robots can perform these tasks. But at a cost: there are fewer people employed in manufacturing these days because technological advances have eliminated their jobs.

Technology does more than just help an industry grow, it also affects careers. In some cases, high technologies create jobs or improve conditions for workers. In other cases, high technology can actually take the place of a worker or make it easier for one worker to do the job of several, which can result in fewer jobs in the industry.

Technology in Manufacturing

Have you ever wondered how your parents' car was built? Most likely, it was put together with the help of a **hydraulic arm**, a machine that creates power by pumping water through a confined space. This allows a kind of robot to lift objects much heavier than any person could lift on his or her own. And like a computer, a hydraulic arm is an example of high technology.

The hydraulic arms of a robot are just one example of high technology in manufacturing. **Manufacturing** is the making of things on a large scale using machinery. This means that a factory that produces toys is an example of manufacturing. A plant that produces microchips for computers is also a manufacturer. Usually, the machinery used in manufacturing is a good example of technology. This can be heavy machinery, like the hydraulic arm used in making cars.

This car assembly plant uses advanced technology to produce goods. How can learning about computers help you pursue a career in manufacturing?

High-Tech Careers

Do you know someone who works in one of these high-tech jobs?

✔ Computer chip manufacturers work in completely dust-free environments where they wear special white full-body suits and masks.

✔ In the finance industry, programmers write complex computer puter *algorithms*—computer programs made especially for problem solving—that help bankers and stockbrokers make important financial decisions.

✔ In the alternative energy industry, technicians climb hundreds of feet in the air to maintain and fix the machinery of giant wind turbines that generate energy for nearby cities or towns.

✔ In the entertainment industry, digital graphic artists use specialized computer programs to create animated features or enhance live-action movies with CGI, or "computer generated images."

✔ In medicine, surgeons use special joysticks to control miniature robotic fingers with special surgical tools to operate on sensitive areas of the body that are out of reach of human hands.

Can you think of other careers that make use of high technology?

It can also be small and highly specialized technology, like the tools used to make computer chips. Technology in manufacturing comes in a large range of sizes and applications:

- Giant cranes make modern skyscrapers.
- Robotic hydraulic arms manufacture cars.
- Computers control automated assembly lines.
- Just-in-time computerized inventory control means that companies don't have to waste time and money on inventory that just sits around in the warehouse. Instead, these systems order inventory at the exact right time and only for the exact right amount.
- High-powered lasers etch and cut giant slabs of metal used in oil tankers.
- Microscopic computer chips are configured using microscopic tools.

Keep this in mind: Even if a computer controls an assembly line, or if a robotic hydraulic arm can put together a car, none of these examples of technology in manufacturing work without skilled people controlling, programming, troubleshooting, and operating them.

Myth Computers are replacing people in the workplace.

Truth The workplace will always need skilled workers; computers are only as powerful as the people using them.

Technology in E-Commerce

"Electronic commerce," or **e-commerce**, is the buying and selling of goods or services over the Internet. Some popular examples of e-commerce include: online banking, online retail sales, online travel resources, online ticket auctions, and so on. For instance, it used to be that in order to buy shoes, you needed to go into a store and try them on. Nowadays, you can buy shoes online and return them overnight if they do not fit. Likewise, buying tickets, books, or movies was only possible from inside the stores that sold them. But now, you can buy all of those things from your living room. And banking is never easier than when it is done on the Internet.

Many technological breakthroughs had to happen before widespread e-commerce became possible. Mathematicians invented encryption software to keep our credit card numbers safe. Systems engineers invented computerized inventory systems to keep track of products and orders. Even the hand-held scanner that your local package delivery employee uses is important for e-commerce because it allows businesses to instantly see that you have received your products.

How does e-commerce affect careers? Unfortunately, the convenience of online shopping has caused some small businesses to close. As a result, there may be fewer career opportunities in retail. However, there may be increased opportunities at online retail businesses.

Careers in Information Technology (IT) Systems

Information technology systems, or **IT**, is a term used to describe computer and telecommunication systems that transmit and store information. Almost every business has an IT department, responsible for managing the data and computer systems that keep the company running. For example, the IT department makes sure that the human resources department has the tools it needs to keep track of employees, benefits, and training. It makes sure that the accounts receivable department can track and record payments.

What are some requirements and rewards of a career as a computer service technician?

A career working in an IT department would involve setting up, maintaining, and repairing the computer systems of an organization. Since virtually every large company uses some kind of computer system, there is great demand for skilled IT technicians. IT departments are usually separate from the rest of a company, since their employees are specifically dedicated to maintaining a computer **infrastructure**, which is the computer systems, networking, and other devices. Because many computer systems are similar, the skills you might develop in the IT department of one company are the same skills you would need in another. This means that if you decide on a career in IT, you can potentially work for many different companies. A good word for this is **fungible**, which means being able to trade one item or quality for another, interchangeably—as in, "your skills as an IT technician are fungible."

As an IT technician, you might:

- Resolve a problem with a company's e-mail system.
- Configure a wireless Internet network for thousands of employees.
- Recommend the purchase of certain computer hardware, such as servers, printers, or consoles.
- Do regular maintenance and backup on servers.
- Install programs, or write programs for specific uses.
- Maintain a company Web site.
- Deal with security issues, such as attempts to hack into company files and viruses.
- Train company personnel on IT issues.

IT technicians are usually highly respected, because they have knowledge of complex technologies. Most other members of a company generally do not have this knowledge, so the importance of having skilled IT technicians in any company makes a career in IT desirable.

Does Everyone Use Technology?

Even if you don't have a cell phone or computer, you use technology if you ride a bus, drive a car, or even write with a ballpoint pen. What if you don't want to use technology?

★ *You would have to forgo using any indoor plumbing or electricity in your home.*

★ *You could only go places that are within walking distance—unless you have a horse.*

★ *You could communicate with people only if you are face to face.*

★ *You could sit and tell stories with friends and family for entertainment.*

Can you think of other ways you could remove technology from your life?

Career Profile: Information Technology Manager

Information Technology Career Cluster

What skills and interests do you think would help someone succeed in a career as an information technology manager?

Job Summary

Working as an IT manager requires knowledge of both technology and business practices. The duties of an IT manager will vary depending on the work environment, company size, experience level, and other factors.

However, you can expect to deal with a wide range of issues. You must have excellent people skills, an understanding of business management, and technical expertise in some aspects of computer infrastructure. All of these skills are equally important. Technical expertise is often gained through previous professional experience, especially working in a related information technology position.

Use the Internet, a library, or career center to research the responsibilities, education and training requirements, and salary range of an information technology manager. Prepare a presentation about the career.

Career Trend

Careers in IT are becoming more desirable as we rely on technology in the workplace more. In economic terms, this means that a career in IT may pay more now than ever before.

Technology in Agriculture

Believe it or not, technology has had a big impact on our food over the past fifty years. Prior to the 1950s, American farms were usually small, family-owned operations that could not output a large number of crops per year. However, important advances in agricultural technologies have enabled large-scale farm practices. Technologies such as modern fertilizers, pesticides, genetically engineered seeds, and farm equipment have shaped what we eat. The food available to us in grocery stores depends on what our farmers produce. And what our farmers produce is affected by the kinds of technologies available to them. Here are some examples of how farm technology has affected the foods you eat:

Technology affects our food supply in many ways. How would a farmer plant and harvest crops without technology?

- Science has allowed food manufacturers to process corn into a cheap sweetener, known as high-fructose corn syrup, which is used in thousands of different products, from soft drinks to soups.

- Large-scale harvesting of wheat is possible with modern tractors, which can turn thousands of pounds of wheat into usable crops per day. This has increased the American output of wheat by a huge amount, making America one of the biggest wheat producers in the world. This means that the foods you eat that contain wheat, like bread and cereal, are all very inexpensive.

- Industrial meat production has increased the amount of beef, pork, and poultry we can produce per day. This makes meat less costly.

Tech Connect

Technology has helped make food supplies around the world safer. Refrigeration keeps food fresher longer. Water purification systems kill harmful germs and insects. **Irradiation** exposes fruit and other food products to a source of radiation that kills bacteria and insects.

Technology has also caused safety problems. Chemicals used in farming to promote growth or kill insects can get into the water supply or into the meat and plants we eat. Bacteria such as salmonella can transfer from farms that produce meat, eggs, and poultry into the water supply and contaminate plants that we eat.

Use the Internet or your library to research technologies that protect or harm the food supply. Pick one, and then create a presentation about it to show your class.

Using Technology in Your Career

Can you think of a job that does not require you to use technology? Phone systems, computers, printers, copy machines, and fax machines are standard equipment in almost every office. Even entry-level positions that require no previous work experience often require some use of technology. Employees at every level, and even job applicants, may be expected to make oral presentations using appropriate technology, such as presentation software, projectors, and monitors. Your best bet in preparing for almost any career is to become familiar with technology so that you can become an expert when the time comes.

Technology for the Entrepreneur

For many years, entrepreneurs have started successful businesses—and sometimes kept them running for a long time—from inside their own homes. It has become even easier as technology has improved and come down in cost. Home computers, all-in-one printers that scan, copy, and fax as well as print, software for scheduling and data management, Internet access, and mobile phones allow entrepreneurs to function as if they had an office and staff—even when they are home alone working at the kitchen table. What technologies do you think an entrepreneur might find most useful when starting a business?

Laptops That Make a Difference

One Laptop per Child is a nonprofit organization with a goal of creating educational opportunities for the world's poorest children by giving them laptop computers they can use to access and share information.

To achieve its goal, the organization developed a low-cost, rugged laptop called the XO. It has built-in wireless communications and a screen that can be read even in bright sunlight—so it can be used outdoors. The battery lasts about six hours, or 24 hours for reading only, and features 2,000 recharge cycles, which is about four times more cycles than the typical laptop.

Use the Internet to learn more about One Laptop per Child and the XO. Why does the XO need to be rugged? How can providing children with laptops help solve problems associated with poverty and the lack of education? Can you think of other solutions that might work? Use presentation software to make a short presentation about the topic. Keep a bibliography listing the sources you consult in your research. Then, share your presentation with the class.

Career Tips

You can make your Internet searches more powerful by using what's called **Boolean operators**, special words or quotes that make your searches more specific. For instance, putting quotes around a phrase searches for that phrase exactly, instead of just searching for sites that contain at least one of the words from the phrase.

Transferrable Computer Skills

Do you type? Have you ever used the Internet to search for information? If so, then you already possess several basic, transferable skills that you will need to interact with technology in your career. And the best part is, since technology is everywhere in the world of work, if you have one or more of these basic skills, you can take them wherever you go.

What are basic skills you need when you start almost any career? You need to be able to:

- Run and use various computer programs and take measures to secure your data.
- Type on a computer keyboard without making many mistakes.
- Access the Internet and move from one Web site to another.
- Use a search engine to do basic Internet research.
- Log on to work groups and use collaboration technology.
- Write and send e-mail.

Some of these skills might seem easy to you. Others might sound difficult or new. Either way, these are all skills worth having. The important thing to know is that you might be asked to use these or other basic, transferrable skills at some point in your career. Even if you do not know how to do some of these things, you can learn. Being flexible, open-minded, and willing to learn new skills is the key to success.

CAREER COUNSEL

21st Century Learning

Jessica is a recent retiree who worked her career in a textile manufacturing plant. She has never used a computer, but since most of her extended family uses social networking sites, she would like to be able to connect with them over the Internet. There are computers at her local library, but she doesn't know how to use them. Her local community center offers classes, but she is embarrassed to admit she needs lessons.

Jessica feels that it's too late to learn and that she will be doomed to computer illiteracy forever.

What steps can Jessica take to become comfortable using a computer? Should she try to learn, or is it too late? How can she overcome her fear and embarrassment? Use your 21st Century Skills such as setting goals, managing resources, and recognizing the impact of technology to help Jessica. Write an ending to the story. Read it to the class, or form a small group and present it as a skit.

Using Communications Devices

Do you text? Can you send e-mail? Do you have a phone that can do either of these things? If yes, you are like millions of Americans who own a smartphone. If not, chances are, you will one day soon.

Cell phones make it possible to communicate with others from almost anywhere in the world. At the same time cell phones were becoming an important means of communication, **personal digital assistants (PDAs)** were developed to help people track their appointments and business contacts. Gradually, cell phones added more features like those of PDAs, and PDAs added wireless communications and calling capability. The combination of features from both types of devices resulted in smartphones.

Why is it important for construction workers to have cell phones?

Besides being able to communicate with your parents and friends, what is the value of having one of these devices?

- Employees with smartphones can be reached by e-mail, the communications medium of choice, at any time of day. In fact, many companies give their employees a smartphone for free. But with this free device comes the understanding that employees are reachable for work even when they are at home. Therefore, they can work longer hours.

- "Texting," sending small written messages from one cell phone to another, can be an efficient way to communicate. Sending a text instead of placing a call can save both the caller and the recipient time. And in some businesses, getting information from one person to another quickly is essential to making money.

- Employees can access the Internet to look up information when they are not at their office.

A **smartphone** is a cell phone that is also a miniature computer. What makes a smartphone "smart"? How can it help you in your career?

Using Computer Applications

A **computer application**, or **program**, is a tool you use to perform specific tasks on a computer. Employers look for employees who have experience using computer applications. Some of the most common and useful include:

- Modern word processors allow you to quickly edit and rearrange your written thoughts. If you are unhappy about a certain paragraph, you can move it somewhere else in the document or delete it entirely. Word processors also automatically check your spelling, so there are fewer mistakes in your work.
- Presentation software can produce slide shows that are very helpful for presenting certain types of information. The slide shows can be displayed on a monitor, over the Internet, or by connecting the computer to a projector.
- Spreadsheets can take tens of millions of pieces of data, usually in the form of numbers, and organize them quickly. This is why many companies use spreadsheet software to contain and manipulate their data.
- Web browsers are becoming so important to modern computing that some computers are nothing but a Web browser! This means that the files normally on a computer—your documents, photos, music, etc.—are all stored online and accessed through the Internet. This type of computing is called **cloud computing**.
- A database management program helps you organize and manipulate data.
- A Web design program helps you create Web pages.
- A graphic design program lets you create sophisticated graphics or repair or manipulate photographs.
- A page layout program lets you place type and graphics on pages for publications.

You can use a spreadsheet to store data, create graphs, and perform calculations. How many careers can you think of where spreadsheet skills might be necessary?

Understanding Mobile Apps Is Important, Too!

*An **app** is a computer program designed for use on a mobile device. There are literally millions of apps available for download. They enable you to do many things directly from your smartphone or other mobile device, such as play games, transfer bank funds, or purchase tickets to a concert.*

Many companies have found apps to be useful tools for communicating with customers and promoting their products and services. Employees, too, use apps to schedule appointments, track inventory, and manage information.

What type of apps do you think an employee could use in any career?

Case Study

Bethany really wanted a new bicycle. But her allowance was too small to pay for it, and her parents did not want to buy it for her. So she decided to try to get a part-time job after school. She felt she had some good job skills, including familiarity with using a computer, phone skills, and typing skills.

Bethany's parents were too busy at work to take her around to any local shops that might be interested in hiring her part-time. She felt unsure walking around by herself. A friend recommended that Bethany use the Internet to find out more information before going to look for a part-time job, but Bethany had never done a job search online before. She had heard of some Web sites where people got jobs, but never used one herself.

- How can Bethany use technology to help her find work?
- What Internet sites could help Bethany find part-time, after-school work?
- What kind of job might Bethany look for?
- How can Bethany find out how much money she might expect her job to pay her?

Answer It!

1. What is technology?
2. How has technology changed the way your generation might plan for and settle on a career?
3. According to the U.S. Bureau of Labor Statistics, how many careers will a person from your generation probably have in a lifetime?
4. How has videoconferencing affected the way companies conduct meetings?
5. What are the three reasons any business would use any type of technology?
6. Name at least three examples of manufacturing.
7. Why is it that technology will never replace people?
8. What would a career in an IT department involve?
9. Why do many companies give their employees smartphones for free?
10. What reason do companies have for using spreadsheet software?

Web Extra

Many informational Web sites provide reliable and accurate information for job seekers, including monster.com, careerbuilder.com, and payscale.com.

Use the Internet to locate some sites that you think might be useful for your peers. Using a spreadsheet application, make a directory of the sites. Include the Web site address, a brief description, and whether the site is free or has a fee. Post the directory on your school Web site or in your school career center or library.

Write Now

Applying for jobs online can be more convenient than doing it in person. You can mail hundreds of resumes and cover letters with the click of a button. This would take far more work when done by hand. But, with this convenience comes a downside. Just as easy as it is to send e-mail to a potential employer, it is also just as easy for a potential employer to ignore you.

Write an essay about the benefits and drawbacks of applying for jobs over the Internet. Include information about why the Internet can make it both easier and harder to get a job. Then, think of ways you can avoid being ignored and how you can make your job application stand out. Talk with people who have experience—your parents, siblings, a teacher, or librarian. Use examples that you find online, if possible. In your essay, provide examples from the Career Portfolio that you are working on.

COLLEGE-READY PRACTICES

Most people think that speaking in front of an audience is scary. College-ready individuals know that being prepared to deliver an oral presentation live shows you are responsible and capable. Being prepared includes knowing how to use technology appropriately to support your presentation. Depending on the situation, you may have to make a marketing presentation in front of a customer, conduct a training seminar over the Internet, or explain a process while on a videoconference.

Select a career in a cluster that interests you. Conduct research on how technology is used in that career, or how technology has impacted that career. Prepare an oral presentation about your findings. Use appropriate technology to prepare and deliver your presentation.

Career-Ready Practices

Information technology has a significant impact on consumers. Advertising on television and the Internet influences our buying decisions, access to online shopping affects how and where we shop, and access to information helps us compare and evaluate products and stores before we shop. To remain competitive and be successful, businesses have had to adjust to these changing consumer buying habits. For example, many companies now have Web sites—in addition to bricks-and-mortar locations—where consumers can buy products and services.

Career-ready individuals recognize that using and maximizing technology can enhance workplace productivity and give their employer a competitive edge. They equip themselves with basic technology skills and embrace new skills that will help them accomplish their tasks and meet the needs of their customers.

Imagine you work in the IT department of a large corporation. Working in teams of four or five, research the influence technology has on products and services, and how it impacts consumer decision making. Then explore how these changes affect the use of technology in the workplace. Using a presentation program, organize the information into a presentation that you can deliver to the class.

Career Portfolio

Many computer applications are common in business these days. Mostly, this software is related to everyday productivity. This includes software that helps people write, organize data, and present information. Knowing how to use these types of software is usually essential for even entry-level jobs. Being an expert at one or more of these applications is almost always a good skill. And since the software is used nearly everywhere, knowing how to use it is a transferable skill.

Your Career Portfolio is a place to showcase the basic, transferrable skills that you possess. For instance, have you used word processing software to create letters or reports or spreadsheet software for analyzing numerical information? Make a list for your Career Portfolio. Include any of these pieces of productivity software (or others) that you have used. Briefly describe your familiarity with each as "Basic," "Intermediate," or "Expert." Include samples of your work.

Career Development

Part III

In This Part . . .

Chapter 12 ■ Career Planning

Chapter 13 ■ Managing a Job Search

Chapter 14 ■ Getting Started in Your Career

Chapter 15 ■ Being Productive in Your Career

Chapter 16 ■ Living a Healthy and Balanced Life

12 Career Planning

Skills in This Chapter . . .

- **Planning a Career**
- **Managing a Career Self-Assessment**
- **Analyzing Career Planning Resources**

GETTING STARTED

Planning for a career is a job in itself. It takes time, energy, and careful management. So why do it? Putting effort into career planning can help you set realistic and attainable goals for education. It can help you identify your strengths and weaknesses, so you focus your resources on finding a career that you will enjoy. Spending time exploring career opportunities can also be fun and exciting, because you experience new situations and activities. Career planning allows you to gather the information you need to make healthy decisions and take control of your future.

➤ Working in small groups, take three minutes to make a list of the most exciting and interesting careers you can think of. Next to each one, write whether or not it is a career you might realistically be able to attain. Why or why not?

But I Haven't Chosen a Career!

How can you use career-related goals to help identify a career that matches your interests and abilities? What if you are unsure about the career path you will follow?

★ Set goals for academic achievement.

★ Set goals for volunteering.

★ Set goals for researching different types of careers.

★ Set goals for building skills and abilities.

★ Set goals for earning and saving money.

Can you think of career-related goals you can set right now? Make a list and discuss it with the class.

How can planning your career help you take control of your future?

Planning a Career

You might be wondering why you should care about planning your career now. You might be thinking that you have time for that in the future. In truth, the sooner you start thinking about your future options, the easier it will be to make those options attainable.

Planning a career involves assessing your career-related skills and interests and then setting goals. It also involves exploring the career clusters, pathways, and jobs so you recognize the possibilities. Finally, it involves analyzing career resources so you can recognize opportunities that can help you achieve your goals.

Steps for Planning a Career

A career, if it is the right one for you, can be very rewarding. The following steps outline the basic process of planning a career.

1. *Assess your interests, skills, values, and abilities.* What are your talents and interests? What do you value? What rewards do you want from a career?

2. *Research the big picture.* Which job areas are the fastest-growing in your area? What kind of education and training will workers need for these jobs?

3. *Explore career options.* What types of careers are there? What are the requirements for different careers? What are the day-to-day responsibilities?

4. *Create a career plan.* What are your career goals? How will you get the education and training you need to achieve those goals? How long will it take?

5. *Get hired.* Are you employable? Are there opportunities for you to work right now that might build your skills and provide you with valuable experience?

6. *Be successful.* Do you have the skills you need to keep up with your work responsibilities? Do you have the personal qualities other people admire and respect? Do you know how to make decisions, set goals, and solve problems?

7. *Continue planning.* Are you satisfied in your current position? Are there new goals you might want to achieve? How can you continue to grow and develop new skills and experience?

Mapping out a Career Plan

A **career plan** is a map that shows you the way to achieve your career goals. You use it to identify your skills and interests, to find a career or career cluster that suits your skills and interests, and to determine the type of education and training you will need to succeed.

A career plan is similar to an academic plan. In fact, you can't really have one without the other.

- Your academic plan should include goals for developing career-related skills and abilities.
- Your career plan should include goals for achieving academic requirements.

As with an academic plan, your career plan should be specific, realistic, and attainable. It should include a long-range career goal as well as a series of short-term milestones that will help you get there. Every once in a while, you should revisit your career plan and make adjustments or corrections. Your goals might change as you develop new interests and abilities, so your career plan will change, too.

What to Include in a Career Plan

There are different ways to develop a career plan. Your teacher or counselor may have a form for you to use. It may be part of your academic plan or a separate document. Most career plans include the following information:

- A career self-assessment, including a prioritized list of skills, interests, and abilities
- An analysis of careers or career clusters that match personal skills, interests, and abilities
- An academic plan aligned with a selected career or career cluster
- An analysis of the outlook for the selected career or career cluster
- An assessment of the financial resources required to achieve career goals

Myth Once you make a career plan, you are stuck with it for life.

Truth A career plan is flexible. Most people revise and refine their career plans many times in their lives. Your plan might change because your interests or abilities change, or you might discover new opportunities or resources.

Surveys show that high pay does not necessarily lead to job satisfaction. While it is important to consider salary when choosing a career, it is also important to look for a job that you enjoy.

I Hate This Job!

Problems can come from choosing a career without careful planning. Just because a career is right for someone else doesn't mean it is right for you. Here are some of the reasons a job might work out—or not.

- 👎 You don't have the skills or abilities.
- 👎 You think it is boring.
- 👎 You think it does not pay enough.
- 👍 You spend time looking for a career that matches your interests.
- 👍 You succeed because of your education.

What are some reasons people work in careers they do not enjoy? What steps can you take to find a career you do enjoy?

College Entrance Exams

In the United States, most colleges and universities consider a wide range of factors when assessing applicants, including scores from standardized tests such as the Scholastic Aptitude Test (SAT) and ACT. In other countries, standardized exams may be the only factor that determines whether a student is accepted.

In South Korea, for example, all students interested in attending a university must take a day-long national college entrance exam. That adds up to more than half a million students taking the same test at the same time. The results determine which school each student will be qualified to attend, which directly affects his or her ability to obtain a job.

Brazil also bases university admission solely on a student's score on a national standardized test. Of course, some countries make it much easier to achieve higher education. In Belgium, the only requirement for enrolling in a public university is a high school diploma.

Do you think it is fair to judge a college applicant on standardized test scores alone? Why do you think most American colleges and universities consider factors such as high school grades, teacher recommendations, and a personal essay? Use the Internet or your library to research the college or university entrance requirements in the United States and in another country. Write an essay or create a chart comparing the two systems.

NUMBERS GAME

The ability to add fractions is important in many careers. Chefs use fractions when developing recipes. Carpenters use fractions to measure lumber and other materials. All workers use fractions to measure time and money.

To add fractions that have the same denominator, simply add the numerators, leaving the denominator unchanged. So, $3/8 + 3/8 = 6/8$ (which can be simplified to $3/4$).

To add fractions that have different denominators, such as $3/8$ and $3/4$, you must rewrite the fractions using the **least common denominator** (LCD). The LCD is the smallest number that two different denominators can divide into evenly. In other words, it is a multiple of all the denominators.

To find the LCD, write down the multiples of one denominator, and then write down the multiples of the next denominator. (That means multiplying the denominator by one and writing down the product, multiplying it by two and writing down the product, multiplying by three, etc.) Then, circle the numbers that are on both lists, and pick the smallest.

So, the first step is to find the LCD for $3/8$ and $3/4$. List the denominators and circle the ones they have in common:

⑧, ⑯, 24, 32, …   4, ⑧, 12, ⑯, …

Next, rewrite the factions using the LCD, which in this case is 8 (multiply the numerator and the denominator by the same multiple):

$3/8 = 3/8$ $3/4 = 6/8$

Finally, add the fractions:

$3/8 + 6/8 = 9/8$, which can be written as the mixed number $1\,1/8$.

You do the math: A tailor has $3/4$ of a yard left on one bolt of fabric, and $2/3$ of a yard left on another bolt. How many total yards does he have?

Managing a Career Self-Assessment

Recall that a self-assessment process helps you identify the things you enjoy and do well. When you focus the self-assessment process on career-related skills, interests, values, and abilities, it is called a **career self-assessment**. A career self-assessment can help you choose a career path that will bring you satisfaction.

A career self-assessment can be hard, but rewarding. At this point, how do you identify your career-related abilities? How do you apply your values to a work environment? To perform a career self-assessment, you use a process to examine your strengths and weaknesses honestly and objectively.

Developing a Career Self-Assessment Worksheet

There are many ways to assess your career interests, skills, and abilities. One way is to use a six-step process to help you complete an objective career self-assessment. The result will be a worksheet that you can use to set academic and career goals. You can develop the worksheet on your own, but it may be easier if you work with an adult such as a teacher or counselor. The adult can help you be objective and honest about your skills, interests, and abilities.

1. List your two favorite school subjects. Usually—but not always—these are the subjects you do well in. They might be core subjects, such as science or social studies, or they may be electives, such as technology education or art.

2. List at least four specific skills you have acquired in your favorite subjects. For example, if you chose Spanish, your skills might include reading or speaking Spanish or understanding Spanish culture, or even cooking Spanish food. If you chose science, your skills might include understanding chemical properties, using a microscope, or valuing peer review.

3. List at least four achievements in your favorite subjects. Your achievements are the things you have earned based on your performance. They might include good grades, awards, acknowledgments, or even an assignment you are proud of. For example, for language arts, you might list a poem you wrote for class. For music, you might list a solo in the fall concert.

Thinking Critically About Yourself

It might be difficult to think critically about your strengths, interests, and abilities, but it is important if you are going to develop an honest career assessment.

✔ You might have trouble identifying your interests. You enjoy riding your bike. Is that an interest that relates to job skills?

✔ You might feel self-conscious talking about your strengths. You are embarrassed when people praise your love of animals or your patience with toddlers.

✔ You might not want to admit that you enjoy a subject such as math, because you worry that your peers will tease you.

Why is it important to work with someone objective when you develop a career self-assessment worksheet?

4. List at least four abilities—the things you do well or your talents. You can list specific abilities, such as playing guitar or repairing machines. You can also list qualities that help you succeed in many areas, such as being a good listener or having good time management skills. (Refer to the table on the following page for categories of abilities.)

5. List at least two interests. Again, you can list specific interests, such as working with animals or photography, or you can list interest areas. (Refer to the table on page 208 for descriptions of six common interest areas.)

6. List at least four work values.

Career Tips

There are many ways to conduct a career-self-assessment. Your teacher or counselor might use a different process or have a form for you to complete. You can try different methods to find one you are comfortable using.

21st Century Learning

Maria is a junior in high school. She plays on the school basketball team. She is a member of the school's chapter of Future Business Leaders of America. She does very well in school and takes advanced-level classes in math, science, and language arts.

Maria works part-time helping her father with his landscape business. Her father wants her to join the business after graduation, and he plans to give her the business when he is ready to retire. Maria enjoys landscaping because it is active, outdoor work, but she is not sure if it is the right career for her. She wonders if she will get bored or miss out on other opportunities if she joins her dad's business right after high school.

What can Maria do now to start planning for her career? What resources can she use to determine whether landscaping is right for her or if she should consider other career opportunities? Use 21st Century Skills such as problem solving, critical thinking, communications, and decision making to help Maria assess her options. Write an ending to the story and read it to the class, or form small groups and present it as a skit.

Identifying Your Abilities

Use the chart below to find the ability (or abilities) that describe you best and the corresponding careers.

Ability	Description	Related Careers
Artistic	If you have artistic ability you are creative and understand artistic ideas.	Fashion designer, graphic designer, interior decorator, or photographer
Clerical	If you have clerical ability you pay attention to detail and do precise, accurate work.	Administrative assistant, bookkeeper, medical, public safety dispatcher, or records technician
Interpersonal	People with this ability relate well to others. They are good communicators, friendly, polite, and adapt well to different situations. Most careers require strong interpersonal ability.	Customer service representative, flight attendant, home health aide, retail salesperson, or social worker
Language	If you have language ability you are an excellent reader, writer, and speaker. You know how to ask questions and listen to answers.	Editor, social worker, speech therapist, or translator
Leadership	People who have leadership ability are able to get others to work together toward a common goal. They also manage resources effectively.	Athletic coach, chef, educational administrator, firefighter, or lawyer
Manual	If you have manual ability you are skilled using your hands, fingers, and eyes, and are good at operating equipment and following directions. This area also includes athletic ability.	Animal caretaker, athlete, auto mechanic, carpenter, or conservation worker
Mathematical/ Numerical	If you have mathematical/numerical ability you are good at solving problems and using math skills.	Financial analyst, network administrator, pharmacist, or software engineer
Musical/ Dramatic	People who have musical ability have a keen ear for sound. People who have dramatic ability are skilled at acting.	Actor, director, music or drama educator, musician, or producer
Organizational	People with organizational ability are good at planning and managing tasks and maintaining records.	Air traffic controller, computer systems analyst, nurse, or travel agent
Persuasive	If you have persuasive ability you can influence others by sharing your own views and opinions.	Advertising executive, lawyer, psychiatrist, or retail salesperson
Scientific	If you have scientific ability you are good at research, logic, and critical thinking.	Agricultural inspector, chemical engineer, dental assistant, dietitian, or physical therapist
Social	People with social ability use logical thinking and effective communication to help others.	Home health aide, recreational worker, school counselor, or social worker
Visual	People with visual—or spatial—ability can see differences in size and shape and understand how objects fit together or come apart.	Architect, carpenter, dentist, graphic designer, landscaper, or mechanical engineer
Technical/ Mechanical	If you have technical/mechanical ability you understand how to set up and operate machines.	Airline pilot, auto mechanic, electrician, or master mechanic

Identifying Your Interests

Use the chart below to find your interest areas and the corresponding careers.

Interest Area	Description	Related Careers
The Arts	If you enjoy music, painting, writing, entertaining others, and are creative and independent, you may be interested in the arts.	Actor, artist, editor, fashion designer, musician, photographer, or reporter
Business	If you enjoy coming up with new ideas, managing projects and other people, and selling things, you may be interested in business.	Bank manager, business executive, entrepreneur, lawyer, or salesperson
Crafts	If you enjoy working with tools and building things, you may be interested in crafts.	Architect, auto mechanic, cook, electrician, fashion designer, or graphic artist
Office Operations	If you enjoy working with words and numbers and if you like being organized and accurate, you may be interested in office operations.	Accountant, administrative assistant, bank teller, computer operator, medical records, tax preparer, or technician
Scientific	If you enjoy math and science, study a lot, work independently, and are curious and creative, you may have scientific interests.	Architect, biologist, computer scientist, economist, mechanical engineer, medical lab technician, pharmacist, physician, or veterinarian
Social	If you enjoy helping others, get along well with other people, have a lot of patience, and enjoy communicating, you may have social interests.	Clinical psychologist, counselor, nurse, police officer, recreation worker, school administrator, social worker, or teacher

What abilities might be useful for a career as a social worker? What interests might a social worker and a teacher have in common?

Completing Your Career Self-Assessment Worksheet

Once you have filled in your career self-assessment worksheet, you can look through the information to find connections that might point to a possible career path. For example, your favorite subject might be history, your abilities might include patience with senior citizens, and you might have an interest in social services. Can you see any connection between them?

For the final step to complete the worksheet, list your top two preferred career clusters—the clusters that best match your skills, interests, and abilities.

Using Your Career Self-Assessment Worksheet

Every career self-assessment worksheet will be unique, because you are unique. You might share interests with a friend, but you might have different abilities or work values. What career path might you consider if your self-assessment profile includes the following?

- Subject: Technology
- Skill: Computer programming
- Achievement: Award for Web page design
- Ability: Artistic
- Interest: Business
- Work value: Independence

You might look at the Information Technology career cluster and the Web and Digital Communications pathway, or you might consider the Business Management & Administration career cluster and the Business Information Management pathway. Can you think of other career opportunities that might appeal to someone with the profile above?

Self-Assessment, a Valuable Tool for Entrepreneur

Entrepreneurs might conduct a self-assessment for reasons other than identifying career opportunities. An entrepreneur can use an honest self-assessment to identify his or her strengths and weaknesses. Then, he or she can use the information to hire employees who are stronger where the entrepreneur may be weak. An entrepreneur might also use a self-assessment to determine if he or she is starting the business at the right time, for the right reasons. In such a situation, the self-assessment might include questions such as, "Is this the right time in my life for this opportunity?" "Am I committed to this venture?" and "Is there another way to achieve my career goals?" How might an entrepreneur and his or her investors benefit from this type of self-assessment?

Analyzing Career Planning Resources

Part of career planning is identifying career opportunities that meet your needs and fit your skills, interests, and abilities. In addition to the resources discussed in Chapter 4, such as the Occupational Outlook Handbook and informational interviews, you can also use **job search resources**—tools designed to help you find **job leads**.

You can use a variety of resources to identify career opportunities. You will probably need to use more than one of the following resources.

- Networking uses people you know to help you find opportunities.
- Online resources let you access information and job listings on the Internet.
- Want ads are a traditional source for job listings.
- Career counselors help you identify jobs that match your skills, interests, and abilities.
- Employment agencies work to match employers with employees.
- Job fairs provide an opportunity to introduce yourself to many different employers.

What Is Networking?

Networking in career planning means sharing information about yourself and your career goals with personal contacts—people you know already or new people you meet in any area of your life. Hopefully, one of the contacts works for a company that is hiring or knows someone at a company that is hiring. The contact recommends you for the position. Employers like to hire people who come with a recommendation from someone they know and trust.

Here's how it works:

1. You see a cousin at a family party and tell him that you are trying to find a job as a paralegal—someone who works in a law office assisting lawyers with legal paperwork.
2. Your cousin is bowling with friends and mentions that you are looking for a position as a paralegal.
3. One of the bowling buddies tells his mother, who is a lawyer, about you.
4. The law firm where she works is looking for a paralegal. She invites you to apply for the job.

Some studies show that nearly 80 percent of all job openings are never advertised.

How Do I Network?

The first step in networking is to tell everyone you know that you are planning your career. Be specific about your career goals. Tell your family, friends, classmates, and teachers. If you volunteer, tell the people at the volunteer organization. If you are a member of a club or organization, tell the other members.

Once you make a contact, it is important to stay in touch through regular calls or e-mails. The first time you call or e-mail a contact, explain who you are and why you are reaching out. Do not ask for a job. Instead, ask for:

- Information about occupations and companies, such as what trends are affecting a certain industry
- Introductions to people who might become part of your network
- Job leads

Remember to give the contact your phone number and e-mail address so he or she can reach you. Always be polite, speak clearly, and say thank you.

How can meeting new people help you find career opportunities?

Money Madne$$

You currently work at the information booth in a local state park. You earn $10.75 an hour. Recently, you met a woman who works as a wilderness first responder—someone who helps people injured while participating in a wilderness activity such as hiking or skiing. The job sounds like something you might be interested in. She suggests you take a wilderness first aid course, which is the first step toward certification. The course costs $175.00. How many hours must you work to save enough money to pay for the course?

Job Search. Career Research. What's the Difference?

You completed a career self-assessment worksheet. You picked two careers that match your interests, skills, and abilities. Why haven't you found a job?

- 👎 You are spending only a couple of hours a week on your job search.
- 👎 You are applying for positions that require more skill and training than you currently have.
- 👎 You do not have personal recommendations for the positions.
- 👎 You are applying for the same job that hundreds of other candidates are applying for, too.
- 👍 You are continuing to spend time building a network of contacts.
- 👍 You keep your job search materials organized to make it easy to apply to any new job opening.
- 👍 You use all available job search resources to increase your chances of finding a job.

Remember, career research helps you identify types of careers that match your skills, interests, and abilities. A job search identifies specific job openings and opportunities for employment. How can you use your career research to help you in your job search?

Organizing Your Contacts

Set up a networking file to keep track of each contact. That way, you will have a record of each communication to help you stay organized and focused. Set up the file using index cards or a computer program. In the file, include:

- The name, occupation, mailing address, phone number, and e-mail address of each contact.
- A reminder of how you know the contact. Is it a personal friend? Did you meet through someone else? Did you meet through a club or organization?
- The date and time of each phone call, message, or other type of communication with the contact
- Notes about each conversation or exchange with the contact.

Using Online Resources

The Internet is a great tool for finding a career. You can even use it to make contacts for networking. Some of the more effective online resources include:

- *Company Web sites.* You can learn a lot about a company from its Web site, including what they do, the backgrounds of the people who work there, and who to contact in each department. Most sites also have a page listing job openings, with information on how to apply. Even if there are no openings that interest you, you can contact the human resources department to try to set up an informational interview.
- *Government sites.* Like corporations, government agencies list information and job openings. There are also government Web sites that provide job listings.
- *Industry sites.* Many industries and industry associations have Web sites that list job opportunities. For example, if you are interested in a career as a social worker, you could look for positions on socialworkjobbank.com.
- *Online job agencies.* Companies such as Monster.com or CareerBuilder.com list job openings and let you upload your resume for employers to look at. You can search these sites for jobs in a field or career that interests you. Some of these sites charge fees.
- *Social networking sites.* You can use social networking sites such as Facebook or LinkedIn to meet contacts and learn about jobs. There are groups for people in certain careers or who work for specific companies. Employers join these sites, as well. They look for potential employees based on the personal profile you create. That's an important reason for posting only information that would be appropriate for a potential employer to see!

Using a Career Center or Agency

A career center is an excellent place to start planning your career. Your school might have a career center that you can use free of charge. Career centers have job listings, research resources, and counselors who will help you identify jobs that match your skills and interests. They can also introduce you to former students who are now employed—giving you more opportunities for networking.

Employment offices are similar to career centers. Some are sponsored by the state or local government. They provide job search resources and assistance free of charge.

Private employment agencies charge a fee to match employees with employers. Sometimes you pay the fee, and sometimes the employer pays the fee. Sometimes you pay even if you don't find a job. They all have different policies, so be sure to ask before you sign a contract.

Does your school have a career center? What resources does it have that you might use?

Career Profile: Human Resources Manager

Business Management & Administration Career Cluster

Job Summary

Human resources managers supervise the staff in a human resources department—or a personnel department. They oversee all personnel-related activities. For example, people who work in human resources review job applications, advise employees on work- and career-related issues, administer compensation and benefits policies, manage training for new and current employees, represent the company at career fairs and other events, and manage employee records.

The role of a human resources manager often depends on the size of the organization. If the company is small, he or she may be responsible for all aspects of human resources work. In a larger company, he or she may specialize in one human resources activity, such as employment and placement, compensation and benefits, training and development, or labor relations.

Use the Internet, a library, or career center to research the responsibilities, education and training requirements, and salary range of a human resources manager in your state. Write a job description or want ad for the position.

Government Job Search Resources

Some Web sites that link to government job search resources include:

✔ usajobs.gov
✔ governmentjobs.com
✔ www.jobsfed.com
✔ www.studentjobs.gov

Temporary employment agencies are hired by a company to fill temporary jobs. For example, a company might need temporary data entry help to complete a large project or help to fill in for a worker on vacation or on a leave of absence. Temporary jobs can lead to a permanent position. They are also a good way to build skills, gain experience, and meet people in the industry.

Career Fairs

An event such as a career or job fair is a great opportunity to meet representatives from many different companies. Companies set up booths where you can talk to representatives to learn more about the business, types of careers, and even available positions. Some companies use career fairs to collect resumes and set up interviews; some conduct interviews on the spot.

What might be some benefits and drawbacks of attending a job fair that focuses on a specific industry?

Although many career fairs are general, which means there are companies from many different industries, there is a growing trend toward career fairs that focus on a specific industry. For example, at a high technology career fair, you can meet representatives from companies in the high technology industry. At a health sciences career fair, you can meet representatives from companies in the health sciences industry. Attending a focused career fair may be useful if you know the industry or cluster that interests you, but you may miss the opportunity to learn about other industries or types of jobs.

When you attend a career fair, assume you are being evaluated as a potential employee. You may be interested in collecting information for career planning, but the company representatives are looking for people to hire. Dress appropriately, be confident and outgoing, and be prepared to ask and answer questions.

What About Want Advertisements?

Another resource that can help you identify career opportunities are help wanted ads. **Help wanted ads**—which are sometimes called classified ads—are job listings for specific positions. They are printed in newspapers and magazines and posted online at newspaper and magazine Web sites. You can use the help wanted ads to find a specific job, but also to learn about career outlooks in your area. For example, you can identify companies that are hiring, industries that appear to have many openings—or very few—and salary ranges for listed jobs.

Usually, there are many ads listed alphabetically. They may be organized into general categories, such as Medical, Professional, and Education. For example, if you are looking for a teaching position, you would look at the listings under Education.

Help wanted ad listings usually include a job title and a very brief description of the responsibilities and experience required. Sometimes they include information about wages and hours. There may be a phone number or e-mail address to call for more information or to apply, or a mailing address where you can send a resume.

What can you learn about career opportunities from a help wanted ad?

Help Wanted Shorthand

Help wanted ads must convey a lot of information in a small amount of space. Here are some common want ad abbreviations.

- ✔ Appt — Appointment
- ✔ Avail — Available
- ✔ Eves — Evenings
- ✔ Exp — Experience
- ✔ F/T — Full-time
- ✔ H.S. Grad — High school graduate
- ✔ Hr — Hour
- ✔ Immed — Immediately
- ✔ M-F — Monday through Friday
- ✔ Nec — Necessary
- ✔ P/T — Part-time
- ✔ Perm — Permanent
- ✔ Pref — Preferred
- ✔ Refs — References
- ✔ Resp — Responsible
- ✔ Wkdays — Weekdays
- ✔ Wkends — Weekends

Tech Connect

Printed newspapers once graced the front porch of nearly every home in America. Now, easy access to the Internet and 24-hour news cycles on television have helped to push newspapers online. While printed editions are still available, you can access and read articles and features in papers around the world using the Internet.

The online editions make it easy to read want ads in any area of the country—and in other countries, too. When you are ready to apply for jobs, you can use search tools to identify the ones that meet your criteria. You might also be able to upload your resume and apply for available positions.

Does your local newspaper have an online edition? If so, can you access its help wanted ads online? Use the Internet or your library to find out. Then, use a word processing program or a desktop publishing program to create a flyer advertising the benefits of using online newspaper help wanted ads to identify career opportunities.

Case Study

Marcus is a shy, quiet young man who does well in math and science. In fact, he won the school science fair and has the highest average in advanced algebra. He does not play any sports, but he sings in the school choir, is a member of the drama club, and plays on the school chess team.

Marcus filled out a career self-assessment profile in school. He selected language arts and music as his best subjects. Among his skills he listed "strong communications" and "excellent singer." He included musical/dramatic, language, and social for his abilities. His achievements included second place in the school spelling bee and chorus in the school musical. He listed his interests as social and the arts. He selected entertainment and music as his preferred career groups. His work values included independence, creativity, and working with people.

- Do you think Marcus was honest and accurate in his self-assessment?
- Do you think Marcus selected career groups that match his values, interests, and abilities?
- If you were Marcus's career counselor, what would you recommend he consider?

Answer It!

1. What is the first step in the basic process of planning a career?
2. What is the purpose of a career plan?
3. Why is it important to revisit your career plan every once in a while?
4. List five things to include in a career plan.
5. Why might it be easier to work with an adult when you develop a career-assessment worksheet?
6. What is the final step to complete your career self-assessment worksheet?
7. List at least five resources for identifying career opportunities.
8. What is one reason networking is an excellent way to find career opportunities?
9. Why is it important to keep track of your networking contacts?
10. Why is it important to dress appropriately when you attend a job fair?

Web Extra

Bls.gov/k12/ is an official U.S. government Web site that provides links to careers that relate to your interests and presents fun facts about the economy and jobs. Use the Student Resources link to explore the careers that are related to your interests. Use the information you find to develop a list of potential careers, and create a table for each that identifies the average salary, education requirements, work experience requirements, and the job outlook.

Write Now

Use the six-step process to create a career self-assessment worksheet for yourself. Be honest and accurate. Work with a partner, teacher, or counselor, if necessary. Analyze the worksheet and look for connections that point to a possible career path.

Complete the worksheet by listing two career clusters that interest you, based on the career assessment worksheet. Select a pathway in each cluster, and research the job outlook for each. Write a paragraph explaining how you might explore the opportunities in each pathway.

COLLEGE-READY PRACTICES

A career assessment helps identify a career path that fits your interests and abilities. College-ready individuals recognize it also helps identify the type of academic and educational opportunities you should pursue.

If you have not already done so, complete a written or online career assessment (texascollegeandcareer.org, educationplanner.org careerkey.org, bls.gov/k12). With a partner, analyze the career assessment, and then discuss the results with a teacher, parent, guardian, or school counselor to determine the career cluster and pathway that might fit you best. Make a chart listing the academic requirements for at least three careers in that cluster and pathway. Write a paragraph explaining what you can do now to start achieving the academic requirements you will need for the career of your choice. Read the paragraph out loud to your partner or to the class.

Career-Ready Practices

As you have learned, networking is a valuable career-planning tool that allows you to share information about yourself and your career goals with personal contacts. Career-ready individuals understand that in order to build a network of quality contacts—people who are eager to help you meet your career goals—they must demonstrate employability characteristics and skills such as leadership, flexibility, cooperation, and a positive attitude. They exhibit integrity and use a variety of means to positively influence others. They are honest and ethical in the roles they have in all areas of their lives.

Write a short paper that explains to a potential contact why he would want to include you in his network of personal contacts. Summarize your employability characteristics and skills and provide examples of how you have had a positive impact on others.

Career Portfolio

Your career portfolio should include your career plan. Working with your teacher, a counselor, or a family member, develop a career plan that starts now. The plan may be part of the academic plan you developed in Chapter 7, or it may be a separate document. If your school has a form for the plan, use it. Otherwise, develop the plan on a sheet of paper. Be sure to state a realistic and attainable long-term career goal that matches your skills, interests, and abilities. Also include short-term education and career goals that map out the steps you will take to achieve your ultimate career goal. Add the plan to your career portfolio along with your career self-assessment.

13 Managing a Job Search

Skills in This Chapter . . .

- **Creating Job Search Materials**
- **Preparing a Resume**
- **Applying for a Job**
- **Interviewing for a Job**
- **Evaluating a Job Offer**

GETTING STARTED

Searching for a job is hard work. You will need all of your 21st Century process skills. For example:

- Communications skills to write job search materials, describe your strengths, and convey your interests nonverbally
- Decision-making skills to identify job opportunities and to accept or turn down a job offer
- Problem-solving skills to negotiate employment needs and improve your search materials or interviewing technique

A successful job search depends on being organized and thorough, as well as knowing how to use all available resources—in other words, on being an effective manager.

➤ As a class, brainstorm resources you think would be helpful for managing a job search. Be creative. For example, in addition to a computer with Internet access, you might think of a relative with a business that needs employees or a new outfit to wear on an interview. After you compile the list, discuss how you could effectively use each resource in a job search.

227

Creating Job Search Materials

Employers often consider tens or even hundreds of candidates for every job opening. Sometimes they shuffle through stacks of letters or field multiple phone calls. How can you make a positive impression and show you are serious and qualified for the job?

You can prepare professional, accurate job search materials, including a resume, cover letter, and list of references.

- Your resume may be your most important job-search document. It is a written summary of your work-related skills, experience, and education.
- A cover letter is sent along with your resume. It introduces you to a potential employer and highlights the qualities that make you suitable for the position you want.
- A list of references includes the names and contact information of people who know you and your qualifications, and who are willing to speak about you to potential employers.

Why Do I Need a Cover Letter?

When you include a cover letter with your resume, you make a better impression on the employer than if you send the resume alone. Including a cover letter shows that you have taken the time to match your qualifications with a specific job.

A cover letter should be short and to the point. Direct it to the person who is responsible for hiring. If you do not know the person's name, title, and address, call the company and ask.

You can customize the cover letter to the job opening you are applying for. Use it to tell the employer why you are interested in that job and why you are qualified to do the work.

Career Tips

Even when you send a resume electronically, it is important to send a cover letter. For example, you can write the cover letter in an e-mail message and include the resume as an attachment.

When you write a cover letter, ask yourself these questions:

- Have I identified the job title for the job I want?
- Have I stated where I learned about the position?
- Have I listed the skills I have that qualify me for the position?
- Have I thanked the reader for his or her time and consideration?
- Have I corrected all spelling and grammatical errors?

Listing References

A **reference** is someone who knows you well and is willing to speak to employers about your qualifications. A good reference knows your positive work qualities and values, understands your abilities and interests, and can describe how you behave in a work environment. People such as relatives and friends should not be used as references because they may not be objective about you and your strengths and weaknesses.

Create a reference list to include in your career portfolio, and make sure you have it available to give to employers when they ask. Put your name and contact information at the top, and use the heading "References."

Select at least three people, and ask them for permission to use them as a reference. On the list, include each person's full name, occupation, mailing address, e-mail address, and daytime telephone number, including area code.

October 3, 2015

Ms. Helen White
MNO company
4671 Highland Avenue
Palatine, IL 60067

Dear Ms. White,

I am writing in response to the job posting for a human resources assistant that I saw on your Web site. I am enclosing my resume for your review.

For the past two years, I have worked as an account assistant at Employ Action, LLC., an employment agency. In that time, I have learned a great deal about employment options and opportunities. I am a hard worker who is looking for new challenges.

I would welcome the opportunity to speak with you personally about the human resources assistant position. You can contact me by phone or e-mail using the information below.

Thank you for your time and consideration,

Sincerely,

Thomas Leslie

Thomas Leslie
2323 McCloud Street
Palatine, IL 60067
555-555-5555
thomasleslie@mail.net

What message do you think spelling errors in a cover letter sends about your qualifications as an employee?

A good reference can speak objectively about your strengths and weaknesses. Can you think of three people you can include on your reference list?

Choosing a Reference

People who make good references include:

- ✔ Former or current employers
- ✔ Coaches or club advisors
- ✔ Teachers

People who do not make good references include:

- ✔ Your parents
- ✔ Any relative
- ✔ Friends

Do I Need a Portfolio?

Everyone looking for work should keep a career portfolio for storing documents such as a resume, cover letter, personal data sheet, and list of references. You can use a career portfolio as a resource when you are applying for jobs or going to an interview. For example, if you have your reference list with you when you are filling out a job application, you do not have to worry about remembering names and contact information.

If you are looking for a position in a creative career, such as photography, graphic design, or journalism, a portfolio that includes samples of current work is an important part of your job search materials. It lets you show potential employers your abilities and achievements.

- For a job as a reporter, you would want to have a portfolio of published writing samples—clips from newspapers, magazines, newsletters, or printouts from a Web site. If you do not have published samples, you can include samples from your school work.
- For a graphic designer, you would want to have a portfolio of materials you have designed, such as logos, business cards, or stationery.

It is a good idea to have the original sample in the portfolio, if possible, and to have copies available to give to potential employers. Artists sometimes create slides of their artwork to include in a portfolio.

Money Madne$$

Can you save money and time by submitting your resume and cover letter electronically? How much does it cost to send a first-class letter using the U.S. Postal Service? How many resumes are you going to send out?

Say it costs 52 cents for each resume, and you plan to send out 20 resumes. How much will it cost? What if you send out 55 resumes? Write a paragraph explaining why the cost of mailing a resume and cover letter is worth it, or not.

Recent research indicates that the average length of a job search is now more than seven months! In a down economy with a high unemployment rate, it can take longer than that. Knowing that it can take a long time to find a job helps keep you from becoming discouraged, or giving up.

Managing Your Job Search Materials

Keeping your job search materials organized in a folder or binder will help you follow up every possibility. When you are actively looking for work at many companies, it is easy to forget who you spoke to and even what you spoke about. An employer might not look favorably on someone who repeats the same conversation or cannot remember who referred her in the first place.

- Keep a to-do list of tasks you want to accomplish each day, such as people you want to contact, resumes you have to send out, and thank-you notes to write. Cross off each item you complete, and add new items as they come up.
- Contact some people in your network every day. Make brief phone calls, or send brief e-mail messages to let them know you are looking for work, and to ask for assistance finding job opportunities.
- Follow up on all leads. Keep a record of the people you contact, including phone numbers, e-mail addresses, and mailing addresses. Include the dates and times, the method of communication, and the result. Did they invite you in for an interview? Did they refer you to someone else?
- Set up folders for storing documents that relate to your job search. Use the folders to keep track of information you send to each contact or potential employer and the response you get back. The folder might include copies of the cover letter and samples of your work that you sent. It might also include notes you took during a phone call or interview, a brochure about the company, and a printout of an e-mail message you received.

Why is it important to keep your job search materials organized?

Typo!

Even the best candidate might not get an interview if there's a spelling error in the cover letter or resume. What can you do to avoid that problem?

- 👎 Assume your word processor's spell check feature will catch all errors.
- 👎 Hit the Send button without previewing the documents.
- 👎 Proofread your documents when you are tired.
- 👍 Use a spelling checker, but also proofread it yourself.
- 👍 Ask two other people to proofread your documents.
- 👍 Preview all documents before sending them electronically.

What could you do if you discover a typo after you have already sent your resume to a potential employer?

Career Tips

Some experts recommend that you begin an active job search six to nine months before the date you hope to start working.

If you treat your job search like a job itself, you know you are actively working to achieve your career goals. How much time would you spend working each day if you had the job you want? Try to spend that time working on your job search.

You might not be earning an income, but you can meet your responsibilities by developing your employability skills, creating and improving job search materials, and exploring resources that might lead to job prospects—the possibility of employment.

21st Century Learning

Is It Too Good to Be True?

Do not fall for employment scams, such as a job ad that promises income of thousands of dollars a week and requires no experience or education. If a job description sounds too good to be true, it probably is. Be wary of any job that:

✔ Asks you to pay for something up front

✔ Asks for a scan of your passport or other personal information

✔ Offers a part-time, work-from-home job that earns a large salary

✔ Includes money transfers or money transactions in the job description

✔ Comes from an agency representing a cruise ship

Jane is looking for a summer job. She is considering a career in hospitality, so she would like to work in a hotel. Glenna is also looking for a summer job. She is considering a career in law, so she would like to work in a law office.

Jane goes online and starts looking through want ads. She sends out her resume to a few hotels in the area, but she does not hear back from any. She goes to the library and looks through the want ads in newspapers and magazines. She sends out more resumes and still gets no response. She is beginning to think she will never find a job.

Glenna also looks online and in newspapers and magazines. She makes an appointment with a career counselor at school, who helps her compile a list of law offices in the community. The counselor gives Glenna the name and phone number of a lawyer she knows. Glenna makes time one afternoon to personally drop off her resume at a few of the local law offices. She calls the lawyer and asks if she can come in for an informational interview. She feels that her chances of finding a job she wants are pretty good.

What is different about the way the two girls are approaching their job searches? Which girl do you think has a better chance of finding a summer job? Use 21st Century Skills to answer these questions. Write an ending to the story. Read it to the class, or form a small group and present it as a skit.

Tech Connect

Most word processing programs such as Microsoft Office Word come with resume templates or let you download resume templates from a Web site. A template is a form, or sample document, that provides text and formatting to help you create standard documents. A resume template helps you quickly create a professional-looking resume.

When you use a resume template, you create the document, then replace the sample text with text customized for your own experience. The template prompts you to insert specific information, such as your name and address, and indicates where you should type your work and education experience. The template also includes formatting, such as lists, bullets, and fonts. Some have graphics, such as lines or shapes, or prompt you to insert your own graphics.

The benefits of using a resume template are obvious—a template saves time and makes it easy to produce a neat, complete resume. There are drawbacks, too. For example, other people might use the same template.

Explore the resume templates available with your word processing program at home or at school. Select a template and use it to create your resume.

Preparing a Resume

A **resume** is a document that provides a snapshot image of your qualifications. It summarizes you, your skills, and your abilities. It is a statement of who you are, what you have done in your life, and what you hope to do next. Your resume may be the first communication between you and a potential employer. You will make a positive impression if your resume is:

- Neatly printed on white paper
- True and accurate
- Free of any typographical, grammatical, or spelling errors
- Direct and to the point

You want your resume to describe you in a way that makes the employer want to meet you. A well-written resume will help you to get an interview.

How Should I Format My Resume?

There are many ways to organize or format a resume. Most word processing programs come with resume **templates**—sample documents. You can also find sample resume designs in books and on the Internet.

Choose a format that highlights your experience and skills so they stand out to someone who might just glance at the resume quickly. You may also want to consider these tips:

- Make it easy to read. Leave space between lines so it is not crowded or overloaded.
- Use one, easy-to-read font, and apply different font styles and sizes for emphasis.
- Bullets are effective for making lines of text stand out.
- Use proper spelling, punctuation, and grammar.
- Keep it to one page, if possible; two pages at the most. (If you use two pages, be sure to put your name in the header or footer on page 2, in case it becomes separated from page 1.)

Sometimes you will mail your resume in an envelope with a cover letter. Sometimes you will send it electronically by e-mail. Make sure it looks professional when it is printed, as well as when you view it on a computer.

What Do I Include on My Resume?

A typical resume has four main sections:

- *Contact information.* Include your full name, address, telephone number, and e-mail address at the top of the page. This gives the employer all the information necessary to get in touch with you.

- *Objective.* An objective describes your career goal. It should be short and clear. You can have a general, long-term goal as your objective, or you can customize your resume for a specific position by using a short-term goal. For example, a general objective might be, "To work in fashion sales." A customized objective might be, "To work as a part-time sales associate in a fashion clothing store."

- *Education.* If you have little work experience, it is important to highlight your education. List the name, city, and state of every school where you earned a degree, starting with high school (or the last school you attended). Include the years you attended the school, the degree you earned, and any special courses or certificates that relate to your career. You might want to include your grade point average, as well.

- *Work experience.* Include all full-time and part-time jobs, internships, apprenticeships, and volunteer experience, starting with the most recent experience first. Include employment dates, job titles, company names, city, and state. Briefly describe your main responsibility, and list your duties and accomplishments. Be specific. If you have work experience, you might want to put this section before Education.

Clyde Duggan
2556 Granger Avenue
Palatine, IL 60067
(555) 555-5555
clydeduggan@mail.net

Objective	To work as a part-time sales associate in a fashion clothing store.
Education	Currently enrolled in Wickham High School; expect to graduate in June 2017.
Work Experience	
9/10–present	Fashion consultant, HH Shelter for Women and Children, Palatine, IL
	• Select and coordinate outfits for women preparing for job interviews.
5/09	Fashion show organizer, Wickham High School, Palatine, IL
	• Proposed, planned, and managed fashion show of student designs to raise funds for the HH Shelter for Women and Children.
Skills	• Sewing
	• Clothing coordination and fit
	• Basic math skills
	• Fluent in Spanish
Extracurricular Activities	
	Family, Career and Community Leaders of America (FCCLA); member since 2014; chapter treasurer 2015–present

---

Why do you think it is important to include co-curricular activities on your resume?

You may also include a Skills section for listing jobs skills such as your ability to use computers or speak a foreign language. You might also want to include a Personal Information section that lists co-curricular activities, awards, and honors. You can include all clubs and organizations of which you are a member, and things you do in your school or community. Remember to update your resume on a regular basis, so employers will see the most current and accurate information.

Use Action Words!

Action words are verbs that describe your actions in a way that makes them stand out to the reader. When you use action words to describe your responsibilities and skills in a cover letter or on your resume, it will bring your actions to life.

- Instead of *Made lunch*, try *Cooked healthy lunches for 50 people.*
- Instead of *Filed papers*, try *Organized customer files alphabetically.*
- Instead of *Took pictures*, try *Photographed school events and functions for use on Web site and in newspaper.*
- Instead of *Know CPR*, try *Certified by the American Red Cross in cardiopulmonary resuscitation (CPR).*

Applying for a Job

When you find a job you think you want, you apply for it. Applying means that you present yourself as a **candidate**—possible employee—for the position. A successful job application leads to an invitation for a job interview.

Applying usually starts with sending in your resume and a cover letter. Sometimes you can improve your chances of landing an interview by applying in person.

- In a small company, that means speaking directly with the owner or manager. For example, if you are applying for a position in a local restaurant, you would speak with the manager.
- In a large company, applying in person might mean going to the human resources department. Employees in the human resources department often meet job applicants first to decide whether they meet the qualifications for the position.

Be sure to bring a few copies of your resume and reference list to hand out when you apply in person. Also, ask for business cards from the people you meet so you know how to contact them.

Tips for Applying

When you apply for a job, your short-term goal is to get invited for an interview. There are a few things you can do to improve your chances and to make your application stand out from the crowd.

- Use a professional e-mail address. A cutesy address such as suzieq, lonewolf, daredevil, or wildgirl828 is not likely to impress a potential employer.
- Even when you are asked to fill out a standard job application form, submit a resume as well.
- Be polite and professional to everyone you meet. Whether you apply in person or are speaking on the phone, be courteous and respectful to everyone.
- When you submit a resume online, ask whether you must use a specific format. Some companies want resumes in plain text format, others accept files created using a word processing program.
- Customize your resume for a specific job. You can customize the objective, or rearrange your experience and skills to highlight the information you think is most appropriate for the position.

Other Action Words to Consider

Use a thesaurus—reference book of definitions and synonyms—to find action words to replace common verbs, such as go, have, or get. Some attention-getting action words for resumes include:

✔ Created
✔ Operated
✔ Supervised
✔ Prepared
✔ Built
✔ Produced
✔ Published
✔ Planned
✔ Coached
✔ Improved

What resources might help make filling out a job application easier?

Practice Makes Perfect!

When you apply for a job, many employers expect you to complete the application on the spot. Providing accurate information and writing legibly can mean the difference between your application making the cut or ending up in the reject pile.

That's why you will find it useful to fill out a practice application before you actually start applying for jobs. Your school's career center or counseling office may have sample applications on file. You can also find practice applications online. Some Web sites that provide applications are jobsearch.about.com and careerchoices.com/lounge/files/jobapplication.pdf.

With your instructor's permission, download and complete a sample job application. Then exchange it with a classmate and make suggestions on how it could be improved.

Filling out a Job Application

A **job application** is a standard form you will fill out when you apply for a job. You might fill it out in person when you visit a potential employer, or you might fill it out online. It requires a lot of the same information that you put on your resume, such as your contact information, as well as details about your education and work experience. (It may ask for your Social Security number, but you can withhold such information until after you receive a job offer.)

Filling out an application form may seem simple, but a lot of people make mistakes or forget important information. A messy or incomplete job application will not make a positive impression on the employer.

- Read the form before you start filling it out.
- Follow all instructions.
- Be truthful and accurate.
- Write neatly.
- Enter N/A for not applicable if there is a question that does not apply to you.
- Check your spelling and grammar.
- If you make a mistake, ask for a new form and start again.

Use a Personal Information Card

You might find it helpful to bring a personal information card with you when you apply for a job. A **personal information card** is an index card—or other piece of paper—on which you write or print the information you might need to fill out a job application accurately. You may even be able to store the information in your cell phone or other handheld device.

To create a personal information card, simply write the information you will need, including:

- The names, addresses, and contact information of your current and previous employers, including dates and the amount you were paid
- The names and addresses of schools where you earned a diploma or degree, including dates
- The names, addresses, and contact information of your references

Career Tips

You can use the documents you have in your career portfolio to help fill out a job application. Your resume includes facts you will need about your previous employers. Your reference list includes the names and contact information of your references.

Interviewing for a Job

What happens when an employer reads your cover letter and resume and thinks you have the qualifications that she is looking for? She invites you for a job interview. A **job interview** is a formal meeting between a job seeker and a potential employer—the interviewer. The job interview helps you and the interviewer make important decisions regarding the position at stake.

- The interviewer decides if the job seeker is the best person for the position.
- You decide if the position is one you really want.

Both you and the interviewer can use the job interview to get to know each other. You learn information that you cannot learn from a cover letter or resume. For example, the interviewer learns whether or not you make eye contact. You learn if people at the company are friendly. A job interview is successful if you convince the interviewer to offer you the job. It is also successful if you learn that the position would not be right for you.

Preparing for a Job Interview

A test is always easier if you are prepared. A job interview is like a test—if you pass, you are offered the job. Use these four steps to prepare for a job interview.

1. Research the company or organization where you are going for the interview. Talk to someone who works there. Visit the company's Web site.
2. Make a list of questions an interviewer might ask you. Common questions include, "Tell me about yourself," "Why do you want to work here?" "Do you have the skills to get the work done?" and "Why should I hire you?"
3. Prepare answers to the questions. Be specific. Emphasize your strengths, skills, and abilities. Explain how you solved a problem, made an important decision, or showed responsibility. Mention your goals, and briefly explain how you plan to achieve them.
4. Make a list of five to ten questions you can ask the interviewer. Ask about the company, the work environment, and the position. Common questions include, "What kinds of projects or tasks will I be responsible for?" "Is there opportunity for advancement?" "What are the hours?" "What is the salary range?" and "When will you make a hiring decision?"

How can practicing for an interview help you perform better in an actual interview?

Practicing for a Job Interview

A job interview is stressful. You are trying to make a good impression. You want to look and sound your best. You want the interviewer to like you and to respect you. Practicing for the interview by rehearsing your behavior and answers to questions helps give you confidence.

Working with a partner is probably the best way to practice. You can take turns being the interviewer and the job seeker. If you are alone, practice in front of a mirror. If possible, record your practice so you can watch yourself.

Making the Most of the Job Interview

Many interviews are 10–15 minutes long. How can you best use that time to get a job offer?

- Arrive ten minutes early.
- Introduce yourself to the receptionist, and explain who you are there to meet.
- Be polite and respectful to everyone you meet.
- Shake hands with your interviewer when you arrive and before you leave.
- Listen carefully, using positive body language. For example, smile and lean forward slightly when the interviewer is talking.
- Use proper English when you speak; no slang.
- Avoid chewing gum.
- Turn off your cell phone. If you forget and it rings or vibrates, apologize and ignore it or turn it off without checking to see who called.
- At the end of the interview, shake hands again, and thank the interviewer. Ask for a business card, so you have the interviewer's contact information.

Tips for Practicing for a Job Interview

Practicing your interview will help you feel comfortable and prepared. Use the time to rehearse your words and actions for the actual interview. Here are some tips for practicing.

- ✔ Be truthful.
- ✔ Pronounce your words in a strong, clear voice.
- ✔ Keep your answers brief and to the point.
- ✔ Use positive nonverbal communication, such as eye contact, relaxed arms, and good posture.
- ✔ Dress as you would for an actual interview.
- ✔ Avoid fidgeting or playing with your hair.
- ✔ Ask someone to critique your interviewing skills and use his or her comments to improve your technique.

Studies show that interviewers judge job seekers first by appearance, next by behavior, and third by what they say. So, having a neat appearance, positive attitude, and showing confidence all help you make a good first impression

Managing a Job Search ■ Chapter 13

Telephone Interviews

Many companies use telephone interviews to *screen*—make a first decision about—potential employees. They might be interested in you based on your cover letter and resume, but want to see how you present yourself during a conversation before they invite you for a face-to-face interview.

- Some employers will call or e-mail to schedule a telephone interview. Then, you can prepare for the call the same way you would prepare for a face-to-face interview.
- Some employers will call out of the blue. This lets them learn how well you handle stress and if you are able to communicate effectively without preparation.

You can prepare for surprise calls by practicing with a friend or relative. You can also make sure you are always prepared to talk about a company you apply to, the position, and yourself.

Why do you think some companies screen applicants by phone before meeting them?

Write a Thank-You Note

The interview is over. Now what? Start by writing a thank-you note. A thank-you note reminds the interviewer that you are serious about wanting the job.

You use a thank-you note to restate your interest in the job and your qualifications and to thank the interviewer for spending time with you. Refer to something specific that you discussed during the interview. Address the note to the person who interviewed you—that's why you asked for the business card! Send the note within 24 hours of the interview.

Follow Up

During the interview, you asked how long it would be before the interviewer made a decision. If you do not hear from the interviewer within that timeframe, you can call or send an e-mail. Ask if he or she has made a decision, and say again that you are interested in the position.

Sometimes, one interview leads to another. The first interviewer might like you for the job, but not be authorized to make hiring decisions. Or, he or she wants to see how other employees react to you. Being called back for a second interview is a good thing. It means you passed the first test.

Looking Your Best for an Interview

Some tips for presenting a professional appearance:

✔ Wash, dry, and comb your hair.

✔ Trim and clean your fingernails.

✔ Use deodorant.

✔ Brush your teeth.

✔ Avoid products such as perfume or body spray that have a strong odor.

✔ Wear small, neat jewelry.

✔ Wear clean, neat clothes that are appropriate for the work environment.

✔ Avoid bright or contrasting colors that might distract the employer.

✔ Avoid revealing too much skin.

✔ Arrive early enough so you have time to visit the restroom to check your appearance and freshen up.

Myth All interviewers ask the same questions.

Truth Most interviewers ask questions that relate to the job opening, the company, and the individual candidate. A good interviewer will know how to ask questions that encourage you to describe your own strengths and weaknesses.

What can you learn from a job rejection that might help you get an offer in the future?

When You are the Interviewer

When you are a manager, you may find yourself on the other side of the table, interviewing job applicants. You may be the one to write the job description, determine the compensation, and advertise the position. You must evaluate applicants based on the cover letter and resume they submit and invite the most promising candidates in for an interview. How would you go about hiring employees? What questions would you ask in an interview? What challenges might you encounter, and how would you deal with them?

What If You Get the Job?

If you are offered the job, you have reason to celebrate. But, before you accept the offer, make sure you have all the information you need to make the best decision. You should not have to accept or reject the offer immediately. If possible, ask to talk to someone in the human resources department who can explain the company policies to you. For example, you might want to ask:

- What is the salary?
- What benefits—health insurance, vacation time—come with the position?
- When does the job start?
- What are the hours?
- When does the company need to know your decision?

When you have all the information you need, use the decision-making process to decide whether to accept the position or not.

- If you accept the offer, thank the employer and ask when and where you should report to work. You may have to sign a formal letter of acceptance, or write a letter of intent, which states that you are accepting the position.
- If you reject the offer, you should still write a letter of intent, thanking the employer and stating that you are not accepting the position.

What If You Do Not Get the Job?

If you do not get a job offer, it is alright to be disappointed. It will be very helpful if the interviewer can explain why you were not selected for the job. When you are feeling calm:

- Call the interviewer and politely ask what you could have done differently to get the job.
- Listen carefully and take notes. The information will help you succeed at your next interview.

Managing a Job Search ■ Chapter 13

- Say thank you at the end.
- Ask the interviewer to keep you in mind for other opportunities.

Some interviewers will not want to talk to you. Do not be discouraged. Remind yourself that the interviewer thought highly enough of you to invite you for an interview. Sooner or later, you will find the right position, and you will get the job.

Career Profile: Image Consultant

Human Services Career Cluster

Job Summary

An image consultant is someone who advises clients on factors that contribute to making a positive first impression. That includes personal appearance, fashion, and style as well as effective verbal and nonverbal communication and etiquette.

Clients may be either individuals or corporations looking to change or enhance their image. For example, image consultants might run training sessions to teach corporate employees how to dress for business meetings or casual Fridays or how to provide customer service.

Image consultants are usually self-employed. They often study human ecology, psychology, and sociology in college, as well as public speaking and general business management. They also benefit from training and certification programs offered by professional organizations.

Use the Internet, a library, or career center to research the responsibilities, education and training requirements, and salary range of an image consultant. Create an advertisement for an image consultant, including information about the services the consultant provides and how clients will benefit from them.

I Thought It Went Well, But...

You showed up on time. Your hair was neat. Your tie was clean. But you didn't get the job. You asked the interviewer why. She said you came across as uninterested. What actions might affect an interviewer's opinion of you?

👎 You don't shake hands.

👎 Your cell phone buzzes during the interview.

👎 You are chewing gum.

👎 You don't make eye contact.

👎 You ask no questions.

👍 You are knowledgeable about the company.

👍 You ask lots of questions.

👍 You smile and lean forward to listen when the interviewer talks.

👍 You follow up immediately with a thank-you note.

How can you use the information an interviewer gives you about your behavior during an interview to improve your chances of getting a job offer in the future?

What type of skills do you think would help someone succeed as an image consultant?

Evaluating a Job Offer

When you receive the job offer you have been working for, it's time to think about your employment needs. Employment needs range from **compensation**—wages and benefits—to recognition; from training to accommodations for disabilities.

Employers provide compensation to employees in exchange for work. You may be satisfied with the first offer from the employer. It may meet your requirements. However, if you believe that the company could pay more, or offer different benefits or accommodations, knowing how to negotiate for the things you need will help you make sure the exchange is fair.

How Much Does It Pay?

Usually, the most important consideration when evaluating a job offer is the **monetary compensation**, or how much the job pays. Monetary compensation comes in different forms. There is **base pay** which is the hourly wage or annual salary that you earn. Some employees are paid a commission instead of base pay, or in addition to base pay. A **commission** is a payment based on a percentage of sales. Some employees earn a **bonus**, which is a lump sum paid in addition to base pay, usually once a year. Some employees receive tips, or *gratuities*, in addition to base pay. Tips are paid by customers, not the employer.

When you evaluate the salary and other monetary compensation, it helps to be prepared. Do some research on how much comparable jobs pay in the area. The Bureau of Labor Statistics publishes salary information on its Web site (www.bls.gov). You can also find information about salary ranges in classified want ads and other job postings.

It also helps to know how much you need. Make a budget to identify expenses such as food, rent, utilities, and transportation. Then, you know how much income you require to meet those needs.

Can I Count on a Bonus?

Unless a bonus is guaranteed in writing, it is never a sure thing. What if your employer says you may receive an annual bonus?

★ A bonus might be based on your personal performance, such as meeting certain sales goals.

★ You might receive a bonus if the company achieves goals, such as a certain profit level, or landing a major contract.

★ Some bonuses are standard and are a set percentage of your annual base pay.

★ You might receive a bonus for signing a contract or if you refer someone who is hired by your employer.

Do you think it is better to receive a higher salary or to receive an annual bonus?

What Are Benefits?

Benefits are things other than wages that have value. Most companies offer some type of benefits, although smaller companies may not. Some benefits, such as unemployment insurance, are required by law. Some types of benefits include:

- Health insurance
- Life insurance
- Long-term care and disability insurance
- Vacation time
- Holiday time
- Sick time
- Tuition assistance for work-related classes
- Retirement plan
- Stock options

Some companies offer benefits in the form of a **subsidy**, which is a cash payment toward a specific work-related need. For example, a company might offer a child- or elder-care subsidy to use to pay a caregiver or a transportation subsidy to use for travel or parking expenses.

Career Trend

There is a trend toward flexible—or cafeteria style—benefits. With flexible benefits, an employee can choose the benefits that are appropriate. For example, a single adult might choose an education benefit instead of a child-care benefit.

Analyzing Insurance

An important benefit to consider when you evaluate a job offer is the insurance plan that the employer provides. **Insurance** is an investment that protects you financially against everyday risks. It protects you from the costs of an unexpected emergency. Insurance is available to cover almost anything, including your health, car, home, belongings, and even life. If you suffer an accident or loss in any of the covered areas, you can collect money to pay for the loss or to repair the damage.

Employers often offer some types of insurance to employees. Large employers may be able to pay a portion of the premiums or to negotiate lower, group rates.

Job Offer Factors to Consider

Before deciding whether to accept a job offer, consider these factors.

✔ *Responsibilities.* Will you enjoy the work?

✔ *Employer.* Do you respect your supervisor?

✔ *Compensation.* Does the salary meet your needs?

✔ *Benefits.* Are the benefits standard for the industry?

✔ *Co-workers.* Will you get along with the other employees?

✔ *Hours.* Will you be able to balance the time commitment with other responsibilities?

✔ *Location.* Is the position in a convenient location?

Why is medical insurance an important part of a job offer?

Insurance Terminology

Understanding insurance starts with understanding the terminology.

✔ *Insurance policy.* The contract issued by the insurance company.

✔ *Insurance agent.* Someone who sells insurance.

✔ *Premium.* The amount of money paid by the policy holder.

✔ *Claim.* A request for payment to cover a loss.

✔ *Beneficiary.* The person who collects the money if there is a claim.

✔ *Deductible.* A set amount the beneficiary must pay toward a claim before the insurance company pays any money.

Usually, the employer pays a percentage of the cost of the insurance, and the employee pays the rest. When you evaluate benefits, you must carefully read the information about the insurance policies so you understand your portion of the costs, as well as the actual benefits you will receive.

Medical insurance is probably the most important, and every policy is different. For example, your employer might offer a plan that pays a set percentage of the bill—say 80%—leaving you responsible for the rest. Or, you might make a co-payment of $25.00 every time you see a doctor, but the insurance pays the rest. Most medical policies are very specific about what is covered and what is not covered. They also vary in terms of the doctors you can use. For example, a fee-for-service policy lets you choose any doctor, while a preferred-provider policy has a list of doctors from which you can choose.

Other types of insurance often offered by employers include life insurance, which pays money to your family if you die, long-term care insurance, which pays for care if you are unable to care for yourself, and disability insurance, which pays money if you are disabled and cannot work. Both sums are usually a percentage of your annual salary.

How to Negotiate Your Compensation

Do not automatically turn down an offer if the salary and benefits are lower than you expected. You may be able to **negotiate**—discuss options until you reach a compromise—with the employer. Negotiating is a bit like problem solving. The employer makes you an offer. You point out the things about the offer that you think are problems. You work together to find solutions that make you both happy.

For example, you might think the salary is too low. Your employer might not be able to offer you more money, but maybe he can offer a benefit that has value to you, such as free parking or a flexible schedule. Or, he may be able to offer you a raise if you meet expectations for three months.

When you negotiate, always be polite and positive. State your goals clearly so your employer knows what you are trying to achieve. Be prepared to compromise. Remember, you want the job.

How can you use the problem-solving process to negotiate employment needs?

Managing a Job Search ■ Chapter 13

Do You Need Accommodations?

An **accommodation** makes it possible for a disabled employee to perform his or her job responsibilities in a safe and accessible work environment. If you have a disability that might get in the way of your performing a task the same way a nondisabled person would, you may need to negotiate for accommodations when you receive a job offer. For example, if you use a wheelchair to move around, you might not be able to access a storage room that is down a flight of stairs.

By law, employers who are aware of an employee's disability must make accommodations, if the accommodations do not impose an extreme hardship on the business. Tell your employer if you need an accommodation to do your job. Not all disabilities are visible—your employer may not know that you have one.

Examples of common accommodations include:

- Making nonwork areas such as cafeterias and lounges accessible by installing wheelchair ramps or elevators
- Modifying work schedules
- Replacing or modifying equipment
- Changing exams, training materials, or policies to accommodate factors such as reading disabilities
- Providing sign-language interpreters or assistive equipment for the hearing impaired

---

Who Made Your Sneakers?

Laws protect workers in the United States. Workers in other countries might not be as fortunate.

Many of the pieces of clothing sold in the United States are made in Southeast Asia. In countries such as Thailand and Cambodia, workers have very few rights and very few opportunities. Sweatshops—unsafe factories—are common. Children as young as seven years of age may work, and wages for everyone may be very low.

Do you own clothes made in Southeast Asia? How do your own buying habits impact sweatshops in Southeast Asia? How do the sweatshops impact the economy of the United States? Use the Internet or the library to learn more about working conditions in other parts of the world. Write an essay on the topic, "Should We Care About Workers in Other Countries?"

NUMBERS GAME

Some salespeople are paid a commission in addition to or in place of a salary. A **commission** is a payment calculated as a percentage of total sales. Earning a commission encourages sales people to sell more, because the more they sell, the more they earn.

Expensive items such as cars, boats, recreational vehicles, houses, and furniture are commonly sold on commission.

If you are a real estate agent earning a 7% commission, how much would you earn by selling a house for $195,000.00?

Multiply $195,000.00 by 7%:

$195,000.00 × .07 = $13,650.00

What if you sell a house for $242,900.00?

What if you sell cars and earn 8% on your total monthly sales. In one month, you sell one car for $13,955.00, one for $15,495.00, and one for $14,455.00. How much commission will you earn for that month?

Case Study

Claude is hoping to land an internship at a company where he can use his skills in computer programming. He researches companies in his community and identifies five that hire student interns. He picks the one that is closest to his home and calls the human resources department. The assistant tells him to send in a resume and cover letter, along with a school transcript.

Claude wants his application to stand out from those of the other students who apply. He inserts his class photo at the top of his resume. He prints his cover letter on neon yellow paper.

Two weeks after sending in his information, Claude receives a letter telling him that he will not get the internship.

- What do you think Claude did well in his job search?
- What do you think Claude could do to improve his job search?
- What do you think Claude should do next?

Answer It!

1. What is the purpose of a cover letter?
2. Why don't relatives and friends make good references?
3. What are four things you can do to keep your job search materials organized?
4. What is the most important job search document?
5. What are the four main sections of a resume?
6. List four types of information you are likely to need to complete a job application.
7. List four steps for preparing for a job interview.
8. List at least four things you can do to make a good impression at a job interview.
9. List at least six types of benefits.
10. What is the purpose of an accommodation in the workplace?

Web Extra

There are Web sites for teens that provide useful information about managing a job search. For example, snagajob.com has a section on resume-writing tips and links to sample cover letters.

Use the Internet to identify the sites that you think might be the most useful for your peers. Make a directory of the sites, including the Web site address, a brief description, and whether the site is free or has a fee. Post the directory on your school Web site or in your school career center or library.

Managing a Job Search ■ **Chapter 13**

Write Now

The government protects the rights of children as workers. The Fair Labor Standards Act (FLSA) sets the rules for workers under the age of 18. Use the Internet or your library to research the laws that set policies for youth workers in your state. Write an advertisement, flyer, or brochure explaining the laws and providing the address, URL, and phone number for the government agency where people can go for more information.

COLLEGE-READY PRACTICES

Managing the search for where you will attend college is very similar to managing a job search. College-ready individuals realize that the same search strategies that help them focus and prepare for finding a job will help them find a school.

With a partner, prepare for a college admissions interview by researching a school you might be interested in. Based on your research, prepare a list of questions and answers. Take turns acting as the interviewer and the college applicant. If possible, videotape the practice. Analyze the interview carefully. Was your research helpful in answering questions? Can you identify ways to improve your interviewing skills?

Career-Ready Practices

As you have learned, managing a job search can be a complex and time-consuming undertaking. But it all begins with basic research. Career-ready individuals realize that using valid and reliable research strategies will help them focus their job search. It also provides them with valuable information that can help them prepare for interviews and make decisions about prospective employers.

With a partner, write a job description for a specific job at a specific company. Prepare for an interview by researching the company, its competitors, and the industry in general. Based on your research, prepare a list of questions and answers. Take turns acting as the interviewer and the job seeker. If possible, videotape the practice. Analyze the interview carefully. Was your research helpful in answering questions? Can you identify ways to improve your interviewing skills?

Career Portfolio

The most important documents in your career portfolio may be your resume and list of references. Sample cover letters and thank-you letters are also useful. If you have not already done so, prepare these documents for your portfolio.

- Develop a resume. Be complete and accurate. Include an objective specific to the job for which you plan to apply. Use action words to describe your experience and skills. Use formatting that is professional and easy to read.

- Write a cover letter to submit with your resume. Customize it for a specific job that interests you. Keep it brief and to the point.

- Develop a list of references. Try to include at least three people who are willing to speak on your behalf. Ask them for permission to include them on the list.

- Write a sample thank-you letter that you could send after an interview. Use the letter to remind the interviewer about your particular skills and qualifications.

Ask a classmate, teacher, or advisor to review the materials and offer comments and suggestions. Use the comments and suggestions to improve the documents, and then add them to your career portfolio. Periodically, look over the documents and revise them in order to keep them up-to-date.

14 Getting Started in Your Career

Skills in This Chapter . . .

- **Beginning a New Job**
- **Building Work Relationships**
- **Benefiting from a Performance Review**
- **Requesting Additional Education and Training**
- **Obtaining a Raise or Promotion**
- **Making a Career Change**

GETTING STARTED

Starting a new job can be stressful, like starting a new school year. Will you show up on time? Will you make new friends? Will the work be difficult? Like adjusting to a new year or a new school, you will have to meet new responsibilities. Most important, starting a new job requires a positive attitude. Remember that your employer hired you over a lot of other candidates. He or she is confident that you have the skills to be a successful employee; you should be confident, too. To make the first days less stressful, it helps to be prepared by knowing what to expect.

➤ Imagine you are an employer. Make a list of five reasons you would promote someone on the job. Then make a list of five reasons you would fire someone. Share your lists with other members of the class. What did you have in common? What does this tell you about being successful at work?

249

Chapter 14 ■ Getting Started in Your Career

Ask Now

One of the best ways to avoid feeling confused is to speak up and ask questions. Asking questions during orientation can help you to minimize mistakes later on.

✔ Ask questions to get the information you need.

✔ Ask questions to show your employer and co-workers that you are eager to learn.

✔ Ask questions so you don't make mistakes later.

✔ Ask questions about workplace policies, but be sure to read any handbooks or other material you're given.

What questions would you ask of a new employer? If you were the boss, what questions would you expect a new employee to ask you?

Beginning a New Job

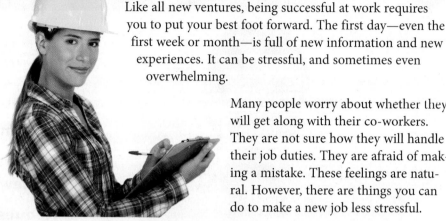

Like all new ventures, being successful at work requires you to put your best foot forward. The first day—even the first week or month—is full of new information and new experiences. It can be stressful, and sometimes even overwhelming.

Many people worry about whether they will get along with their co-workers. They are not sure how they will handle their job duties. They are afraid of making a mistake. These feelings are natural. However, there are things you can do to make a new job less stressful.

■ *Understand exactly what you are supposed to do.* If you are confused about what is expected of you, talk to your manager.

■ *Make a commitment to do your best.* Everyone has a "learning curve," and it may take you a while to learn how to do certain tasks. If you are determined and work hard, your employer will recognize the effort and help you to do your work to the best of your ability.

■ *Get to know your co-workers.* They can answer questions, show you how to do your work more efficiently, and, most important, they can make you feel more comfortable on the job.

What can you expect on your first day of work? Is it normal to be nervous?

You can expect quite a few experiences during your first day and first week at work: forms and rules and tours, to name a few. However, one thing you can always be certain of is that you will learn something new.

Filling out Forms

Your first day will probably include time for filling out forms. One of the first forms you will fill out is a **W-4**—the Employee's Withholding Allowance Certificate. It is a form employees fill out to provide information the employer needs in order to calculate how much money to withhold from wages to pay taxes. You can follow the instructions on the form to fill it out correctly, or ask someone in human resources to help you.

Other forms you may need to fill out include the following:

■ State tax withholding form.

■ Employment Eligibility Verification form, which proves you are a U.S. resident or are authorized to work in the United States.

Getting Started in Your Career ■ Chapter 14

- Authorization to deposit your paycheck directly into your bank account.
- An application for an identification card.
- Benefits selection forms. For example, you may have to make decisions about the type of health insurance plan you want.

Most employers expect you to bring proof of residency (that you are a United States citizen) and all necessary documents with you on the first day—if not before. If your employer does not tell you what to bring, call and ask, so you have the right documents ready. Be prepared to bring the following:

- Your birth certificate or passport, or resident card
- Your Social Security card (If you don't have a Social Security card, you can apply for one by visiting your local Social Security office.)
- A work permit, which is required before anyone under the age of 18 is allowed to work in most jobs, or employment certificate, if necessary

You will spend part of your first day filling out forms. What should you do if you have questions?

Making the Most of Orientation

When you start a new job, you will likely have an orientation your first day. It may be led by someone from human resources or by the manager of your department, or it may be an online or video presentation. Orientation will answer most of your questions and provide the information you need to get started.

At orientation, you are likely to receive an employee handbook. An **employee handbook** describes company policies and procedures, such as how to request vacation time, and the different benefits that are available.

In addition to the orientation, many companies provide new-employee training on the first day, or within the first few days. Some training may be for all new employees. For example, you may receive training about safety procedures, or about harassment policies.

Some training may be specific to your job. For example, if you are a cashier, you may receive training on how to use the cash register. If you are a salesperson, you might learn about the different products available. If you are a cook, you may start learning how to prepare food. Make sure you pay attention during training, and ask questions.

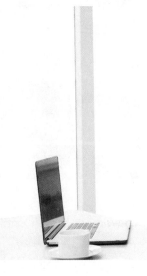

Why should new employees take the time to read their employee handbook?

Making It Through Probation

Some employees have a probation period, which is a set amount of time during which you prove you are right for the job. Probation varies in length depending on the company. It may be a matter of a few weeks or several months. Because they don't expect you to know everything right from the start, employers look for certain behaviors during the probation period.

- 👎 Taking time off or calling in sick
- 👎 Showing up late or leaving early
- 👎 Not following the dress code
- 👎 Failing at basic tasks that you said you were qualified for
- 👍 Making an effort to get along with co-workers
- 👍 Asking questions when appropriate
- 👍 Taking initiative to learn new things
- 👍 Following the rules and regulations

Can you think of other behavior an employer might look for during the probation period?

Making a Good Impression

Recall the importance of making a positive first impression in a job interview? It doesn't stop there. You need to approach your first few days of work the same way you approached the interview that landed you the job. That means staying positive and enthusiastic.

Keep in mind that the person who *hired* you isn't always the person who will be *managing* you. In fact, most of the people you will be working with will probably be meeting you for the first time on your first day. You need to make a good first impression on everybody from day one: your supervisor, your co-workers, and your customers. Some tips to keep in mind:

- *Come to work a little early the first few days.* This will not only help you get adjusted to your surroundings, it will also show that you are eager to do a good job.
- *Dress the part.* Read about the dress code (if there is one) in your handbook. If it's not specific, pay close attention to what your co-workers are wearing and dress similarly. It is better to overdress than to underdress.
- *Be proactive.* Take the initiative to meet your co-workers and learn new job tasks.
- *Keep any negative opinions to yourself.* While other employees may not always seem happy to be at work, avoid passing judgment yourself.

Should you make an effort to introduce yourself to your co-workers or wait until they approach you?

Building Work Relationships

Like any other relationship, the relationships you build at work can affect your on-the-job success, job satisfaction, and overall well-being. Building positive working relationships helps you in the following areas.

- Makes you more productive. Working well with others helps avoid conflict, which usually leads to work being done faster and better.

- Gives you additional opportunities to learn. Your co-workers can be excellent resources for helping you improve your skills.

- Makes work more enjoyable. While work isn't meant to be all fun and games, it can be less stressful and more agreeable if you like the people you work with.

On the other hand, negative working relationships can lead to conflicts, which can decrease productivity, increase stress, and even be the cause of dismissal.

Perhaps the most important thing you can do during your first week of work is to get to know your fellow workers. Be sure to introduce yourself (without being intrusive), keep an open mind, and don't be afraid to ask questions. You might even consider asking co-workers if they'd like to have lunch (or they might ask you). Take advantage of opportunities to help out as well. Co-workers will remember the offer even if they don't take you up on it.

Employers are looking for team players. How much easier is it to complete group projects when you get along with your group members?

NUMBERS GAME

At your job, you may have to choose how much you want to deposit in a retirement plan, such as a 401(k). Many companies offer a matching contribution for retirement plans. They often state the contribution as a ratio, such as 2:1 or 1:1.

A **ratio** is a proportional relationship between two numbers or quantities. For example, if there is one teacher in your class and 24 students, the ratio of teachers to students is 1 to 24, which is written as 1:24.

A 2:1 ratio means that for every $2.00 you contribute, the company will contribute $1.00. A 1:1 ratio means that for every $1.00 you contribute, the company will contribute $1.00.

If you contribute $100.00 a week into your 401(k), and your employer makes a 2:1 matching contribution, how much is deposited into your account each week?

Your contribution:

$100.00

Employer contribution:

$100.00 ÷ 2 = $50.00

Total contribution:

$100.00 + $50.00 = $150.00

How much is contributed each month if your employer makes a 3:1 contribution and you put in $525 a month? How much is deposited each week if you contribute $126.00 at the same 3:1 rate?

Career Trend

You already know that you will hold several different jobs over the course of your career. Did you know that some people choose not to hold a job longer than a few years? These job jumpers aren't interested in moving up the corporate ladder. More often than not, they simply get bored with the work they are doing and want to try something new. Job jumping has become much more expected in today's working world.

Types of Work Relationships

Imagine you are hired as a server at a family restaurant. What work relationships will you build? Naturally you will build relationships with the other servers, the hosts and hostesses, the cooks, and the bussers who clear and set the tables. You will also build a relationship with your manager, and maybe even the restaurant owner. But what about customers? There may be some repeat customers who ask for you specifically because you provide excellent service. You may even have working relationships with people you only see occasionally, such as delivery people or maintenance staff.

Regardless of the person—customer or co-worker, manager or peer—it's important to have a positive attitude. Consider the following tips, no matter whom you are working with:

- Be enthusiastic.
- Don't make excuses.
- Pitch in.
- Adapt to changes quickly.
- Try not to complain.
- Smile!

Also remember that work relationships are just that—*work* relationships. While you may be friends with your co-workers and see them outside of the work setting, while you are on the clock, you should try to keep any personal problems to yourself.

Would you act differently around your co-workers than you would around your supervisor? Would you treat some customers differently than others?

Getting Started in Your Career ■ Chapter 14

Teamwork and Leadership

People who can work successfully as part of a team—whether as a leader or a team member—are valued in the work environment, just as they are valued in school, the community, and at home.

As a team member, you should do the following:

- Use active listening to make sure you understand the needs and opinions of others.
- Show cooperation in order to achieve your team's common goals.
- Compromise when necessary—be sure all parties come away with something positive.
- Complete your tasks and responsibilities on time, and to the best of your ability.

As a leader, you can do the following:

- Help build cooperation among team members.
- Show *initiative*, which means you take charge to accomplish a task that needs to be done.
- Make healthy choices at work.
- Respect the diverse opinions of others while still getting your point across.
- Be assertive, but not aggressive. Don't force other people to follow your lead—encourage and persuade them to.

Remember: No two team members will always agree on everything or be equally skilled at all tasks. Good team members recognize that each person on the team has different ideas and skills. They use these differences to achieve their common goals.

Career Tips

It is important to treat your co-workers with respect. If you are honest and respectful toward others, they are likely to be honest and respectful toward you.

Who Would You Pick?

Unless you are the boss, you don't get to pick your co workers, but you still have to interact with them on a daily basis. But suppose you could pick them? What if you got to choose a group of five people that you would work with at a new job?

★ Would they be close friends or just acquaintances?

★ Would they do what you told them?

★ Would you want them to be smarter, faster, or better workers than you?

★ Would you pick people on the basis of their skills or their personality?

★ Would you hire strangers if you thought they were qualified?

Think about your answers to these questions. Then think about the people who make hiring decisions. Do you think they would make the same choices?

Benefiting from a Performance Review

How do you know when you are doing well in a certain class? You probably have a general idea as the term progresses, but you can't be absolutely certain until your report card comes in. The same is true of the working world. When you are an employee, you receive a job review or **performance review**—a report that rates how well you do your job.

A performance review evaluates you in different areas, depending on your job. For example, if you are a carpenter, it might evaluate you on your ability to use tools or follow safety procedures. If you are a salesperson, your review will likely be based on your sales history, quotas (sales goals) reached, and customer feedback. Regardless of the job, some common areas for evaluation include the following:

- Job performance
- Attendance
- Attitude and behavior
- Ability to communicate with customers
- Relationships with co-workers
- Relationship with manager

A performance review is a resource that helps you understand how your supervisor views your ability to manage your work-related tasks. You can use the information to improve and advance in your career.

What if you received an actual report card every three months at work? How do you think you would do? What kinds of grades would you earn? How is a performance review rating similar to a report card?

Knowing What to Expect

While you are being graded in a way, performance reviews aren't like final exams that you can cram for. However, knowing what you can expect—and when you can expect it—can help you ensure that your work is up to par and that your review will be a good one.

Most companies schedule performance reviews on a regular basis. Your supervisor or a representative from human resources should tell you the schedule, or you might find it in your employee handbook.

Many companies use an assessment form, similar to a report card. Your supervisor will rate your performance in different areas. If you are not sure how performance is rated, ask your supervisor to explain the performance standards ahead of time, so you know what is expected of you.

Some performance reviews will involve your supervisor watching you work—completing specific job tasks while the supervisor observes and takes notes. Other reviews will be more like meetings where you and the supervisor discuss your work more generally.

While your performance review will likely be based on your work as a whole and over a certain period of time, it doesn't hurt to put forth extra effort on the day you are being reviewed—and even a few days before.

If you had to rate yourself as a student, what aspects of your performance would you include? How would you do?

Career Trend

Some companies use peer reviews, in which your co-workers rate your performance, or self-reviews, in which you rate your own performance. This can be harder than you think. Some people are harder on themselves than an employer would ever be.

Money Madne$$

Your performance review is coming up and you want to make sure you've met your annual sales quota. You are required to sell $200,000.00 in digital widgamadigits each year. It's December 30, and so far you've only sold $188,000.00. Widgamadigits cost $100.00 each, though the price is reduced to $75.00 for customers who buy quantities of 100 or more. You are in the process of closing a sale of 18 widgamadigits for one customer and are hoping for one more big sale to put you over the top. How many widgamadigits would you have to sell to one customer to meet your quota?

What the Performance Review Means

As part of your review, you will probably sit down with your supervisor to discuss your scores. Though every review is different, reviews generally evaluate employees in terms of their ability to meet expectations.

- You might receive a score of Meets Standards if you consistently complete your responsibilities on time.
- You might receive a score of Exceeds Standards if you often help others complete their own work.
- You might receive a score of Needs Improvement if you are frequently late for work.

What does it mean if you receive a score of "Needs Improvement" on a performance review? Does your supervisor hate you? Will you lose your job? Don't panic. The information in your performance review is intended to help you succeed at work. Use the time in the meeting to discuss with your supervisor ways you can make the most of your strengths at work, as well as how you can improve in the areas where you may be underperforming.

Use these tips to make the most of your performance review.

- Listen attentively to everything your supervisor says.
- Take note of areas in which your supervisor expects you to improve.
- Be prepared to talk about yourself and your role at work.
- Accept criticism calmly; it is meant to help you improve.
- Ask questions so you fully understand what is expected of you.

Your performance review is also a good time to discuss the goals you have achieved and to set new goals for the coming year. You can discuss things that are on your mind, such as requesting training or asking about opportunities for new responsibilities.

Why is it difficult for people to accept criticism? What can you do to make sure you don't take any criticism too personally?

Requesting a Review

Sometimes, a supervisor does not schedule a performance review. He or she may be busy, or might not enjoy the process. While this may sound like a good thing—like the teacher who forgets to give a quiz—it's not. Passing up a performance review means passing up a chance to improve, to show off your successes, and to ask for more responsibility and possibly more money.

If, for some reason, your supervisor neglects to schedule a review, you will have to speak up. Be polite but assertive. Remind your supervisor that it is time for the review, and that you would welcome information that will help you succeed at your job.

Requesting Additional Education and Training

To keep up with changes in the workplace, you may need to develop new skills. Some careers, such as those in health care, require that you update your certification on a regular basis. For other careers, it is up to you to make sure you stay up to date on changes and advancements.

- A computer technician must keep up with changes in technology.
- A tax accountant is responsible for knowing about changes to tax policy and regulations.
- A travel agent may want to keep up with world events so she can recommend safe vacation spots.

Ask your teacher if he or she has to undergo additional education and training on a regular basis and he or she will likely tell you about conferences, teacher training, and in-service days. Regardless of the profession, you almost always have opportunities for improving your skills and learning more about your job. Seek out additional education and training, and make sure you are prepared for changes and advancement opportunities that come your way.

Career Tips

When a manufacturing plant replaces old equipment with new, updated equipment, it's called retooling. Nowadays, more and more workers are *retooling* as well—replacing old skills with new ones. Community colleges, for example, are experiencing record enrollment from students headed back to school for more training. This not only makes workers better at their jobs, it often leads to better jobs.

Do You Make the Grade?

Here are four reasons employees might not meet performance standards, and actions for improvement.

✔ *Not enough training.* If you do not know the proper procedure, it is difficult to perform up to expectations. Ask for training so that you can improve.

✔ *Distractions.* Problems in the workplace can interfere with performance. For example, you might be bothered by noise, conflict with co-workers or customers, poor lighting, or equipment that does not work. Politely discuss the problems with your supervisor.

✔ *Lack of motivation or encouragement.* You may be bored or tired, or you may feel as if your work is not valued by your supervisor or co-workers. It is hard to be positive if you are not motivated. Discuss your feelings with your supervisor to see what steps you can take to improve conditions.

✔ *Failure to follow rules.* If there are company policies or rules that you haven't complied with, then you should either start complying or look for a different job. If there are policies you don't agree with, you can discuss the matter with your supervisor, provided you have a good reason. Be sure to listen openly to feedback.

What kinds of workers would you expect to stay most up-to-date with their knowledge and skills? Why?

How Can You Develop New Skills?

There are many ways to develop new skills. Co-workers and supervisors often provide on-the-job training simply by showing you how to improve the way you do your regular tasks, or teaching you new skills. For example, a co-worker might train you on a new computer application, teach you how to install a piece of drywall, or simply show you how to make a conference call.

Your company might also offer other opportunities for education and training.

- Specialized courses, such as a seminar about customer service for sales representatives, or a session on how to use Web-based storage for data management employees.

- **Professional development**—training in your chosen career. Teachers, lawyers, and nurses, for example, are usually required to participate in professional development to learn about new trends in their fields or to prepare for certification exams.

- Tuition reimbursement programs. If you take a class, your company may pay the tuition costs. You can take classes at a local college on nights or weekends to earn a degree or certification, or simply to learn material that might help you advance.

Developing new skills prepares you for new responsibilities and helps you achieve career, academic, and life goals. Even if your employer does not offer these opportunities, you can take steps to continue your education and training on your own by taking classes or reading books. This kind of education makes you a more valuable worker and helps you grow as a person.

Volunteering Overseas

One of the best ways to learn new skills and to further your professional development is to volunteer. You have hundreds of options for volunteering in this country, and hundreds more opportunities abroad. Underdeveloped countries especially have a tremendous need for educated, committed volunteers. You could teach English, help build a school, or help run an orphanage.

The benefits of volunteer service overseas are numerous. You'll have a chance to learn a new language, experience a different culture, travel to new places, and develop key career and leadership skills. Volunteering abroad can enhance long-term career prospects whether you want to work for a corporation, a nonprofit organization, or a government agency.

In addition, several organizations, such as the Peace Corps, offer educational benefits, such as college credit or student loan deferment, for service. The Peace Corps can even open doors to graduate school.

Use the Internet to research your options for volunteering in another country. You might start with peacecorps.gov or globalservicecorps.org. Learn what the requirements are and what specific opportunities are available. Pick one program that you would like to teach your class about and develop a three-minute presentation, including photos and contact information, to give to your class.

Finding a Mentor

One of the best ways to develop your skills is to find a mentor. A **mentor** is someone knowledgeable and experienced in your field who is willing to teach you, advise you, and help you reach your goals. While all mentors are teachers of sorts, not all teachers are mentors. A mentor is someone who is willing to give you extra individual attention, to guide you along your career path, and to pass along what he or she has already learned.

Have you ever had someone "take you under their wing" and help you to improve? Have you ever considered being a mentor to somebody else?

Finding a mentor at work can be as easy as asking someone you know and respect at work to train you or help you learn new skills. Consider these other tips when finding and choosing a mentor.

- *Know what you want.* Are you looking for someone to teach you specific skills, to help you make important career decisions, or to just be a sounding board for when you have problems? Know what you'd like your mentor to do for you and be clear about this up front.

- *Explore your network.* Your mentor doesn't have to be your supervisor or another member of your work team or department, though that is often the case. He or she may be a teacher or a counselor, a coach, or even a family member.

- *Show your gratitude.* Most mentors will help you because they honestly want to see you succeed. However, it doesn't hurt to thank them for their efforts and to help them out if ever the occasion arises.

Becoming a Lifelong Learner

Lifelong learning means continually acquiring new knowledge and skills throughout the course of your life. While this might sound like a given, it actually goes against traditional thinking that education ends the moment one graduates from high school or college. The truth is you never stop learning. It's just that nowadays people are making a more concentrated effort to *keep* learning, and not just because their careers demand it.

Education and training are not limited to learning new skills for the workplace. You should consider educational opportunities to enrich your life at home, with friends, and in your community. Studies have shown that people who continue to take an active role in their learning lead happier, even healthier lives. To become a lifelong learner, consider the following.

- Take a class to learn a new hobby, such as Web design or woodworking.
- Join an exercise program, such as karate, golf, or fitness.
- Enroll in a community college to improve your personal finance skills or learn more about computers.

- Join a book club to develop your reading skills and expand your social group.
- Attend a seminar at your bank about starting your own business or investing in the stock market.
- Teach others about something you are good at or knowledgeable about, such as art, sports, or music.

Regardless of what you study, remember that the goal of lifelong learning is not simply to advance your career, but to cultivate your interests and gain new experiences.

Lifelong learning doesn't mean you always have to improve your job skills. What would you learn how to do if you had the time?

Tech Connect

Can you imagine graduating from college without ever leaving your home? Welcome to the world of online education. Many colleges offer strictly online courses that you access over the Internet. The instructor posts video lectures, notes, and assignments on a Web site. You log in and participate as if you were on campus. Sites often have tools that allow students to communicate in real time so they can study together, or work on group assignments. Some of these courses are even free. Though they are generally for no credit, they provide a new way to learn about something that interests you.

Some companies provide online training, too. You might be able to complete a course in ethics online, or use a simulation training program to learn how to use new equipment.

Use the Internet to research the online courses offered by colleges in your area. Do they lead to a degree? What requirements must you meet to enroll? Make a presentation about the benefits and drawbacks of taking a course online. Try to find at least one class that you think you would be interested in taking online.

Obtaining a Raise or Promotion

In school, you are rewarded for all of your studying and hard work with good grades. You may be rewarded for your performance in a sport or club with a ribbon or a trophy. In the working world, your rewards come in the form of raises and promotions.

A **raise** is an increase in pay. A **promotion** is an advance in your career that includes a new job title and additional responsibilities. Both are ways that a company recognizes your achievements.

Earning a raise and/or promotion is an important part of career success and satisfaction. It is proof that your employer appreciates your contributions. In addition to the financial rewards, promotions bring with them new challenges. They also may open up even more rewarding career opportunities down the road.

Myth Companies are required to give employees a raise every year.

Truth Companies understand that the cost of living goes up every year. While most companies try to reward their employees with annual raises, they are not required to do so. Raises are often based on merit. That is, they have to be earned. There are times when employers can't afford to give raises even to the people who deserve them. In the working world, nothing is a given.

The Importance of Advancement

Promotions often come with an increase in pay, but they also come with added responsibilities. You may be asked to take on additional or more challenging work tasks. You may be asked to work different hours, to travel more, or to supervise others. You might also experience more pressure to perform. You might even question if it is worth the raise.

It's important to remember that advancing in your career is about much more than a pay increase. The money is nice, but a promotion offers other benefits.

- Promotions sometimes include added **perks**—which is compensation in addition to wages or salary—and benefits, such as a nicer office, better parking, more vacation time, or better health insurance.
- Advancing in your career provides a boost to your self-esteem.
- Being recognized and rewarded brings career satisfaction.
- Achieving certain goals allows you to set new goals for the future.
- Earning a promotion makes you look more attractive to other employers, which could lead to even better job offers.

Not to mention that new challenges can be exciting. Staying in the same position and doing the same work every day can become tedious. By moving up, you learn new skills and push yourself to grow as a person.

Do you want to be recognized for the work you do? Are the benefits worth the hard work required?

Career Profile: Veterinarian

Health Science Career Cluster

Job Summary

Veterinarians take care of pets and other animals. They treat injuries and illnesses, perform surgery, administer vaccinations, prescribe medicines, and inform animal owners of the proper way to care for their pets. Some are involved in research and work with laboratory animals. Others work for farms, ranches, or zoos. More than half of all veterinarians treat small animals—cats, dogs, birds, and rabbits, with the occasional snake thrown in for good measure.

Becoming a veterinarian is no easy feat, however, requiring much the same educational requirements as a physician. A doctorate of veterinary medicine usually takes four years; that's above and beyond other postsecondary requirements, such as a bachelor's degree. Veterinarians also must be licensed—which requires them to pass an 8-hour examination! Once you become a veterinarian, don't expect an easy job. Unlike human patients, most animal patients have few reservations about biting, clawing, kicking, and hissing.

Use the Internet, a library, or career center to research the responsibilities, education and training requirements, and salary range of a veterinarian. Prepare a presentation about the career.

What skills do you think a veterinarian would need?

How Can You Earn a Raise or a Promotion?

So, how do you achieve a raise or promotion? The best way is to work hard, excel at your job, and show your supervisor that you have the skills, interests, and abilities to take on more responsibility.

How long does it take to earn a raise? The answer depends on your job and on company policies. There are two basic types of raises:

- An *annual raise* is awarded to all employees on a regular basis—usually once a year. It may be based on seniority—how long you have worked for the company—or on how successful the company has been. You will be notified of the increase a few weeks in advance. It is usually a percentage of your current pay.

- A *merit raise* is based on your performance. If you are exceeding expectations, or if you are promoted to a new job or assume new responsibilities, you might earn a merit raise. A merit raise can be awarded at any time.

What does it take to earn a promotion? First, you must prove to your supervisor that you are ready for more responsibility. Then, there must be an opportunity.

- You might earn a promotion if someone with more seniority leaves. For example, your supervisor might retire, or a co-worker with more responsibility might take a job in a different department.

Getting Started in Your Career ■ Chapter 14

- Another way to earn a promotion is if the company grows. Companies that do well might expand into new areas of businesses. They will need employees to take on the new tasks and responsibilities.
- A third way to earn a promotion is to create a new job for yourself. You might see an opportunity for something new in your department. For example, if you work in a restaurant, you might suggest to your manager that you could expand business by offering home delivery. Your manager might put you in charge of the new home delivery service.

Making a Career Change

You've already learned that the average worker changes jobs many times over his or her working career. But how do you know when it is time for a change?

That depends on whether it is a decision you make for yourself, or one that your employer makes for you.

- You might feel that you have reached your potential, and that there is no more room for advancement at your current place of work.
- Your family might move to a different city, making the commute to your current job impossible.
- You might just want to try something new.
- New family or personal responsibilities may require you to take a job with different hours, different pay, or different benefits.

Or, it could be because of one of the following reasons.

- Your employer might not be satisfied with your performance.
- Your job might be eliminated.
- Your company might go out of business or move operations to another country.

Changing jobs is stressful, just like changing schools or starting a new grade, and there is often much more at stake. It means leaving a familiar situation. It means beginning your job search all over again. It may even mean going back to school or learning new skills.

A job change is also an opportunity to think critically about your career and to set new career goals. Looking at a job change as an opportunity instead of a loss can lead to greater career satisfaction and positive well-being.

Going Above and Beyond

Imagine you have been with the same company a few years. Your reviews have been good. You earn an annual raise every year. But you can't seem to earn a promotion. What can you do to get recognized?

- 👎 Start acting as if you were promoted anyway and boss your co-workers around.
- 👎 Be disrespectful to your supervisor.
- 👎 Pass up opportunities for developing new skills.
- 👍 Join a professional organization.
- 👍 Propose new projects or suggest ways to improve productivity.
- 👍 Meet with your supervisor to discuss opportunities in your department, or in other departments.

How could you advance in your career if there are no opportunities for a promotion available?

Is leaving a job ever an easy decision to make? Think of one or two examples.

How to Say Goodbye

Even when you are the one who makes the decision to leave, changing jobs can be confusing and depressing. There are some things you can do to make the process easier for yourself and your employer.

- Giving plenty of notice. **Giving notice** means telling your employer you are leaving. You should be clear about the date you plan to stop working. Most companies ask you to give notice two weeks before your last day. That gives them time to look for your replacement. You may also need to write a **letter of resignation**—a brief, formal, and positive letter that states that you are leaving the job, the date your employment will end, and thanks your employer for the opportunity to have worked there.

- Creating a job folder for your replacement. The folder should contain lists of the responsibilities, procedures, and resources needed for the job. It will help your replacement settle in and become productive.

- Meeting with a human resources representative to discuss your benefits, including what to do with your 401(k) or other retirement account. This may be part of your **exit interview**—a meeting with your supervisor or human resources representative to discuss why you are leaving.

Leaving on good terms is important. You want to maintain positive relationships with your supervisor and co-workers. You may want them as references for future jobs or as contacts for networking, or you may want to work for the company again someday.

Jobs for the Self-Employed

Some jobs have higher proportions of self-employed workers than others. Often this has to do with the kind of work being performed—whether it can be done by an individual or a group, for example, or whether it can be done from home or by telecommuting.

Below are some of the jobs with high percentages of self-employed workers.

- Personal financial advisors
- Chiropractors
- Interior designers
- Creative writers
- Property managers
- Construction managers
- Veterinarians
- Graphic designers
- Lawyers
- Podiatrists
- Real estate agents

Pick one of these jobs and research it. In particular, try to discover why these jobs are more attractive to people interested in self-employment. Write a brief (two- to three-paragraph) report on the job you've choosen and present it to the class.

Getting Started in Your Career ■ Chapter 14

Josh has always wanted to move to California. He has lived in a small town all of his life and is thinking about heading to downtown Los Angeles or San Francisco. He's worried about getting a job there, however. He knows that the cost of living in those cities is high. He has a bachelor's degree in English and has some work experience building Web sites. He also has four years of experience as a waiter and another two years working as a junior copywriter for a local advertising agency.

Use your 21st Century Skills such as problem solving, critical thinking, and decision making to help Josh decide what his options are. What kinds of jobs are plentiful in those cities? What kind of work do you think he could do? How much would he have to earn in order to afford a place to live? Write two endings to Josh's story: an ending in which Josh is successful and an ending in which he isn't. Share your stories with your class.

Career Tips

Just because you have a job doesn't mean you must stop looking for another. Even if you are satisfied with your current position, you might find one that has a similar work environment and responsibilities that offers better pay or benefits. Just don't surf the job boards or want ads while you are at work.

Coping with a Layoff

At some point, through no fault of your own, you may lose your job. You might be working hard and doing everything right and still find yourself let go or laid off. A **layoff** is a job loss caused when a company has no work for certain employees for a period of time. Some layoffs are short—you might work at a factory that receives no orders for new products for six months, but when new orders come in, you are called back to work. Some layoffs are permanent. The factory might shut down and never reopen.

Losing your job can make you angry or depressed. It seems unfair. You lose your income. You lose your daily routine. You may even lose friends. To better cope with a job loss, consider the following.

- Make sure you understand the reason for your termination. Try to get it in writing.

- Ask for letters of recommendation. Most likely these will come from supervisors, though you might ask co-workers for recommendations as well.

21st Century Learning

In Good Times and Bad

Layoffs are more likely to happen when the economy is slow, such as during a recession. To make up for lost profits, companies will look to cut costs by trimming personnel or outsourcing work. There are some jobs and industries, however, that are more naturally resistant to tough economic times. Consider the following careers that tend to have more staying power:

✔ *Nurses*
✔ *Teachers*
✔ *Firefighter*
✔ *Government Employees*
✔ *Veterinarians*
✔ *Accountants*
✔ *Pharmacists*
✔ *Database Administrators*
✔ *Engineers*
✔ *Police Officers*

Have you ever asked anyone for a letter of recommendation? Why do you think employers request them?

- Analyze the terms of your **severance package**—the compensation you receive because you are being terminated. It may include wages, training, career counseling, and temporary continuation of medical insurance.
- File for unemployment benefits as soon as you are eligible. Unemployment benefits include money and free career counseling services that are available to unemployed workers while they are looking for new jobs.
- Check into your health insurance options. If you were insured through a company policy you will need to get your own health insurance.
- Reevaluate your budget. The loss of income may require you to make changes to your spending until you can find a new job.
- Ask for help. Make sure you have a support network in place to help you deal with the stress of losing your job and finding a new one. This same network may be able to help you to find work as well.
- Stay positive. Feelings of anger or sadness are natural, but they can lead to hopelessness or depression. By staying positive you will be more likely to seek out new opportunities and will find work faster.

You've probably heard the phrase, "a blessing in disguise." Many times, a job loss is exactly that. While it is stressful and life-altering, it can also provide the incentive for you to move on to something more rewarding.

Though the name may suggest otherwise, if you are laid off, COBRA can be a good thing. COBRA provides former employees and their families the right to temporary continuation of their health coverage at group rates. COBRA helps make sure that individuals don't have to go without health insurance while they are looking for a new job.

Finding a New Job

The best time to look for a new job is while you still have the old one. If you make the decision to resign, or if you know a layoff is coming, you can update your resume and start your job search before you are unemployed.

The process for finding a new job is basically the same as for finding a first job. Assess your skills, abilities, and interests; research job opportunities; and send out your resume. However, there are a few advantages to finding a second (or third or fourth) job.

- *You have experience.* Make sure your resume highlights your most recent job description. Give specific examples of your successes.
- *You have developed contacts in the industry.* Get in touch with people you met through work, including customers and clients, members of professional organizations, and even your competition. Let them know you are looking for work.
- *You know more about yourself.* You know if the kind of work you have been doing is right for you. You know what it takes to succeed in that industry. You know if you are ready to try something new.

You should be careful about where and when you look for work, however. Surfing job boards for new opportunities from the workstation of your current job could get you laid off faster. If you are thinking about making a career change, it is best to keep that information to yourself—at least while you are at work. Once a new opportunity arises, however, you should discuss it with your current employer. The possibility of your leaving might persuade him or her to make a counter offer—perhaps giving you enough reason to stay.

Taking Time Off

Every day, thousands of people put their careers on hold. Often it's time needed to raise a family or care for relatives. Others take time off to travel, go back to school, or start their own business. Sometimes people just need time away to reevaluate their career choices. At some point, however, many of those same people will want to reenter the workforce. Going back to work after substantial time off isn't always easy. Workers trying to find jobs after taking time off can feel under-skilled and overwhelmed. It can sometimes be even more difficult to reenter the workforce than it was to enter it in the first place. Returning to work after taking time off brings with it several considerations.

- Should you go back to your previous career or at least work in that same industry?
- Do you still have contacts you can count on for job leads and references?
- Do you still have the education and skills required, or will you need to go back to school?
- Have your career values and priorities changed? For example, do you need work that pays more, has more time off, or is less stressful?
- Has the job market changed? Are there new jobs available that you'd consider?

You also will need to consider the time off from an employer's perspective. While it may seem unfair, most employers see a gap in employment history and assume that it means a worker with out-of-date skills and training. For that reason, you'll want to think about how you will approach your time off on your resume or in an interview.

Use a functional type resume that emphasizes your key skills rather than your work history. Use your cover letter as an opportunity to explain the time off and emphasize any non-work related activity you've done to improve your knowledge and skills and keep up with the industry.

In the interview, focus on why you want to return to work rather than why you left in the first place. Emphasize what you hope to accomplish for the future.

Remember that no matter what you did in your time off—whether it involved traveling the world, raising a family, or volunteering for a good cause—it helped you grow as a person and probably improved important skills that you can transfer to your job.

Why is it important to keep in contact with co-workers, clients, and other people you work with even after leaving a job?

A Closed Door or an Open One?

Being unemployed may be an opportunity to expand your knowledge and skills. What if the restaurant you are working for closes permanently?

★ You might decide to open your own sandwich shop.

★ You might retrain in an industry that is growing, such as health care.

★ You might move to a part of the country that has low unemployment.

★ You might take a temporary job while you decide what to do next.

★ You might take time to travel.

★ You might join a not-for-profit organization such as the Peace Corps.

What skills and resources can you use to turn a job loss into an opportunity for advancement?

Case Study

Leah has been working as an assistant in a medical office for three months. At her first performance review, her supervisor praised her for having a perfect attendance record. However, Leah received a score of Needs Improvement for organization, because she often filed patient records incorrectly.

Leah did not like the Needs Improvement score. She interrupted her supervisor to say that she thought it was one of the other assistants who made the errors. She said that filing was too easy, anyway, and that she needed something more challenging to do.

Leah's supervisor replied that if Leah was not happy with her role in the office, maybe the job was not right for her. Leah was stunned. Was she being fired?

- Do you think Leah handled herself appropriately at her performance review?
- What could she have done differently?
- What decisions does Leah have to make now?
- How should she respond to her supervisor?

Answer It!

1. List two things you can do to make a new job less stressful.
2. Name three kinds of forms you might be expected to fill out on your first day at work.
3. What does an employee handbook do?
4. Name some strategies for making a good impression during your first few days at work.
5. List three characteristics of being a good leader.
6. List the six areas that are commonly evaluated on a performance review.
7. What should you do to get the most out of a performance review?
8. What is a tuition-reimbursement program?
9. What is the difference between an annual raise and a merit raise?
10. Why might it be easier to find a second job than a first one?

Web Extra

Each state has its own system for handling unemployment benefits. Use the Internet to research the procedure for your state. You might start at doleta.gov, the site for the U.S. Department of Labor's employment and training administration.

Summarize the information in a report, and include links to important information for your state. Publish the summary on your school Web site, or make it available in your career center or library.

Write Now

Can you think of reasons why someone might become dissatisfied with a job that he or she used to enjoy? Is it because of changing values? Changing goals? Could changes in other areas of life affect his or her job satisfaction?

If you were a career counselor, how would you advise someone who is dissatisfied with work? Write a report describing the person. Include information about the job and the person's career history, such as how long he or she has had the job, what responsibilities there are, and the quality of his or her relationships with co-workers.

Describe the reasons the person is dissatisfied. Explain how you would recommend he or she approach the problem and find solutions.

COLLEGE-READY PRACTICES

Like a performance review, a report card provides valuable feedback on how well you are doing on your schoolwork. College-ready individuals understand that a report card is a resource that you can use to improve and advance in school.

Take a good, objective look at your most recent report card. Use critical-thinking to identify areas where you excel, where you are doing well, and where you can improve. Then, write yourself a performance review. Suggest at least five specific things you could do to improve the results of your next report card.

Career-Ready Practices

Career-ready individuals seek new methods, practices, and ideas from a variety of sources that they can apply to their own workplace. A professional organization is one such resource.

Joining a professional organization provides you with opportunities to learn more about your industry, strengthen your job skills, and expand your network of personal contacts. To an employer, it shows that you are motivated to look beyond the workplace to find new and innovative ideas that can help you and your organization be more productive.

One way that a professional organization helps its members is by distributing information about the industry through Web sites, newsletters, or magazines.

Working in teams of three or four, select an industry and create a two-page publication for its professional organization. Try not to be too general in your choice (choose nursing instead of medicine or marketing instead of business). Use the Internet to research current trends, policies, or events affecting the industry. Give the publication a suitable name, and write three or four articles that the members would find useful. Include pictures if appropriate.

When your project is complete, print it to share with the class, or publish it on your class Web site.

Career Portfolio

A performance review makes an ideal piece for a career portfolio—provided it is a good one, of course. Such reviews showcase specific skills and exhibit your performance on the job in ways that general recommendations or a resume can't. Plus, several performance reviews together can show a history of improvement and personal growth, providing a kind of timeline of your work experience.

Some employers may be reluctant to provide you with a copy of your review for your own records. If that is the case, consider asking your supervisor to type up a summary of the review that you can include in your portfolio in lieu of the actual review.

In addition to the performance review, this chapter discussed several other kinds of experiences and documents that you might want to include in your portfolio. Can you think of any?

15 Being Productive in Your Career

Skills in This Chapter . . .

- **Being a Successful Employee**
- **Applying Time-Management Techniques in the Workplace**
- **Managing Workplace Conflict**
- **Recognizing Your Rights and Responsibilities**

GETTING STARTED

What does it take to succeed? Think about your classes, hobbies, or co-curricular activities. Think about what it takes to get an A on a test, to finish a project, or to cross the finish line first. Like success in school, career success has a lot to do with your ability to adapt to changes, to work well with others, to think clearly and critically, and to manage your resources effectively to continue to learn and grow. Most important, career success requires you to take responsibility for your actions.

▶ Think of a class or an activity (such as a sport or hobby) in which you do well. Make a list of five personal traits you have that help you to succeed in that class or activity. Now think of different work situations or scenarios in which that personal characteristic would come in handy, one scenario for each trait. Pick one of those scenarios and write a paragraph describing how your personal characteristics would help you to excel.

Being a Successful Employee

The qualities that make you successful in other areas of your life make you successful at work, too. The determination required to run a race or work through a complex math problem would come in handy when you need to work extra hours to complete a marketing project or finish your delivery route. The critical thinking skills you use to perform a science experiment would be valuable when you are diagnosing a patient or preparing a legal brief. The communication skills you use to keep two friends from arguing would also be helpful when you are dealing with an angry customer insisting on a refund.

Think about someone you know who is good at his or her job. Why do you think he is good at what he does? Is he smarter than others who do similar jobs? Does he have more resources? Or does he simply work harder than the rest?

Successful employees exhibit the traits that employers need most. According to the National Association of Colleges and Employers, employers look for job candidates who are honest and motivated, who have a good work ethic, exhibit dedication and perserverance, and who have strong communication and teamwork skills. Think of just about any job on the market: Who doesn't want a motivated and honest worker who knows how to get along with others and be part of a team? It doesn't really matter *what* you are doing: The skills and qualities that make for a successful employee apply across the board.

What does it take to be the best at something? Do you want to be the best at your job, or do you want to do "just good enough" to get a paycheck?

Being Professional

Have you ever heard someone say that a worker exhibits *professionalism*? Maybe a repair worker calls back two days after fixing your furnace to make sure your house is still warm. Maybe an emergency dispatcher calmly talks a frantic mother through the steps necessary to stop her child's bleeding while waiting for the ambulance. Maybe a busy waiter takes the time to help you select a meal and feel comfortable.

Professionalism is the ability to show respect to everyone around you while you perform your responsibilities as best as you can. It includes a basic set of personal qualities that make an employee successful—no matter what job or career he or she has. The personal qualities that combine to create professionalism include the following:

- Work ethics
- Integrity

Being Productive in Your Career ■ Chapter 15

- Dedication
- Perserverance
- Courtesy to all
- Honesty
- Dependability
- Responsibility

When you show professionalism, you earn the trust and respect of your supervisor, co-workers, and customers. You contribute to your own well-being, because you can feel good about the way you treat others and the way you meet your work responsibilities. Generally, when you feel good about the work you do, it means you are doing good work.

Is "being a professional" the same as "showing professionalism"? Can anyone show professionalism, regardless of the kind of work he or she does?

In today's workplaces, people change jobs often, and companies have trouble hanging on to qualified workers. That's why employers value *loyalty* almost as much as hard work, honesty, and dependability. Many employers still reward loyalty with raises, promotions, bonuses, and other incentives.

Being a Problem Solver

While many of the characteristics that make you successful at school will help you to succeed at work, you will still run into problems and obstacles that you have not encountered before. Businesses and organizations want workers who can take the initiative and solve problems as they arise. In fact, the ability to solve problems is consistently ranked by employers as one of the most important skills.

Recall that a problem is a difficulty you must resolve before you can make progress. You are likely to encounter a wide variety of problems depending on your work role.

- Should you give a customer a discount in order to make the sale?
- Should you buy cheaper materials in order to finish a job under budget?
- What do you do if your computer crashes in the middle of a multimedia presentation?
- Do you know the right person to speak to about ordering supplies?
- What should you do if a co-worker wants to gossip about your supervisor?

Meeting Your Employer's Expectations

Every employer will have different expectations, just as every job will come with a different set of duties. There are some expectations that almost every employer can agree on.

✔ Work hard for a full day.
✔ Follow the rules.
✔ Be honest.
✔ Get along with others.
✔ Respect people's differences.
✔ Show up every day on time.
✔ Meet deadlines.
✔ Stay positive.
✔ Avoid mistakes.
✔ Accept responsibility.

Can you think of other expectations that employers have? Have you ever had trouble meeting any of these expectations at work or at school?

Someone You Can Count On

Regardless of the job, employers need their workers to be reliable. If an employer can't count on you to show up and do your work to the best of your ability, he or she will find someone else who will.

👎 Showing up late day after day

👎 Taking too many sick or personal days

👎 Taking longer than allowed breaks and lunches

👎 Missing deadlines

👍 Completing work accurately and on time

👍 Calling ahead if you are going to be late

👍 Scheduling days off well in advance

👍 Working overtime if required

👍 Helping out co-workers when they need it

Are you reliable? Can your friends and family members count on you? Is there anything you could do to be more dependable?

Because the kinds of problems you'll face will vary, it's important to have a process for resolving them. You can use your problem-solving skills at work the same way you use them in other areas of your life.

- Identify the problem.
- Consider all possible solutions.
- Identify the consequences of each solution.
- Select the best solution.
- Make and implement a plan of action.

Taking responsibility for your problems and working to find solutions shows you are independent and capable. That, in turn, can lead to increased responsibility and more recognition.

Can you think of a single job in which workers *aren't* expected to solve problems?

Being Ethical at Work

Recall that *ethics* are a set of beliefs about what is right and what is wrong. **Work ethics** are beliefs and behaviors about what is right and wrong in a work environment. When you behave ethically at work, others will trust and respect you.

Most work ethics are similar to principles from other areas of life that you apply to your workplace.

- Always try to do your best.
- Respect the authority of your supervisor.
- Respect your co-workers.
- Respect company property.

Not everyone behaves ethically at work. When you are in a situation where others are acting in a way that you know is unethical, you will have to use your decision-making skills to choose how to respond.

Being Productive in Your Career ■ Chapter 15

- You might see someone taking office supplies home for personal use, which is stealing.
- You might hear co-workers insulting a customer or manager, which is disrespectful.
- Someone might leave early and ask you to lie about it to your supervisor, which is dishonest.

Of course, ethics extends beyond the actions and behaviors of individual workers. It applies to companies and organizations as a whole.

- You might work for a company that dumps chemical waste into a nearby river.
- You might work for an employer who neglects safety procedures in order to cut costs.
- You might work for an organization that doesn't report all of the taxes it owes.

It's important to remember that being ethical means standing up for what you believe, even if that goes against the actions and beliefs of others—even your employer.

What would you do if you caught someone stealing at work?

NUMBERS GAME

Imagine you work at a factory. At a recent meeting, your supervisor happily noted that work-related injuries were down 20% from the previous month. During the current month there had only been 16 injuries. In the month prior, there were 20.

A percentage *increase* represents how much *more* of something there is compared to a previous number or amount. A percentage *decrease* represents how much *less* of something there is.

In the example above, there were four fewer injuries in the current month compared to the previous month. To find the percentage *decrease*, first find the *difference* of the two numbers. Then divide that answer by the *higher* number. Finally, multiply that number by 100.

20 − 16 = 4

4 ÷ 20 = .20

.20 × 100 = 20%

To find the percentage *increase*, divide the difference of the two numbers by the *lower* number. Then multiply the answer by 100.

Now imagine it is a month later and your supervisor sadly reports that the number of injuries rose from 16 to 24. What is the percentage increase?

Applying Time-Management Techniques in the Workplace

You arrive at your work as a marketing assistant at 8:50 a.m. You look at your to-do list. It has ten tasks on it. You get started on task one. The phone rings. Your supervisor asks you to reschedule an appointment. Then, a co-worker needs help preparing a mailing that has to go out today. Your sister calls. She's in the area and wants to meet for lunch. After lunch, a customer calls to inquire about a new product. Then, your supervisor wants you to sit in on a meeting. You look at your watch and it is 5:05. You look at your to-do list. The day is over, and you are only halfway through item one.

Time management is a critical skill for succeeding at work. Employers want you to make the most of your time—to them, time is money. To achieve your career goals, you will have to manage your time effectively and *set priorities*—decide which tasks must be completed first.

Should you work on the project worth half your grade due next week or finish your math homework due tomorrow? What time management skills do you use to succeed at school?

Tech Connect

Sometimes it seems everyone needs a personal assistant . . . or at least a digital version of one. A personal digital assistant (PDA) is a handheld computer, sometimes called a palmtop computer. Most PDAs are smartphones: mobile phones capable of browsing the Web, creating and storing documents, and acting as portable multimedia players. This technology allows users to stay connected to almost everyone from almost anywhere.

But how smart is a smartphone? While the applications available can certainly make you a more productive worker, they can also be a distraction. While your smartphone may alert you to your upcoming meeting, it also tempts you to play games instead of paying attention to the latest sales figures. Technology is only as valuable as the person using it.

Imagine you are the manager of information technology at a small business. Department managers are all issued company smartphones. You are responsible for developing a list of applications that each manager should download to his or her phone. Research business-oriented applications that would help a department manager work more productively. Create a table that describes each application and why you would recommend it. Share your findings with the class.

Employers expect you to respect your time at work. That means showing up on time, taking breaks at scheduled times, meeting deadlines, and taking care of personal business on your own time. Making the most of your time demonstrates to your supervisor that your are dedicated and take your responsibilities seriously.

Career Tips

School, practice, clubs, meets, homework . . . not to mention your social life—knowing how to manage your time is a skill you should start developing immediately. Think about how busy you are now. Now imagine balancing your career with the rest of your adult life.

Myth Americans work too hard.

Truth It is all relative. Compared to other countries, Americans do work longer hours than average. According to the Organization for Economic Cooperation and Development, U.S. workers work an average of 1,790 hours per year. That puts them 10th in the rankings. At the top: the Mexicans, who work 2,226 hours per year, on average.

Tools for Managing Time at Work

You can use many of the same time-management tools you use at home and at school to manage your time at work. Schedules and goal-setting can keep you focused on the tasks at hand. Computer programs can help you organize and meet your responsibilities. Set realistic and attainable goals using daily, weekly, and monthly schedules.

- Different schedules can help you identify tasks that must be accomplished within a specific timeframe. For example, you may have to call a customer today, but you may have to submit a report sometime this week. You may have to request vacation time before the first of next month, and you may have to attend a safety training session before Wednesday.
- Make use of the calendar program on your computer or handheld device. With programs such as Microsoft Office Outlook or Google Calendar, you can enter schedules, phone calls, and appointments. Use the tasks list feature to record and prioritize the things you need to accomplish. Set the program to display a message or make a sound to remind you of deadlines.

Have advances in technology made it easier or harder for people to manage their time?

> ### Did I Wake You?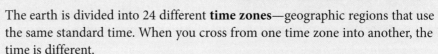
>
> The earth is divided into 24 different **time zones**—geographic regions that use the same standard time. When you cross from one time zone into another, the time is different.
>
> When you work in an office that does business with people living in different time zones, it is important to know the **local time**—time in the time zone where the other people are located.
>
> For example, if you work in Miami, Florida, you are in the U.S. Eastern Time Zone. The local time for a client in London, England, is five hours later, in the Western European Time Zone. When it is 9:00 a.m. in Miami, it is 2:00 p.m. in London. While that time difference is certainly manageable, think about a business call from Miami to Japan, which is *13* hours later.
>
> Can you think of jobs that might have problems resulting from differing time zones? Write about a work problem caused by time zones. Be specific about the job, the problem, and the time zones. Describe how you would solve the problem.

Tips for Managing Time at Work

The most important way to manage your time at work is to *prioritize*.

- Identify the tasks that must be done immediately and finish them first.
- Identify tasks that have multiple steps and plan when steps will be completed.
- Use your goal-setting skills to help you decide how to break larger projects into smaller chunks.

Something as simple as a daily to-do list may be all you need to keep your work priorities in order. However, it doesn't hurt to check in with your supervisor from time to time and make sure that what you are working on is the top priority.

Would your friends be happy if you said you'd meet them at 4:00 but you didn't show up until 4:45? Employers and co-workers expect you to honor and respect time in the workplace, too.

Following are some other tips for managing your time.

- *Give no for an answer.* Some people may ask for too much of your time. They may expect you to take on more responsibility than you can handle. It is okay to say no. Be polite and respectful, but explain that your schedule is full.
- *Ask for help.* If you are having trouble completing tasks that are part of your assigned responsibilities, you will need to find a way to get them done. Ask your supervisor, a co-worker, or someone in human resources to help you learn how to organize your time, or find ways to be more efficient so you can get more work done.

- *Notify your supervisor if you are late.* Most people try to be **punctual**—on time. They don't intend to be late for work or meetings. But things come up to delay them. Your morning bus is late because of traffic or you receive an unexpected phone call just as you are heading for a conference. When possible, call to notify your employer that you will be late, or request the time off for unplanned absences.

Career Profile: Occupational Therapist

Health Science Career Cluster

Job Summary

Occupational therapists help patients to develop, recover, or maintain their daily living and work skills. Often they work with individuals who suffer from a mental, physical, developmental, or emotional disability. The goal of all therapists is to help their clients lead independent, productive, and satisfying lives.

Occupational therapists help clients perform daily activities, engage in rehabilitative exercise, and improve their mental functioning. They work in rehabilitation centers, schools, hospitals, and other healthcare settings, though many of them also work out of patients' homes. Being an occupational therapist requires a college degree and a license. Employment of occupational therapists is expected to grow much faster than most other occupations, especially for those treating the elderly.

Use the Internet, a library, or career center to research the responsibilities, education and training requirements, and salary range of an occupational therapist. Prepare a presentation about the career.

Occupational therapists do more than just help people get back to work—they help people of all ages improve their quality of life. What kinds of skills are necessary for this type of work?

Managing Workplace Conflict

No matter how well you get along with your friends or how much you love your family, you are bound to disagree with them every now and then. It is no different at work. If anything, the pressure to succeed can make workplace conflicts even more intense.

Recall that conflict is a disagreement between two or more people who have different ideas. Conflict at work can interfere with your career goals. It can make you angry and cause you to resent the other people at work. If it is left unresolved, conflict can make it difficult or even impossible to successfully meet your responsibilities.

Some workplace conflicts are small, such as a disagreement about who gets to use the printer first or who put a file in the wrong drawer. Others are more significant, such as a disagreement over who will be the manager of a new project or who will get a promotion.

As in other areas of your life, managing workplace conflict does not always mean eliminating the conflict completely. It means that you are able to recognize what is causing the conflict and that you can cope with it in an honest and respectful way.

Valuing Differences

You will interact with people with diverse backgrounds and experiences in your workplace. You are also likely to encounter people from different cultures. Sometimes, these differences can cause conflict.

- You might think the food a co-worker eats in the lunchroom smells bad, even though it is common in his or her native country.
- You might be frustrated trying to understand an order placed by a customer who speaks English as a second language.
- You might become impatient waiting for a disabled co-worker to complete a task you know you could do faster.

Understanding the differences between your co-workers makes it easier to communicate, which, in turn, helps you to find common bonds. Though you may have different cultures, beliefs, backgrounds, or skills, you are all working toward the same goals: to get your work done and help your organization succeed. At work, focusing on the common goals you and your co-workers share will help you see past the differences to resolve conflicts.

Communications Review

Knowing how to use communication skills to manage workplace conflict will make you a valuable member of your work team. Remember some of these important steps for effective communication.

✔ *Be clear. Use words and body language that the other person can understand.*

✔ *Be personal. Address the other person by his or her name or title. Use "I" statements to show that you take responsibility for your role in the conflict.*

✔ *Be positive. State your message in positive terms directed at how to achieve your common goal.*

✔ *Get to the point. Explain why you feel or think this way.*

✔ *Use active listening to be sure you hear the response.*

✔ *Think before you respond. Use critical thinking instead of emotions.*

The ability to communicate effectively is a transferable skill that you will use no matter what direction your career takes.

Breaking Down Workplace Barriers

Workplace barriers are anything that keeps you from doing your job to the best of your ability. Such barriers often *cause* conflict, but they may be caused *by* conflict, too.

- You might not understand your responsibilities.
- You might not feel that you have received adequate training.
- You might think a team member is not pulling his or her weight on the job.
- You might find your supervisor is not helpful or supportive.

Effective communication is one of the best ways to avoid barriers at work. Be respectful and polite. Show that you value the opinions of others, and present your point of view in a clear and positive way.

You can also knock down a lot of barriers by maintaining a positive attitude. A negative attitude can quickly cause conflict and result in the loss of support from the rest of your team.

How does respecting cultural and social differences help you avoid conflict at work?

Peer Pressure at Work

You are probably aware of the power of peer pressure among your friends at school. Co-workers, customers, and supervisors can influence your thoughts and actions as well. Is peer pressure at work positive or negative?

- Co-workers convince you to leave work early to go to the movies.
- Your supervisor asks you to tell an angry customer that he is out when you know he is in the office.
- A vendor says she will pay you cash on the side if you buy office supplies from her.
- Co-workers encourage you to exercise during lunch.
- Your supervisor invites you to join a professional organization that can help your career.
- Team members convince you to work overtime to finish a project ahead of schedule.

How can recognizing peer pressure at work help you make healthy decisions and achieve your career goals?

Can't We All Get Along?

You respect the opinions of all of your co-workers, make an effort to get to know them, and hope that you get along with them all. But sometimes there is simply a personality clash. What if you just don't like someone you work with?

★ You could avoid working with that individual whenever possible.

★ You could have a talk with the co-worker and find ways to cooperate and compromise.

★ You could discuss any problems you are having with your supervisor.

★ You could ask for a different schedule or a transfer to a different department.

What are the consequences of each of these actions? How can unhealthy relationships with your co-workers interfere with your career success?

Recognizing Your Rights and Responsibilities

What would you do if an employer did not pay you for two months? What if an employer expected you to use a dirty bathroom, or work in a building without smoke detectors? What would you do if an employer promoted someone less qualified instead of you simply because that person was the opposite gender?

As a worker in the United States, you have rights that are protected by law. Federal and state agencies are responsible for making sure your employer treats you fairly and obeys the law. Your employer is responsible for the following:

- Paying you on time
- Providing you with safe working conditions
- Allowing you to leave to care for your family
- Preventing discrimination against you

Understanding your rights as a worker will help you make decisions about where to work and what type of career to choose Moreover, it can help you to decide if an employer *is* treating you fairly and what steps you can take if the answer is no.

Overcoming Discrimination and Harassment

Have you ever felt that you were treated differently from others because of a quality or characteristic that you cannot control? Maybe girls did not want you walking home with them because you are a boy. Maybe a teacher did not choose you to narrate a class presentation because you have an accent. Perhaps you weren't picked to be part of a club because you are in a wheelchair. If you thought the treatment was unfair, you were right. In fact, it was discrimination. **Discrimination** is unfair treatment of a person or group based on age, gender, race, religion, or disability. Discrimination is usually because of **prejudice**, or negative opinions that are not based on fact.

Sometimes employers or co-workers harass others. **Harassment** is unwanted, repeated behavior or communication that bothers, annoys, frightens, or stresses another person. Like discrimination, harassment at work is against the law.

Learning how to cope with and overcome discrimination and harassment can help you achieve success at work. The workplace will be a more productive environment if all employees are happy and comfortable. You can encourage a more productive environment by developing qualities such as tolerance, cooperation, respect, and understanding.

Employers are required to keep your work environment safe, but what can you do to stay safe on the job?

You have the right to be treated fairly in the workplace, and to be safe from harassment. The **Civil Rights Act of 1964** states that employers may not use race, skin color, religion, sex, or national origin as a reason to promote, not promote, hire, or fire an employee. Similarly, the **Equal Pay Act of 1963** prohibits employers from paying workers less simply because of their gender.

Newer laws such as the **Americans with Disabilities Act (ADA)** make it illegal to discriminate based on age or physical disability. States also have anti-discrimination laws, some of which are stronger than the federal laws.

The **Family and Medical Leave Act (FMLA)** provides certain employees with up to 12 weeks of unpaid, job-protected leave per year. The leave must be used for family or medical reasons, such as the birth of a new baby or to care for an ailing parent. The law also requires that the employee's health benefits continue during the leave.

If you believe you are facing discrimination or harassment at work you should do the following.

- Keep a careful written record of every incident. That means writing down what people say or do, including the date, time, and names of all the other people who were there. If you receive written threats or offensive documents such as letters, texts, or e-mails, save them and print them.
- Get a copy of your company's policy on discrimination and harassment. It may be online or in the employee handbook.
- Follow the procedure in the company policy for reporting illegal behavior. If there is no procedure, meet with someone in human resources to report it.
- If the company is not responsive, you can report directly to the **Equal Employment Opportunity Commission (EEOC),** the federal agency responsible for investigating charges of discrimination against employers.

Dealing with discrimination or harassment at work can cause stress and unhappiness. How would this affect an employee's attitude toward work?

Privacy in the Workplace

Would you be angry if you found out your supervisor was reading e-mails from your friends that you received at work? Maybe, but your supervisor did not break the law. Employers have the right to check on their employees to make sure they are being honest and productive. They may:

✔ Read e-mail messages and personal computer files stored at work

✔ Listen to telephone conversations that take place at work

✔ Monitor Internet usage, including tracking the sites an employee visits while at work

✔ Video employees using hidden cameras, if they suspect them of illegal or unethical behavior

You can avoid any problems by assuming that your employer is able to read or hear everything that passes through your computer and telephone at work.

21st Century Learning

Eduardo works at a restaurant, handling take-out orders over the phone. He works hard. He takes extra shifts when his manager needs him. He covers for co-workers who want time off. He likes the work environment. He gets along well with his co-workers and managers.

When a waiter quits, her job becomes available. Eduardo applies. He has the necessary qualifications. He is familiar with the restaurant, the menu, and the procedures. Someone who had never worked in the restaurant is offered the job.

Eduardo is one of the few Latino people who work at the restaurant. He thinks he did not get the job because of his cultural background. Do you agree with Eduardo? Are there other reasons he might not have been hired?

Use your 21st Century Skills such as decision making, goal setting, and problem solving to help Eduardo decide what to do next. Write an ending to the story. Read it to the class, or form a small group and present it as a skit.

Understanding Unions and Trade Organizations

A trade or labor union is an organization of people who do similar work. Trade unions exist for many kinds of workers. The purpose of **trade unions** is to improve wages and working conditions for workers. While every union is different, all unions share some characteristics.

- Unions hold elections to choose representatives.
- Representatives negotiate for workers' contracts.
- They speak up for workers' rights, such as safe working conditions.

Becoming a member of a trade union means that you will have to pay union **dues**—fees for belonging to that union. The dues are used to support the trade union and the work it does. Many teachers belong to unions. Ask your teacher what teachers' unions do and what teachers must do to belong.

Some businesses are "union shops." If you work in a union shop, you *must* join the trade union soon after you start working. In states that have right-to-work laws, workers generally do not have to belong to trade unions.

Why do you think an organization of many workers may be better able to negotiate than one worker alone?

Career FACT: In 2014, a little more than 11% of all wage and salary workers belonged to unions.

Understanding Professional Organizations

Like trade and labor unions, a **professional organization** is an association of people who are all employed in the same field or industry. Usually, the goals of a professional organization include providing education and training and acting as a source of information about the industry. It is similar to a student organization, such as Future Business Leaders of America.

Being a member of a professional organization can help you stay up-to-date on issues affecting the industry, as well as establish contacts with potential customers, suppliers, and employers. There are professional organizations for just about every career you can imagine—sometimes more than one. Benefits of joining a professional organization include:

- *Opportunities to meet people in the industry.* You can increase the number of contacts you have for networking and make friends who share similar interests. This can be especially valuable for getting advice on job and advancement opportunities in your area of interest.
- *Access to the most current news and trends affecting the industry.* Professional organizations publish Web sites and newsletters to make sure members have the latest information.
- *Access to training.* Organizations hold meetings, sponsor speakers and seminars, and offer classes for members.

Joining a professional organization helps you expand your career opportunities and improve your career-related skills. It also shows your employer that you are serious about your career.

Most Dangerous Careers

According to the Bureau of Labor Statistics, the following are some of the most dangerous jobs in America:

✔ *Fishers*
✔ *Loggers*
✔ *Pilots and flight engineers*
✔ *Iron and steel workers*
✔ *Roofers*

Why do you think people choose to work in these dangerous jobs? Are the benefits worth the risks?

Staying Safe at Work

Is the work you do safe? Do only the most dangerous jobs—such as roofer or fisher—require safety procedures? Consider the following situations.

- You are the co-pilot on a jet liner, and the pilot has been flying for 42 hours straight and is noticeably fatigued.
- You work at an autobody repair shop, and your co-worker is trying to use an acetylene torch to weld two pieces of metal together.
- You work for a manufacturing facility testing toxic chemicals, and your supervisor insists that you don't need a respirator because you will only be exposed for a couple of minutes.
- Your company is moving to a new office building, and the owner is looking to cut costs by not installing an emergency sprinkler system.

Part of your rights as a worker is the right to safe working conditions. There are numerous laws and regulations in place to ensure that organizations provide a safe working environment. In addition, most businesses have specific rules and procedures for maintaining workplace safety, whether it's how to correctly operate a machine or how to safely evacuate a burning building.

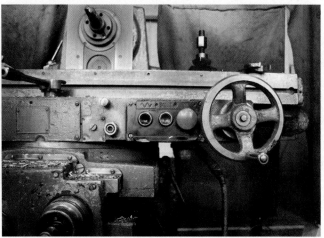

Is the work we do now more or less dangerous than the work our parents, grandparents, or great-grandparents did? What factors have made work safer?

Money Madne$$

You work for a construction company and are injured on the job when the scaffolding collapses and you break your leg. Your employer provides you with disability insurance equivalent to full-time pay for the period that you will be unable to work. Assuming that it will take six weeks until you can return to work and that you normally work a 40-hour week at $18.50 per hour, how much will the company compensate you for by the end?

Being Productive in Your Career ■ Chapter 15

In 1970, the federal government passed a law called the **Occupational Safety and Health Act**. This law requires all employers to provide a safe and healthful workplace. Workers must be provided with safe equipment, protective clothing when needed, and education about safety practices. The **Occupational Safety and Health Administration (OSHA)** was formed to inspect companies and enforce safety laws. Even so, more than 4,000 Americans die from on-the-job accidents every year.

As a worker under the Occupational Safety and Health Act, you have the following rights and responsibilities.

- *Right to know.* You have the right to know about hazards in your workplace, as well as the right to training to learn how to identify workplace hazards and what to do if there is an incident.
- *Right to refuse unsafe work.* If you have reasonable grounds to believe the work you do or the piece of equipment you use is unsafe, you can stop work immediately. You cannot be laid off, suspended, or penalized for refusing unsafe work if you follow the proper procedures.
- *Responsibility to follow safety rules.* It is your employer's responsibility to teach you the safety rules; it is your responsibility to follow the rules.
- *Responsibility to ask for training.* If you feel that you need more training than your employer provides, it is your responsibility to ask for it.
- *Responsibility to speak up.* It is your responsibility to report incidents and unsafe work practices as well as unsafe conditions.

Part of your responsibility as a worker is to make sure that you keep your work environment safe for yourself and for others. You have to take some responsibility for your own safety. If you are a new worker, enroll in any safety classes offered by the company. If you work around chemicals, poisons, or dangerous machinery, ask how they can be handled safely. If a company offers you hard hats or other safety equipment, use them. Read your manuals and handbook, and don't take shortcuts if it means endangering your safety or the safety of customers or co-workers. Remember—failure to follow safety guidelines is cause for dismissal.

What do you do to make sure you stay safe at work, at school, or when exercising or playing sports?

Career Tips

Those interested in a career in a growing industry should consider workplace safety. Some occupations within this area include safety training specialists, environmental health specialists, emergency management specialists, health safety technicians, and safety inspectors.

Putting Others at Risk

You have the right to a safe workplace, but you also have responsibilities to do your part to keep yourself, co-workers, and clients safe. How might your actions affect the safety of others?

👎 You fail to report a worn-out electrical cord.

👎 You do not request training on a new piece of equipment.

👎 You do not report that a new co-worker does not know how to use the equipment.

👍 You participate in your company's safety committee.

👍 You request safety inspections.

👍 You keep your safety training current.

What careers do you think have the riskiest work environments? Does risky work appeal to you?

Case Study

Lucy and Amanda work at an afterschool recreation program for elementary school kids in their town. The program uses the gym at the middle school. The girls give the kids snacks and help them do their homework. Then they play games using the middle school gym equipment. At the end of the day, Lucy and Amanda are responsible for making sure all the equipment is locked up in the storage room.

One day, Amanda takes a soccer ball home with her instead of locking it in the closet. Lucy notices but doesn't say anything. She hopes Amanda will bring it back the next day. Lucy never sees the soccer ball again.

- What do you think of Amanda's actions?
- Can you identify a problem or challenge facing Lucy?
- What do you think Lucy should do next? How should she respond to her supervisor?

Answer It!

1. Name three characteristics of professionalism.
2. List five ways you can meet any employer's expectations.
3. What is the key to managing your time at work?
4. What communication skills can help you to manage workplace conflict?
5. What responsibilities do employers have to their workers?
6. True or false: Employers have the right to record employees' phone conversations and read their work e-mails.
7. What federal law protects individuals from workplace discrimination?
8. What steps should you take if you are facing harassment or discrimination at work?
9. What do trade unions do?
10. What are you entitled to under the Occupational Safety and Health Act?

Web Extra

The nonprofit organization Workplace Fairness provides workers with information about many issues, including employee rights. Use its Web site—workplacefairness.org—to learn about issues facing workers today.

Pick an issue that you find interesting, and write a summary explaining how it affects workers, why you think it is important, and how workers can deal with it.

Being Productive in Your Career ■ Chapter 15

Write Now

Do you think there are similarities between your home environment, school environment, and the environment you might encounter at work? Are there differences? Are your roles the same or different? Do you need similar skills to succeed in each environment? Are the kinds of relationships you develop the same?

Write an essay comparing and contrasting the environments of your home and school with the environment you might find at work. Be sure to include what impact the differences might have on your behavior.

COLLEGE-READY PRACTICES

College-ready individuals know that the characteristics that are necessary for professional success are also necessary for academic success. That includes work ethics, integrity, dedication, perseverance, time-management, and the ability to interact with a diverse population.

Using effective time-management strategies, make a chart with at least five columns. In the top row list characteristics that are necessary for both professional and academic success. In the rows below, enter at least three examples of ways you can demonstrate or build each characteristic. Explain your chart to a partner or to the class.

Career-Ready Practices

Students have the right to be free of harassment and discrimination in school, just like employees in the workplace. As a class, discuss the ways harassment and discrimination occur in school and brainstorm ideas for how to eliminate it.

Career-ready individuals look for ways to contribute to the betterment of their families, schools, community, and workplace. They are responsible and eager to participate in activities that serve the greater good. In groups of three or four, brainstorm ideas for an anti-harassment and discrimination theme and slogan. As a group, create a poster that educates people about harassment or discrimination using the theme and slogan you select. Display the posters in your classroom or on a bulletin board in a common area of the school. Notify your local media about the display. They may want to publish examples of the posters in print or online.

Career Portfolio

When most people think of references or letters of recommendation, they immediately look to their supervisor—and for good reason. Your supervisor is generally in the best position to evaluate your overall performance. However, you shouldn't overlook the opinions of your co-workers and peers, either. Sometimes they are in an even better position to discuss some of the skills and personal attributes most valued in the workplace—the ability to communicate effectively, to work as part of a team, and to manage conflicts, to name a few. Because supervisors aren't always around to watch your every move, it is sometimes just as valuable to get the opinions and feedback of the people you work with most closely.

Consider asking one of your co-workers or fellow students—someone whose opinion you value—to write a letter of recommendation for you to put into your portfolio. Stress that you would like him or her to comment on what it is like to be your co-worker. As a gesture of gratitude, offer to write him or her a recommendation in return.

16 Living a Healthy and Balanced Life

Skills in This Chapter . . .

- **Analyzing Lifestyle Choices**
- **Recognizing the Importance of Lifelong Learning**
- **Living in Your Community**
- **Staying Healthy**

GETTING STARTED

How much time do you want to spend at work or at home? Do you need time to volunteer? When will you see your friends? Do you want to raise a family someday? When do you hope to retire? Though it may seem too early to even think about some of these questions, it is important to start learning how to balance your work with the rest of your life. Balancing work, family, and friends means making the right choices for your education, career, and lifestyle. It means staying active and involved and making time for what's important to you. Finding a balance between all of the demands and desires in your life will lead to happiness and better health.

➤ Imagine you have five hours to do whatever you please. No work. No studying. No bedroom to clean or laundry to put away. What would you do? Would you spend time with friends, with your family, or by yourself? What activities would make you happiest? What activities would enhance your self esteem? Make a detailed schedule of how you would spend that five hours and then share it with one other person in the class. What does your schedule say about your priorities and your ability to balance your responsibilities?

Analyzing Lifestyle Choices

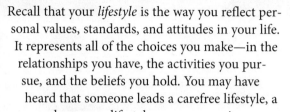

Recall that your *lifestyle* is the way you reflect personal values, standards, and attitudes in your life. It represents all of the choices you make—in the relationships you have, the activities you pursue, and the beliefs you hold. You may have heard that someone leads a carefree lifestyle, a dangerous lifestyle, or an expensive lifestyle. In fact, your lifestyle can seldom be encompassed in just one word, and it changes as you change and grow.

The choices you make—in your career, your relationships, your education, and how you spend your free time—will help dictate your lifestyle and vice versa. The goal is to choose a healthy lifestyle that helps you to meet your goals, fulfill your needs, and find happiness.

Recall that wellness is when all of the areas of your life are working together to make you healthy, happy, and confident. Wellness is not an accident. It happens when you make healthy choices about the way you live your life. In order to stay healthy and happy you will need to learn how to balance all the parts of your life.

Balancing Roles

Part of achieving your lifestyle goals includes balancing the roles you play in all five areas of your life—family, work, school, peers, and community.

- When you spend all of your time on schoolwork, you have no time to volunteer in your community.
- When you spend all of your time at your job, you have no time left to spend with your family.

Of course, achieving balance doesn't mean spending the same amount of time in each role. You might feel balanced spending 45% of your time at school; 35% with your family; 15% at work; and 5% volunteering. Your friend might feel balanced spending 45% at school; 25% with family; 20% at work; and 10% volunteering.

Living a Healthy and Balanced Life ■ Chapter 16

- Staying balanced also means being flexible. When the demands increase in one role, you may have to adjust the balance temporarily.
- If you have to study for midterm exams, you may not be able to take an extra shift at work.
- If your parents expect you to visit your grandmother, you may not be able to hang out with your friends.

Your overall wellness depends on your ability to meet your responsibilities in all of your roles.

Balancing Relationships

Your overall wellness also depends on your ability to balance your relationships. That means knowing how to make space for people and responsibilities without giving up the things that are important to you.

When you achieve balanced relationships, you are more likely to feel confident and happy. You are able to spend time with the people who are important to you, while still meeting your responsibilities in other areas. When you are unable to achieve balance, you may become tired, stressed, and unsatisfied. You may underperform in all areas and have trouble achieving your goals.

Tech Connect

Balancing work, family, and friends can be difficult. Often there isn't enough time in the day to talk to everyone you care about. You might have trouble staying in touch with people who live in a different community, city, state, or even a different country.

Thankfully, the Internet provides many ways for people to meet and socialize online.

- Social networking sites let you post information and pictures about yourself and your friends. You can join groups, exchange messages, and stay in touch.
- Chat rooms let you discuss topics with people who have similar interests.
- Text messaging keeps you in constant contact with friends and family.
- RSS feeds (a form of Web publishing that is easily updated and downloaded) let you broadcast your thoughts and actions, as well as follow the thoughts and actions of others.

Pick one means of socializing online and research it. Ask your parents if you can set up an account, if required. As you explore what the technology has to offer, consider these questions: How could you use this technology to balance your work, family, and friends? How could you use it to maintain your relationships? When does social networking become a problem? Write a brief report of your findings and present it to the class.

Signs of an Unhealthy Relationship

Relationships that are full of conflict or are not satisfying are unhealthy, or **dysfunctional**—not working correctly. Ways to tell if a relationship is unhealthy include:

✔ You are unhappy in the relationship.

✔ You feel you cannot be yourself with that person.

✔ You are afraid the other person will end the relationship.

✔ You are afraid the other person will hurt you—physically or emotionally.

✔ You do not trust the other person.

✔ You are dishonest.

✔ You cannot or will not talk about your feelings.

✔ You do not respect the other person.

Is the solution to a dysfunctional relationship always to end it? Are there positive ways to "fix" unhealthy relationships?

Have you ever struggled to balance your schoolwork with your social life? Do friends ever make meeting your responsibilities harder?

Think of all of the relationships you have. Parents, siblings, close friends, casual friends, extended family, teachers, teammates—these are all people who demand varying amounts of your time and attention. You have a relationship with anyone you interact with, in all five areas of your life—home, school, peers, work, and community.

Relationships are important in every role of your life and at every stage of your life. They help you share and exchange resources and achieve common goals. Often, relationships motivate us to make the choices we do. Think of the parent who works two jobs to support his or her family, or the friends who apply to the same college so that they can be together.

Money Madne$$

Recently, your friend spent a Sunday afternoon helping you study for a social studies test when she could have been at the movies. You decide to show your appreciation by baking her chocolate chip brownies. You are not sure if you should buy a box of brownie mix or bake the brownies from scratch.

You already have butter, sugar, flour, and salt. You would need to buy the chocolate ($4.85), eggs ($2.25), vanilla extract ($4.25), and chocolate chips ($3.75). The mix would cost $3.49. Which would cost less? Is cost the only factor? Which option would you choose, and why?

Career Choices and a Balanced Lifestyle

Previously, you considered how your choice of lifestyle might impact your career options. It's important to keep those same considerations in mind as you think about your personal well-being and the need for balance. Consider the following careers and the lifestyles and demands that go with them.

- An attorney works 80 hours a week. She earns enough money to host her extended family for one month each summer at her vacation beach home.

Events That Can Change Career Plans

Plan for the possible and be prepared for the unexpected! Flexibility, problem solving, and good decision-making skills will help you cope with events that might disrupt your career plans, including:

✔ *Marriage*

✔ *Parenthood*

✔ *Divorce*

✔ *Sickness or disability*

✔ *Layoffs*

✔ *Relocation to a new city*

Can you think of other events that might alter your career plans?

Living a Healthy and Balanced Life ■ Chapter 16

- A chef works nights and weekends, but he is available during the day to volunteer at his child's school.
- A truck driver is frequently out of town, but he calls home every night to talk to his family before bed.

Your career impacts the way you perform your responsibilities in other areas of your life. The choices you make in your career affect your relationships with your family, friends, and community. Using your values and standards to make choices will help you keep the areas of your life in balance.

- Do you pursue a high-paying career because it enables you to provide your family with the things they need and want, and so you can make cash contributions to charities?
- Do you work in human services because you enjoy helping others and think it sets an important example for your family?
- Do you go into medicine not necessarily because you want to help people, but because you like the challenge of the profession and the prestige that comes with being a doctor?
- Do you follow in your parents' footsteps because you think it would make them proud and you hope one day to inherit the family business?

Your values and standards are likely to change over time. You may benefit from reconsidering your choices on a regular basis to make sure they still support your values.

Do you think these people all value the same things in their work? Why or why not?

Career Tips

Finding time for leisure activities may seem difficult. Use goal-setting strategies to incorporate them into your daily schedule and time-management strategies to make sure you stick to your goals.

Balancing Work and Leisure

One hundred years ago, it was common for people in the United States to work 6 days a week, 12 hours each day. They had very little time for **recreation** or **leisure**—fun—activities outside of work. Now, most people value leisure activities and vacation as a necessary part of balancing work with the other areas of their lives.

Leisure activities contribute to your well-being in many ways.

- They improve your emotional well-being by helping you cope with stress.
- They improve your social well-being by giving you time with family and friends.
- Some forms improve your physical well-being by providing exercise.
- They improve your personal well-being by helping you develop new skills and interests.

Leisure activities can range from reading a book, to playing sports, to taking a hike. The activities that are most satisfying match your interests and values. You might want to volunteer for the Special Olympics, go birdwatching, or attend a concert. Leisure is an opportunity to relax and enjoy yourself, and to take time off from your work responsibilities. Often this allows you to go back to work refreshed and rejuvenated—or energized.

How can having fun outside of work help you to be a better worker?

Myth Balancing work and family is only a women's issue.

Truth Once, traditional roles meant men were responsible for earning money and women were responsible for taking care of the home and family. Now, the responsibilities are more evenly shared, and both men and women are concerned with balancing work and family.

Vacation Around the World

On average, Americans receive 13 days of vacation per year. This amount varies dramatically from one occupation to the next, of course, and with how long you work for a company. Those starting out with a company are likely to have only one or two weeks of paid vacation per year, but after ten years of service, that may increase to three or four weeks.

But how does that compare to other countries? Not so well, according to the World Tourism Organization. Workers in Italy and France enjoy more than 35 days of paid time off per year. Those in Canada, Japan, Korea, and the United Kingdom are entitled to 25 days or more per year. And in Austria, workers are legally entitled to a minimum of 22 vacation days plus 13 public holidays. If you had more than a month of vacation time a year, how would you spend it?

Thinking About the Future

When you become an adult, your roles and responsibilities will change. Some common grownup responsibilities include:

- Voting
- Furthering your education at a college or trade school
- Serving in the military
- Holding a full-time job
- Getting married
- Having children
- Making large purchases such as a home or car
- Making financial investments

The more roles and responsibilities you take on, the more difficult it may be to balance them.

One of the most common lifestyle-changing events is marriage. About 75 percent of the population in the United States will get married. **Marriage** is an institution that grants a couple a unique set of legal, social, and religious rights and responsibilities. Adding the role of spouse often requires a rebalancing of goals and priorities. When people marry, they agree to *share* their lives socially, financially, and spiritually. The relationship between spouses can lead to a change in career plans.

- You might follow your spouse to a different part of the country to pursue a job opportunity.
- You might have to quit your job to take care of your spouse if he or she should become ill.
- You might take a job that requires less travel so that you can have more time to spend together.

Like marriage, the decision to have children is an important one that will shape the rest of your life in significant ways. Whether or not you are raising children could determine things such as:

- *The job you have.* Caring for your children is your first priority, so you might not be able to take a job that involves a lot of travel or long hours.

NUMBERS GAME

How many hours, on average, do you spend a week on homework? How many hours do you spend playing video games? How many hours do you spend chatting with your friends on the phone or online?

An **average** is the sum of two or more quantities divided by the number of quantities. For example, you can calculate the average number of hours you sleep in a week by adding the hours slept each day and dividing by 7.

For example, say you got 7 hours of sleep on Monday, Tuesday, Thursday, and Saturday, 8 hours on Wednesday and Sunday, and 5 hours on Friday (sleepover). How much sleep did you get daily, on average?

1. First, add up the hours:
   $7 + 7 + 7 + 7 + 8 + 8 + 5 = 49$
2. Next, divide the sum by the number of days in the week:
   $49 \div 7 = 7$ hours of sleep per day, on average.

Now imagine that your parents are worried about the amount of time you spend in front of the television. You figure you watch about 1½ hours of television (counting movies and video games) Monday through Friday, about 2 hours on Saturday, and 1 hour on Sunday. About how many hours do you spend in front of the television on average per day?

Do you think you might want to get married someday? How might that decision change your career plans?

Does having children always mean putting careers on hold?

- *Your vacation plans.* Rather than a romantic getaway, you might opt for a vacation at a family amusement park.
- *Your home environment.* A home with small children is often noisy and chaotic, with toys everywhere.

When you have children, your time, money, energy, and relationships are not just your own; you must use them in a responsible way for the benefit of your children. That's why it's important not to enter into parenthood lightly—to make sure that you want to have children, and to have them only when you are emotionally, physically, and financially prepared to take good care of them. That way you can keep your career and family life in balance.

Career Profile: Computer Software Engineer

Information Technology Career Cluster

Job Summary

If you enjoy playing video games in your free time, you might consider pursuing a career as a computer software engineer. These workers research and design computer programs, creating applications to meet specific user needs. They might design software to help run a factory, to keep track of hospital patient files, or to create 3-D designs. They could create software for guiding missiles or saving an arcade princess. While they often have strong programming skills, software engineers are more concerned with creating and designing new applications rather than writing the actual code that makes those applications run. As such, their jobs require a balance of artistic vision, imagination, and technical skills.

Because computers are used in nearly every facet of life, the need for software engineers is great, and average earnings are relatively high. In fact, the profession is expected to grow much faster than average compared to other occupations over the next ten years. While there are a few examples of computer whiz kids who make their fortunes right out of high school, most computer software engineers have a bachelor's degree or higher.

Use the Internet, a library, or career center to research the responsibilities, education and training requirements, and salary range of a computer software engineer. Prepare a presentation about the career.

Recognizing the Importance of Lifelong Learning

Lifelong learning is more than just developing new work skills and maintaining old ones. It means being aware of changes that occur around you and adapting your roles and responsibilities as necessary. Setting goals for lifelong learning can help you achieve your career goals and find career satisfaction. It can also help you balance your responsibilities. Consider these common ways of being a lifelong learner.

- *Continuing your education* will contribute to your self-esteem and improve your qualifications for achieving career success. It may also help you adapt to changing roles in your family. For example, you may shift from a dependent (someone your parents take care of) to a provider (someone who takes care of others).
- *Developing hobbies and interests* will help you maintain your role as a friend. You will meet new peers and find new ways to enjoy time with old friends.
- *Participating in your community* can help you to learn about social and cultural issues that are important to you and those around you, making you a more informed citizen and an active, contributing member of society.

Most important, being a lifelong learner means always being open to new experiences and seeing them as opportunities for personal and professional growth. As your roles and responsibilities grow, you must grow with them in order to stay balanced.

Studies have shown that people with four-year college degrees earn, on the average, over a million dollars more in their lifetime than people without one. Continuing your education can lead to all kinds of rewards.

No More School?

People living in the United States are lucky to have free education provided for every child. Not all children across the world are so fortunate. Imagine that all schools in the United States were suddenly shut down and students and their families were now responsible for educating themselves. What if you had to take full responsibility for your own education?

★ Would you watch more educational programming on television?

★ Would you read more newspapers or watch the news more often?

★ Would you spend more time at the library or visit more museums?

★ Would you try to learn another language or try to study in another country?

★ Would you read more books?

★ Would you form a group of people interested in learning about a certain subject and find ways to explore it?

Think about all of the ways you might educate yourself. Can you use these methods to learn about subjects not taught in school?

Staying Informed

Part of being a lifelong learner is staying in touch with the world around you. That means being aware of what is happening at home and at work, in your neighborhood and in your community, in your city or town, in your state, in the country in which you live, and in the world at large. By taking the time each day to read news online or watch it on television, you can keep up with current events and stay informed. By doing so, you will be prepared to identify trends that might affect your career, family, or community.

Some of the ways that you can stay informed and even get involved include:

- Reading newspapers, news magazines, and Web sites that publish international and national news stories, so you will know what the key issues are.
- Sharing with your friends and family what you have learned about the issues, to raise their awareness and to discuss the issues with others.
- Keeping up on legislation that will affect you and your community.
- Volunteering for organizations that support your goals, values, and beliefs.

Why is it important to stay informed about what is going on in the world? How can this help you become a well-rounded person?

Rights Around the World

Citizens of the United States have specific rights defined by the Constitution, such as:

- Freedom to speak about controversial issues
- Ability to vote in elections, for a voice in government
- Freedom to travel within the country and outside it
- Freedom to assemble in groups without government interference

Do other countries afford their citizens the same or similar rights?

Pick one country that you would like to learn more about. Use the Internet or library resources to research the basic rights of that country's citizens. How do they compare to the rights of U.S. citizens? Are there laws or rights that you don't agree with? Are there laws or rights that you think all countries should adopt?

Learning Outside of School

We go to school in order to learn (for the most part). But just as much learning happens outside of school. Think about how much you learn simply by reading books, surfing the Web, traveling to new places, or just talking to your friends. Part of being a lifelong learner is recognizing the many opportunities for learning and growth that exist outside of a traditional school environment.

Of course, there are more formal learning opportunities that exist outside of school as well. Your workplace may offer workshops and training courses, and you may also take advantage of volunteer opportunities. You may take on an **apprenticeship**—a paid training program in a given trade—or an **internship**—an often unpaid, short-term job designed to provide valuable work experience. If you treat your life as an ongoing learning experience, you will soon find educational opportunities everywhere.

Take a moment and think of three or four things you learned yesterday. Did you learn them *all* at school?

21st Century Learning

Hasan looks forward to Sundays. He and his family have always spent Sundays at his grandmother's house. His aunts, uncles, and cousins often come, too. There may be 20 relatives there—eating dinner, playing games, and having a good time.

Hasan is considering a career as a personal trainer. He has a part-time job at a fitness club. He collects towels from the locker room, wipes down equipment, and keeps the floors clean and dry. He usually works on Tuesdays and Thursdays from 3:30 until 6:00, and on Saturdays from 6:00 a.m. until noon.

An instructor offers Hasan the opportunity to help out with a weekly class. She needs someone to demonstrate proper form and to show people the right way to use the equipment. Hasan is very excited, until she explains that the class is on Sundays.

Hasan is faced with a decision that impacts his roles at work and at home. Using your 21st Century Skills such as decision making, goal setting, and problem solving write an ending to the story. Read it to the class, or form a small group and present it as a skit.

Living in Your Community

A community is a group of people who have a common goal. The people might be physically near each other, living in the same neighborhood, or they might be spread out all over the world, connected to each other through the Internet. The people might be very similar to one another, or might have very little in common other than the one interest that draws them together.

Most communities are held together by a shared sense of purpose. That purpose can either be to explore or protect a personal interest (such as a club based around a certain hobby), or to work for a larger cause that the members believe in (such as world peace). A community can have a personal scope or a local, regional, national, or global one.

Examples of communities include:

- Residential neighborhoods
- Small towns
- Religious groups
- Charitable organizations
- Political action groups
- Volunteers for a cause
- Internet message boards and chat rooms
- Girl Scout and Boy Scout troops
- Music, dance, or art-based groups
- Friends who share common interests
- Schools

Recognizing the communities that you are a part of (or that you would like to become a part of) and the roles you play in them is an important step in achieving balance and wellness. Moreover, being involved in your community can lead to increased career opportunities, higher self-esteem, and greater personal growth.

Thank You, Kind Citizen

You may not realize it, but every day you are confronted with opportunities to be a good (or bad) citizen.

- 👎 Throwing trash on the ground
- 👎 Disobeying traffic laws
- 👎 Ignoring or pretending you don't see someone getting bullied
- 👎 Leaving a shopping cart in the middle of the parking lot
- 👍 Giving a dollar to charity
- 👍 Planting a tree
- 👍 Volunteering to help an elderly person with his or her groceries or yard work
- 👍 Writing your congressperson about an issue that concerns you

What motivates you to be a good citizen? What could you do to take more responsibility for your community?

Being a Responsible Citizen

As a citizen of your local community, your state, and the United States, you have certain rights and responsibilities. The service that you owe to such communities is sometimes called **civic responsibility**.

Some civic responsibilities are dictated by law, such as paying taxes, obeying traffic regulations, or serving on a jury. Others, like voting, are not required but are still part of being a good citizen. You can do many things to go the extra mile as a citizen, actively working to make your world a better place. These can include volunteering for worthy causes, serving in local leadership positions such as those on a school or zoning board, picking up litter along the road, or donating money or food to local food banks.

Serving your community is not just a nice thing to do; it can also positively impact your own life. Anything you can do to make your community more safe and pleasant will also make your own life more safe and pleasant within that community.

Volunteering

Volunteering—doing work without pay—is a great way to contribute to the communities you care about and to gain valuable work experience. You have probably done a lot of volunteer activities in your life so far, perhaps without realizing it. For example, when a teacher asked for someone to run an errand or hand out supplies, have you ever raised your hand? That's volunteering for your classroom community.

Volunteering is part of being a good citizen. When everyone pitches in to help, tasks get done more quickly and easily, freeing up the entire group to focus on what's important.

Volunteering also gives you a chance to express your values—in other words, to show what is important to you. For example, if you are passionate about improving the quality of life for children, you might volunteer at an afterschool reading program; if your primary interest is the environment, you might volunteer for a conservation group that plants trees.

What volunteer opportunities are available in your community? How much time do you think students your age should spend volunteering?

What Volunteers Do

Nonprofit organizations use volunteers for almost every aspect of their operations, so there is sure to be something that matches your interests and abilities. For example, volunteers can:

✔ Staff information booths at fairs and trade shows

✔ Tutor children in reading, math, and other subjects

✔ Help elderly or disabled people with daily activities

✔ Do physical labor such as cleaning, yard work, or even building houses

✔ Create flyers, brochures, and other printed materials

✔ Provide childcare

✔ Collect and distribute donations of food, clothing, or money

✔ Give presentations and speeches that promote the group

✔ Entertain groups of people with music, dance, acting, or other creative arts

✔ Pick up trash or gather materials for recycling

✔ Help take care of animals at shelters

Pick a nonprofit organization that interests you and research what volunteers in that organization do. Share this information with your class and consider taking part in that organization's mission in your free time.

Taking Care of the Environment

As it turns out, you are actually part of a community of some 7.1 billion members and counting. The planet we live on represents its own community and requires responsible actions in order to preserve it. The environment is the natural world—the soil, the water, the air, and the plant and animal life. Volunteering for organizations that promote a healthy environment can pay off in creating a world for yourself—and others—where plants and animals flourish, resources are used responsibly, and the air, land, and seas are free of pollutants.

Some common environmental concerns facing our global community include:

- Global warming awareness and prevention
- Planting trees to re-grow forests that have been cut down
- Setting limits on how much pollution companies can release into the air and water
- Protecting the habitats of endangered species
- Working with auto makers to reduce air pollution from vehicles, and supporting the development of alternative fuels
- Cleaning up landfills, dumps, roadsides, and other areas and preventing harmful substances from entering water supplies
- Promoting the use of recycling programs and recycled paper and plastics

There are many simple things you can do to help preserve the environment in your community and beyond, and most of them don't require a great deal of time or money. Here are some ideas.

- Get literature from local conservation groups to learn what environmental programs are active in your area and how you can participate.
- Volunteer after school and on weekends for organizations that do local environmental cleanup activities such as picking up trash along a river or road.
- If your school or religious institution does not recycle, do research to find out what the costs and benefits would be, and present that information to decision-makers in the organization.
- Encourage family and friends who drive automobiles to consider fuel-efficient models such as gas/electric hybrids.
- Conserve electricity at home by shutting off lights and turning off appliances.
- Go green by bringing your own reusable bags when you shop.

By taking care of the environment, you are not only helping to ensure your own personal wellness, but the health and wellness of billions of others.

Who is responsible for taking care of the Earth? Can the actions of one person really make a difference?

Tech Connect

Almost all organizations have a Web site that publishes contact information, a mission statement, and a schedule of activities and events. Some sites also list requests for volunteer help and application forms for getting started with them. You may need some training or skills development, such as first aid, CPR, child care, or computer training, in order to participate in some programs.

Here are a few sites of popular nonprofit organizations that encourage community involvement.

- American Red Cross: redcross.org
- Habitat for Humanity International: habitat.org
- Heifer International: heifer.org
- Greenpeace: greenpeace.org/usa
- The Nature Conservancy: nature.org
- The Humane Society: humanesociety.org

Pick an organization from the list above or one of your own choosing and research the volunteer opportunities it has available, the skills or training required, and the time commitment involved. Present your findings to the class.

Staying Healthy

Think about a time when you didn't feel well. What did you do? Did you spend as much time with your friends? Did you go to work or struggle to finish school work? Did you feel like doing much of *anything*?

In order to keep all parts of your life in balance, you need to stay balanced yourself. That means looking out for your own health and well-being. When you are healthy, you have more energy to meet your responsibilities and achieve your goals. Good health leads to high self-esteem and a positive attitude.

Staying healthy means keeping both your body and mind in top working order. It means staying in touch with your feelings and understanding how the choices you make affect how you feel. Your personal health is made up of three parts, all interconnected.

- *Physical health.* Being physically fit makes you alert and energetic. You can focus when you are at school or work and have lots of energy for enjoying your friends and other activities.

- *Emotional health.* Being emotionally healthy makes you confident and enthusiastic. It helps you make the most of your relationships and find satisfaction in your activities.
- *Mental health.* Being mentally healthy means you can think clearly and critically, set realistic goals, and make healthy choices.

We all have different ways of staying healthy. However, there are a few things everyone can do in order to stay physically, emotionally, and mentally fit.

- Get enough sleep.
- Eat right.
- Get regular exercise.
- Maintain high self-esteem.
- Engage in hobbies and leisure activities.
- Have supportive friends and family members.
- Keep a positive attitude.

The most important thing is to be aware of what works for you. What makes you feel strong, energetic, and motivated? Once you know the answer, you can make the right choices to ensure your well-being.

Stress Management

Stress is the way your body reacts to a difficult or demanding situation. You might feel stress in both good situations and in bad. For example:

- You might feel stress at a birthday party because you have to talk to lots of people and make sure everyone is having a good time.
- You might feel stress when you start a new job or meet new people.
- You might feel stress when your teacher gives a pop-quiz.
- You might feel stress the first time you get behind the wheel of a car.

Any event or situation that is unfamiliar or demanding can be a **stressor**—a cause of stress. Stress can come from serious, life-changing events, such as a death in the family, divorce, a lost job, or a move to a new town. It can also come from small, everyday events, such as missing your ride or losing your lunch money. Some of the most common stressors come from within, from low self-esteem, for example, or from an overwhelming pressure to succeed.

Stress can have positive or negative results. Think about a competition. You might try harder when you are in a real game than when you are in practice. It might bring out the best in you. Your friend might hate competition. It might give him a stomach ache.

Living a Healthy and Balanced Life ■ Chapter 16

Some stress is normal, but too much stress can cause problems and interfere with your overall wellness. It is unrealistic to think you can remove stress from your life, but you can use problem-solving skills to manage it. The key is to identify the source of the stress and then try to deal with *that* problem directly.

However, sometimes you cannot solve the problem causing the stress. You cannot stop your co-workers from fighting. You cannot cure a relative's cancer. **Coping skills** help you deal with or overcome problems and difficulties that you might not be able to solve. Here are some coping skills you can use to relieve stress and make yourself feel better.

- Relax. Sit or lie down in a quiet place. Breathe deeply.
- Do something you enjoy with people you like.
- Stretch your muscles to relieve tension, or be physically active.
- **Prioritize** your responsibilities.
- Accept what you cannot change.
- Try to stay positive.
- Express your feelings. Write down your thoughts in a journal or blog or talk to someone you trust.
- Talk with someone you trust or respect as a professional.
- Take steps to avoid situations that you know will cause you stress.

Not everyone handles stress the same way. What do you do when you get stressed?

Signs of Stress

When you experience stress, your body responds as if you are in danger—your heart beats faster, you breathe faster, and you might have a burst of energy. Following are some other signs of stress.

- ✔ Sadness or depression
- ✔ Mood swings
- ✔ Anger
- ✔ Headaches
- ✔ Upset stomach
- ✔ Trouble sleeping
- ✔ Loss of appetite

Since stress is a common occurrence, how do you know when it becomes dangerous? How much stress is too much?

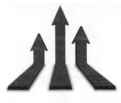

Career Trend

Employers are recognizing the impact of stress on their employees. Stressed employees work less, get sick more often, have strained working relationships, and make more mistakes. For this reason, many employers take steps to help their workers relieve stress, such as having a gym available for employees to exercise or providing free back massages once a week.

Nutrition

What did you eat for breakfast this morning? Was it the well-balanced meal you always see pictured on cereal advertisements, complete with fresh fruit, orange juice, and lowfat yogurt? Or was it a toasted pastry and a quick gulp of milk? Maybe it was just a granola bar on the bus.

Every day, you make choices that affect your physical, mental, and emotional well-being simply by eating the foods you do. Each choice has consequences. Some are positive, such as the health benefits of choosing an apple instead of a bag of potato chips. Some are negative, such as purposely starving oneself in order to maintain an unhealthy body image. It is important to think about your food choices and their impact. Doing so can help you make healthy decisions about what—and how much—to eat.

To make healthy choices, you need to know what your body needs. **Diet** is the food you eat. **Nutrition** is a science that studies the way the food you eat nourishes your body. When you eat a healthy diet and have good nutrition, your body is getting what it needs to work at its best level. Different foods contribute different **nutrients**—the parts of food that your body requires. Recognizing that food is a resource you can use to stay healthy and happy can help you choose the best foods and food combinations and maintain a healthy diet.

So what is a healthy diet? A healthy diet is one that gives your body balanced nutrition. Like a car, if you give your body the right fuel, it will run great and last a long time. If you give it the wrong fuel, it might break down. To achieve a healthy diet, you should eat a balanced variety of foods from different food groups.

Do your friends have any influence on what foods you eat? How do you decide what is best for you?

Other basic guidelines include:

- Eating a variety of fruits and vegetables, including dark green vegetables, yellow vegetables, and beans
- Eating foods that are rich in calcium, such as low-fat milk, cheese, and yogurt
- Eating whole grains
- Eating lean meats, fish, poultry, and nuts
- Limiting the amount of fats, salts, and sugars that you eat
- Drinking plenty of water

Living a Healthy and Balanced Life ■ Chapter 16

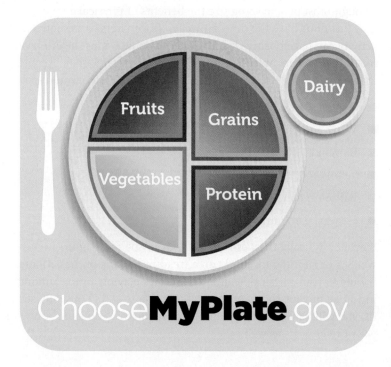

A healthy diet also balances the amount of food you take in with your physical activity. Think of *food* as energy that comes into your body, and *activity* as energy that goes out. For that reason, people with very active lifestyles, such as athletes, generally need to eat more. A healthy goal is to have the same amount of energy in as energy out. That way you can maintain a healthy weight and obtain all the nutrients your body needs to function properly.

Physical Activity

Physical activity is one of the best things you can do to promote your own wellness. It keeps your heart and other muscles strong, increases your energy level, improves self-esteem, and—best of all—it's fun.

Physical activity is also beneficial because it:

- Increases your mental focus
- Helps prevent diseases, such as heart disease, osteoporosis, and breast cancer
- Improves digestion
- Improves endurance
- Improves circulation and helps reduce blood pressure
- Enhances coordination and balance
- Helps to reduce stress

The United States Department of Agriculture (USDA) developed MyPlate to provide the information you need to choose a healthy diet. Check it out at choosemyplate.gov.

Force of Habit

You can tell a lot about a person from his or her eating habits. Postive habits contribute to health and wellness. Negative habits do not.

- 👎 Eating fast food every day for breakfast or lunch
- 👎 Regularly skipping meals
- 👎 Snacking heavily in front of the television
- 👎 Eating foods high in sugar or fat
- 👎 Eating because you are stressed
- 👍 Eating moderate portions
- 👍 Having regular meals with your family
- 👍 Taking part in food preparation
- 👍 Keeping fruits, vegetables, and nuts handy for snacks
- 👍 Taking your time while eating

Recognizing your own eating habits can help you understand why and how you make the choices you do. How might you adjust your eating habits to stay healthy?

312　　　　　　　　　　　　　　　　　　　　　Chapter 16 ■ Living a Healthy and Balanced Life

Energy Expenditure

A *calorie* is a unit of food energy. You can use online calculators to look up how many calories you burn doing different activities, such as running, biking, or walking. They take into consideration your age, height, and weight. Here are some standard numbers for common activities. The value is the number of calories burned per minute, for a person who weighs 120 pounds.

✔ Sitting still　　　　1.4
✔ Walking slowly　　2.2
✔ Walking briskly　　4.8
✔ Bicycling　　　　　3.8
✔ Running slowly　　7.6
✔ Running moderately　9.5
✔ Running fast　　　13.2
✔ Yoga　　　　　　　2.4

How many calories do you take in during a day? How many do you burn?

Do you exercise with your family? What activities could you do as a family to stay active?

Though most people recognize the benefits of physical activity, they complain they just don't have the time. However, researchers have shown that any physical activity can have positive health benefits. In fact, most health professionals recommend that you get at least two to three hours of moderate exercise a week (less than half an hour per day). Moderate physical activity would include:

- Walking briskly (3 miles per hour or faster)
- Water aerobics
- Bicycling
- Tennis
- Ballroom dancing
- General gardening

Staying physically fit doesn't have to mean spending hours in the gym lifting weights or running for hours on a treadmill. Many of the activities you enjoy contribute to your physical health, whether it's playing a sport, hiking in a park, dancing alone in your room, or just walking to a friend's house. By staying physically fit, you will have the energy and focus required to do your best work and meet your career goals as well.

 Some medical professionals believe that obesity is the number one killer in the United States. It is a contributing factor in a number of health conditions from diabetes to heart disease. Even more striking is that the cure for most obese Americans has been around for centuries: a healthy diet and daily exercise. That can be difficult in the modern workplace where many jobs involve sitting at a desk for hours on end. What can you do to stay healthy if your job keeps you chair-bound all day?

Avoiding Risky Behaviors

Living a balanced and healthy lifestyle means making positive choices every day. Of course there are times when you will make choices that are not so healthy—staying up late on a school night talking to a friend or eating a whole pint of ice-cream in one sitting. While maybe not the smartest decisions, such choices are ultimately not dangerous.

However, you will likely be confronted with choices that are dangerous. Any activity that puts your health and wellness in danger is a **risky behavior**. When you make a choice to participate in risky behavior—and it is your choice—you take a chance of harming yourself and others.

Perhaps the most common risky behavior that teens must face today is substance abuse. **Substance abuse** is when you eat, drink, smoke, or otherwise use substances—such as food, drugs, or alcohol—that are bad for your health and wellness. The substances can be legal—such as medicines you can buy at the supermarket. They can also be illegal—such as cocaine, marijuana, ecstasy, or even alcohol and tobacco products if you are underage.

Abused substances usually change the way you think and feel. They confuse your brain so you cannot tell right from wrong. You don't remember what's important to you. Things might seem to move faster or slower. You might feel light-headed. You might lose control over your body.

Most people abuse substances because they like the way those substances make them feel. It's fun and social, and everyone seems to have a good time. Some people abuse substances because of **peer pressure**—when your peers and friends influence you to do something. Some people start because they are curious, or because they believe it helps relieve stress.

Regardless of the reason, it's important to realize that substance abuse is a choice with potentially devastating consequences. You can be fired from a job for substance abuse, and many employers require mandatory drug tests. A documented history of substance abuse can negatively impact your ability to even get a job. Even just showing up for an interview smelling like smoke or alcohol might influence an employer's decision on hiring you. The employer might conclude that someone who drinks or smokes might not make the best, most responsible decisions.

Substance abuse can cause you to hurt yourself or someone else. Over the long term, you can even become addicted to the substance. An **addiction** is a need to continue using the substance even though it has negative consequences. The best way to avoid substance abuse is **abstinence**—which means never using the substance at all.

What would you do if you were invited to a party where you know there will be illegal substances?

A Deadly Habit

According to the Centers for Disease Control, smoking is bad for almost every organ in your body.

✔ It causes cancer in many parts of the body, including the lungs, mouth, bladder, and kidneys.

✔ It raises the risk of heart disease and stroke significantly.

✔ It can lead to emphysema, bronchitis, and many other lung problems.

✔ It can cause infertility or lead to problems during pregnancy and at birth.

As a result, more and more businesses, not to mention state and local governments, have instituted smoking bans. In fact, 28 states and Washington, D.C. have bans on smoking in enclosed public places.

Case Study

Thomas's friends invited him to a party at the house of another student he doesn't know very well. Thomas's parents trust him to make healthy choices and gave him permission to go. Later, he learned that the parents of the student throwing the party would be out of town that weekend. He also heard someone say that there would be alcohol. Of course he knows that underage drinking is illegal, but he has already told his friends that he would go. He feels like they would be disappointed in him if he backed out now—that they would somehow think less of him.

- Should Thomas go to the party?
- Should he tell his parents about the situation?
- Is it okay for him to go the party as long as he promises not to drink any alcohol?
- What should he say to his friends if he decides not to go?

Answer It!

1. Explain the importance of a healthy lifestyle.
2. List the five major roles you must balance in your life.
3. List three characteristics of a healthy relationship.
4. What common life events can alter career plans?
5. How do leisure activities contribute to your well-being?
6. What are important factors to consider when deciding whether or not to have children?
7. True or false: Taking college courses is the only way to be a lifelong learner.
8. List five examples of communities you might belong to.
9. List at least five coping skills you can use to relieve stress.
10. What is the difference between diet and nutrition?

Web Extra

You can use the Internet to identify ways to become involved in your community. Most cities and towns have Web sites and social networking communities. Neighbors exchange ideas and discuss issues that concern them. Local organizations look for volunteers. Take the time to explore your community's online resources. When you find an issue or organization that interests you, join the conversation or ask how you can help.

Write Now

Knowing how to express your feelings is an important part of maintaining healthy relationships. Think about someone who means a lot to you. Think about the personal qualities he or she has that you admire and the values that you respect. Write a letter to that person expressing how you feel and why.

COLLEGE-READY PRACTICES

Managing stress is an important part of a healthy lifestyle. College-ready individuals recognize that developing problem-solving and coping skills for managing stress in middle school and high school will help you achieve a healthy lifestyle in college and beyond.

Conduct research to learn about some common stressors for students. Then, write a personal essay about something that causes you stress, and how you use problem-solving and/or coping skills maintain a healthy lifestyle. Read your essay to a partner or to the class.

Career-Ready Practices

Career-ready individuals understand the relationship between personal health and well-being and their performance on the job. They eat healthy, exercise, engage in mental health activities, and avoid risky behaviors.

But finding and keeping a healthy balance can be difficult when you are working hard to reach your career goals. In teams of three or four, use time-management strategies to plan and execute the following project: Create a three-column table with headings for Physical Health, Emotional Health, and Mental Health. In each column, list as many things you can think of that would contribute to that area of your personal health. Use the information to create a presentation explaining the importance of maintaining a healthy lifestyle and share it with your classmates.

Career Portfolio

A record of community service can be a point of distinction for your portfolio, as well as a point of personal pride. Employers look favorably upon candidates with a record of community service and volunteerism, and college admissions boards almost expect to see some record of community involvement in the records of applicants. In addition, many high schools now make community service a graduation requirement.

For young people, community service and volunteer work often act as a substitute for paid work experience. In fact, such service often teaches valuable work skills. You might learn how to measure and cut wood while helping to build houses for Habitat for Humanity, learn how to cook or serve food while working at a soup kitchen, or enhance your leadership skills by volunteering to coach sports for young children.

Because of the increasing emphasis on community service, it's important to include any record of your own service in your portfolio. Include photos documenting your efforts and descriptions of what you did. And don't be afraid to ask the program director for a letter of recommendation.

Part IV
Financial Management

In This Part . . .

Chapter 17 ▪ Personal Money Management

Chapter 18 ▪ Personal Financial Planning

17 Personal Money Management

Skills in This Chapter . . .

- **Why We Use Money**
- **Comparing Financial Needs and Wants**
- **Setting Financial Goals**
- **Managing a Budget**
- **Analyzing Your Paycheck**
- **Choosing a Method of Payment**

GETTING STARTED

There are thousands of different jobs out there, but nearly all of them have one thing in common: a paycheck. What you do with that paycheck is an important part of your ongoing success. Reaching your goals—whether that means owning a house, going to college, or seeing the world—depends a great deal on how well you manage your money. That means creating a personal spending plan and sticking to it. It also means being a smart consumer—understanding where your money comes from and where it goes, and making sure you make the most of the money you earn. Any millionaire will tell you that it's not the size of the paycheck that matters; it's how you use it.

➤ On the front of an index card or a piece of paper, write down one item you own. On the back, write the dollar amount that you think it is worth. Walk around and show your classmates the item you own—without revealing its value. If someone has something you want, see if you can make a trade. When the trading is complete, compare the value of the item you gave with the one you received. Did you get a good deal? As a class, discuss what makes a good deal.

Why We Use Money

What is money? Is it simply paper or coins, stamped with a picture of someone famous and labeled with a number? What is that dollar bill in your pocket really worth? Is the paper itself worth something? We all know that money can be a powerful thing. It buys us the things that we need and want. But where does its power come from?

Money is anything you exchange for goods or services. In the United States—and most of the world—the money we use is cash. Cash is the money made out of paper—dollar bills—and metal—coins. Money could be made out of anything. Native Americans used shells. Maasai tribes still use cattle. In most countries now, money is made out of paper and metal because these materials are readily available, easy to work with, and convenient for people to carry.

In many countries—including ours—cash money is just one way to pay. We transfer electronic funds between accounts using online banking software or debit cards. We use credit cards to buy something and pay for it later. We write **checks**, which are written orders to a bank to transfer funds from our account to someone else's account. In fact, more and more transactions are done electronically—with no exchange of actual bills or coins involved.

Before money, people **bartered**, or traded, to get things they needed or wanted. If a farmer had a bushel of tomatoes and needed a roll of barbed wire to build a fence, he would trade with someone who had a roll of barbed wire and needed tomatoes. If he couldn't find someone who needed tomatoes and had barbed wire, he wouldn't be able to make the trade. To make the exchange easier, people created money. They assigned a specific value to the money, so everyone knew what they had and what it was worth. The farmer could sell his tomatoes to anyone with money, and then use the money to buy the barbed wire.

Money is generally worth more than the paper it is printed on. What is a dollar worth to you?

What Money Is Worth

The **face value**—or **denomination**—of money is the number printed or stamped on it. If you have a bill in your pocket that has a 5 on it, then its face value is 5 dollars. If you have a coin stamped with 25, then its face value is 25 cents. Face value is determined by the **Federal Reserve**, the government agency responsible for creating and tracking all of the money in the United States. The Federal Reserve decides how many $5.00 bills to print and how many quarters to mint (produce). Obviously, money is worth more to exchange than the paper and metal it is made from. You know what you can get in exchange for the money, because prices are set by people with goods or services to sell.

Personal Money Management ■ Chapter 17

The real value of money is what it is worth to you. How will *you* spend it? What do you want to get in exchange? The decisions you make about spending your money show what you value.

Different people will spend $5.00 in different ways. Your mother might buy a loaf of bread and a gallon of milk. Your brother might buy a video game. You might download five songs to your smartphone. Your friend might save it, because she wants to buy a sweater that costs $19.99.

The Power to Purchase

We know that money is worth more than the paper or metal it is printed and stamped on. But people who work with money for a living know that money's *power* to buy goods and services changes almost daily. **Purchasing power** is the amount of goods or services that can be purchased with a unit of **currency**, or type of money. For example, in the 1960s, a dollar could buy you five loaves of bread. Nowadays it probably won't even buy you one loaf of bread. Likewise, in the 1950s, it cost two quarters to see a movie. Today the price of admission to the movie theater is more than $10.00.

Purchasing power is tied to the economy and the value of currency. Prices of goods and services rise and fall depending on how well the economy is doing. Historically, prices have risen over time. Thankfully, incomes tend to rise over time as well. For example, the minimum wage in the 1960s was about a dollar an hour. So while our bread may be more expensive than the bread our grandparents bought, we have more dough now to buy it with.

Do you think your parents or grandparents had an easier or a harder time managing money than you will? Why?

Why Worry About Money?

Some people live "hand-to-mouth," or "paycheck-to-paycheck," which means they spend whatever they earn. These people can have a lot of stress, because they are always worrying about not having enough money. What if you never had to worry about money?

★ You would never have to work.

★ You could buy whatever kind of car you wanted.

★ You could travel the world.

★ You could go to any college or university that accepted you, regardless of cost.

★ You would never have to save for a rainy day.

If you never had to worry about money, would you even need to go to college? Or graduate from high school? What role does money play in how you see your future?

Money Around the World

In the United States, our currency is the dollar. Other countries have different currencies. For example, Mexico uses the peso and India uses the rupee. Some countries call their currency dollars, but these dollars are not the same as U.S. dollars. For example, Canada, Singapore, Australia, and New Zealand all use dollars.

Each type of currency has a different value, and the values change all the time. Today one U.S. dollar might be worth the same as 48 rupees, but tomorrow it might be worth 50 rupees. European countries formed the European Union (EU) and adopted a standard currency—the Euro—to make it easier for people to travel and trade throughout the region.

Pick another country in the world. Use books or the Internet to learn more about its currency. Be sure to find information about the value of that currency compared to the U.S. dollar. Create a brief presentation about the country's currency and present it to the class.

Comparing Financial Needs and Wants

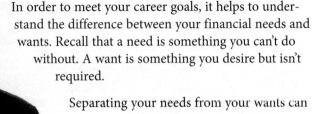

In order to meet your career goals, it helps to understand the difference between your financial needs and wants. Recall that a need is something you can't do without. A want is something you desire but isn't required.

Separating your needs from your wants can be more difficult than it sounds. You may need a car to drive to work, but does that mean you buy a 10-year-old junker, a brand-new fuel-efficient hatchback, or a Ferrari F430 ? Establishing your needs and wants, and learning how to balance the two, is the first step in effectively managing your money.

Establishing Your Needs

Think about what you need to survive. Food, water, and oxygen come to mind. While it's true that these are the absolute basics, your needs actually are much more encompassing. You need a place to live—even if it just means having a roof over your head. You need clothes to keep you warm. Most people need basic utilities such as gas, electricity, and water to cook, clean, and stay warm. While you could go back to the days when you had to dip water from a well, light the house with candles, and cook over an open fire, most people consider a home or apartment with running water, power, and appliances a necessity.

Your **financial needs** and **financial wants** cost money. Your financial needs are the things you must buy in order to survive. Your financial wants are the things you need to maintain a certain lifestyle, or **standard of living**. A standard of living measures how comfortable you are based on the things you own.

If you want a lifestyle, or standard of living, that includes expensive things, such as a big house, fancy car, and designer clothes, then you will spend a lot of money on your wants. You will need to have a career in which you earn a salary. If you want a lifestyle, or standard of living, that includes an apartment, used car, and casual clothes, then you will spend less money on your wants. In other words, your personal budget will reflect the lifestyle, or standard of living, that you choose. When you understand the difference between financial needs and wants, you can make responsible choices about what to buy and when to buy it.

NUMBERS GAME

Meeting your basic needs will actually use up much of your earnings. For example, on average, Americans spend one-fourth of their income on housing. What *percent* of their income is that? When converting a fraction to a percent, divide the numerator of the fraction by the denominator, then multiply by 100.

- 1 ÷ 4 = .25
- .25 × 100 = 25%

You can now multiply total income by the percentage to find out how much money would be spent on housing.

Imagine that you are living on your own and you make $455.00 per week at your job. How much of your paycheck (in dollars) will have to go toward housing?

Personal Money Management ■ Chapter 17

Changing Wants and Needs

Financial needs and wants are different depending on your stage of life.

- Children and teens can spend the money they have on their wants; their parents pay for their needs. Children want small items such as snacks and toys. Teens might want more expensive items, such as electronic gadgets or special clothes.
- Young adults are responsible for their own needs and wants, though they are not usually responsible for the wants and needs of others.
- Adults are responsible for their own needs and wants, as well as for the needs and wants of their family members. They must use their money to provide for the needs of their spouse, children, and sometimes parents.

As you get older you will not only find that your needs and wants change, but that the financial decisions you make become more and more complicated. This will require you to balance your needs and wants more carefully.

Making Financial Decisions

A **financial decision** is a decision about how to manage your money. A basic financial decision might be what you should buy for lunch. A more complicated financial decision might be how much money to save for college. To make healthy financial decisions, you must consider your financial needs and wants, as well as how much money you have available.

You can use the decision-making process for financial decisions. For example, your bicycle might be broken. How can you decide what to do about it?

1. *Identify the decision to be made.* Do you need a new bicycle? Do you need to have your old bicycle repaired?
2. *Consider all possible options.* You could buy a new bike. You could buy a used bike. You could have your old bike fixed. You could use your brother's bike. You don't need a bike.
3. *Identify the pros and cons of each option.* If you buy a new bike, you will spend a lot of money. You will have less money available for other things. It will cost less to buy a used bike, and less than that to have your old bike fixed. If you have your old bike fixed, it might just break again.
4. *Select the best option.* You decide to have your old bike fixed.
5. *Make and implement a plan of action.* You call a few bike shops to find out what they charge for repairs and how long it will take. You select a shop and take your bike there.
6. *Evaluate the decision, process, and outcome.* Your old bike is working fine now. You are happy that you have a bike. You are happy that you spent some money, but not too much.

How have your own financial needs and wants changed over time?

If You Had One Hundred Dollars

People make financial decisions based on how much money they have, what they need, and what they want. What if one hundred dollars was just handed to you?

★ Would you buy something you need?

★ Would you buy something you want?

★ Would you save for something more expensive that you might need or want in the future?

★ Would you buy something for someone else?

★ Would you donate it to charity?

★ Would you ask your parent or guardian to invest it?

How would you spend one hundred dollars?

Not everything you buy will require a thorough six-step decision-making process. Purchasing a bag of chips from the vending matching won't require you to make and implement a detailed plan of action. More-important financial decisions can have long-term consequences, however, expanding or limiting your options for the future. When faced with an important financial decision, consider the following:

- Do I have enough money to pay for it?
- Do I need it?
- Do I need it now?
- Do I want it?
- Why do I want it?
- Are there other things I want more?
- Is there something less expensive that would be as good?
- If I buy it, will I have money left over for something else?

Setting Financial Goals

Financial goals are the plans you have for using your money. Financial goals help direct your actions and guide your decision making. They give you something to work toward. You can use financial goals to help achieve the lifestyle, or standard of living, that you want.

When you set financial goals, you show that you are responsible with money and that you understand the importance of planning for your financial future. There are two main types of financial goals.

- Spending goals are for buying things you need and want.
- Saving goals are for saving money for the future.

Setting financial goals helps you focus on how you want to use the money you have available. You can decide what you need and want now, and what you will need and want in the future. Then, you can make and implement an action plan for achieving the goal. Setting financial goals also helps you focus on how much you must earn to afford your desired lifestyle. The career you choose determines the salary you will earn. So, keep in mind that decisions you make now about your education and career paths will affect your ability to achieve your financial goals. You can use resources such as the Occupational Outlook Handbook to learn salary information for different careers.

Getting Your Priorities in Order

Your parents or guardians might have told you once that money doesn't grow on trees. You have a certain amount available, and once you use it, it is gone until you earn more. This means that if you pay for one thing, you may not have enough money left to pay for something else.

For example, you might have $4.00 in your pocket. You need $1.50 to pay for the bus home from school. You need $3.50 to buy a snack. What happens if you buy the snack before you pay for the bus? When you set financial priorities, you have guidelines to help you make the best decisions about how to manage your money. Are you going to just buy a cheaper snack to tide you over or are you walking home with a full belly and fifty cents in your pocket? Setting priorities means that you make a decision about which goal is more important, and you make sure you achieve that goal first.

Do you plan to go to college? If so, you might want to think about setting some financial goals now.

Short-Term and Long-Term Financial Goals

You can use both short-term and long-term goals to manage your finances. Recall that short-term goals are for the near future, and long-term goals are for the more distant future. You might set short-term financial goals to make sure you have enough money to go to the movies on Saturday night, or to buy a birthday gift for a friend. You might set long-term financial goals so you can buy a car when you get your driver's license, go to college after high school, or have money in a retirement account.

You might feel overwhelmed by the thought of setting financial goals. You probably don't even have very much money of your own right now. Breaking long-term financial goals into a series of intermediate goals makes it easier.

For example, your goal might be to save $140.00 to buy a new digital music player. That seems like a lot of money all at once, but saving $10.00 each month for a year seems much more realistic.

Why is it so much harder to save money than to spend it? Is there anything you can do to make saving easier?

Avoiding the Impulse

Why do grocery stores keep the candy bars and magazines by the checkout? In general, people buy these things on impulse. Usually impulse purchases are wants (not needs), and often it is a better financial decision to avoid them. Why should you buy something?

- 👎 Buy something just because someone else has one.
- 👎 Buy something just because it is on sale.
- 👎 Buy something just because an advertisement told you to.
- 👍 Buy something because you need it.
- 👍 Buy something because you believe it will make you or someone else happy.
- 👍 Buy something that has value to you.

What is the last thing you bought without thinking? Was it worth it? Do you wish you had made a different decision?

Spending Goals and Saving Goals

When you set financial goals, you will probably think first about what you need or want to buy. These are your spending goals. Most of your spending goals right now are short-term, such as an ice cream on the way home from school or a day at an amusement park. You might have some long-term spending goals, such as expensive clothes or video games. Sometimes in order to meet our spending goals, we have to set saving goals.

It's harder to set saving goals. Saving money is not as much fun as spending. It's also hard to think about what you are going to need and want in the future, and how much those needs and wants might cost. Think about your long-term plans and goals.

- Do you have a vacation to save for?
- Will you need to pay for college?
- Will you want to rent an apartment?
- How much will you spend on a car?

Luckily, you don't always have to be specific about what you are saving for. Your goal might be to buy a car—but you don't have to know exactly which car, or what it will cost. Better yet, you can set a goal to save money without having a specific purpose in mind. Your goal might be to save $5.00 each month, knowing that in the future that money will come in handy. This type of saving goal is open-ended and flexible. It gives you the freedom to know that when you need the money, it will be there.

Myth Getting a job with a big salary will make you wealthy for life.

Truth Even movie stars and pro sports players go bankrupt. Wealth is defined by how much money you have accumulated—not how much you earn at work. To be wealthy, you need to save.

Managing a Budget

A **budget** is a plan for spending and saving money. It helps you balance your income and expenses so that you can make sure you don't spend more than you have.

When you make a budget, you keep a record of your income, which is the money that comes in to you, and your expenses, which is the money that goes out. The goal is to balance the budget—which means making sure that income is equal to or greater than expenses.

A budget gives you a clear picture of where your money comes from and where it is going. You can use a budget to keep track of the ways you actually use your money, and to set goals and priorities for how you might use your money in the future. A budget can help you plan, spend, and save so you can achieve the lifestyle, or standard of living, or want.

Personal Money Management ■ Chapter 17

Career Tips

The ability to make, analyze, and manage budgets is a valuable skill, and one that is required in many jobs. Of course, accountants, auditors, and business managers have to be budget experts, but almost anyone who is responsible for watching the "bottom line" needs to know something about budgeting. Roofers have to budget for materials and labor costs. Musicians have to budget to cover the cost of equipment and travel. It's an important transferable skill that will come in handy in the future.

What's in a Budget?

Every budget has two main parts: **income** and **expenses**. All the money that comes in to you is income. Income includes the following.

- Wages and tips you take home from working
- Gifts you receive for birthdays, holidays, and special occasions
- An allowance your parents give you on a regular basis
- Interest that you earn on money in the bank

Expenses are all the ways you use your money. Expenses include the following.

- Purchases, including needs and wants
- Savings
- Charitable donations

You can set up a budget for any length of time, such as a week or a year. Most budgets are set up for a month. A month is convenient because most people pay expenses on a monthly basis, such as rent, phone, and cable bills. They often receive income on a monthly basis, as well.

Do your parents make a budget? Have they ever asked you to give up something because they couldn't afford it?

Fixed or Flexible?

In theory, setting up a budget is easy. You list your income in one column and your expenses in another, and then you subtract your expenses from your income.

In practice, setting up a budget is not quite that easy. Some income and expenses are **fixed**, which means you know what they are and have to pay them each month. Others are **flexible** or **variable**, which means they change from month to month.

- Wages and allowance are fixed income. Gifts and tips are flexible.
- Rent is a fixed expense. Snacks, clothes, and gasoline are flexible.

You can easily enter your fixed income and expenses in a budget, but what do you do about the flexible items? The answer is to **estimate**. Make your best guess based on past knowledge or facts. If you know you spent $25.00 on snacks each of the past two months, you can estimate that you will spend $25.00 this month, too.

The best way to estimate is to look at records of your past income and expenses. For example, a record of how much money you deposited in the bank after your last birthday will give you a good idea of how much you will receive as birthday gifts this year. Receipts from restaurants will give you a good idea of how much you spent last month eating out.

You can use the estimated amounts when you set up your budget, and then go back and adjust them each month by entering the actual amounts you earned or spent.

Career Trend

Financial records used to only be written in ledgers—notebooks full of debits and credits, numbers and dates. As such they weren't always accessible to everyone in a business all the time. Often only the accountant and the business owner knew how a company was doing. Computers and networking technology make it possible for the most current information to be shared by everyone in an organization. The bottom line: More and more people are responsible for watching their own bottom lines, whether they are accountants or not.

Setting Up a Budget

The best way to set up a budget is to use a table or spreadsheet. That way you can keep your numbers in neat, organized columns and rows. You list categories of income and expenses in one column, actual amounts in the next column, and estimated amounts in the third column.

There are four basic steps for setting up a budget.

1. *List the categories or types of your monthly income.* For example, list wages, allowance, and/or gifts. If you know the amounts, enter them in a column to the right of the category. If you don't know the amount, enter an estimate, or guess, in the third column.

Personal Money Management ■ **Chapter 17**

2. *List the categories of your regular expenses.* Include things you spend money on each month, such as transportation to and from school, cell phone service, movies and other entertainment, school supplies, clothing, and food. Don't forget to include a category for saving and one for charity. Also include a "miscellaneous" category that you can use for unexpected or one-time purchases. Again, if you know the actual amounts, go ahead and enter them in the column to the right of the category. If not, enter an estimate.

3. *Balance your budget.* At the end of the month, go back and replace the estimates with the actual amounts that you spent.

4. *Do the math!* Add your total income and your total expenses, then subtract your total expenses from your total income. If the result is zero or more than zero, congratulations! You have a **surplus**, which means you earned more money than you spent. If the result is less than zero, you have a **deficit**, which means you spent more than you earned.

Make Your Budget Work for You

After you set up your budget, you can see exactly how you are spending your money. This might come as a surprise. Did you realize you spent so much money on food and entertainment? Are you shocked by how little you save? To get the most out of your budget, take the time to step back and analyze it. Look at your flexible income and expense categories. Identify areas where you might be able to make changes.

■ *Reconsider your priorities.* You might see that you spend a lot on clothes, but very little on charity. Maybe you could spend $5.00 a month less on jeans, and donate $5.00 more a month to a good cause.

■ *Reevaluate your goals.* You might see that you spend a lot of money on the bus. Could you walk to school, or ride your bike? If so, you could save that money. You could then use that savings to meet a long-term financial goal.

Monthly Budget

	Actual	Estimated
Income		
Wages	$ 150.00	
Allowance	$ 40.00	
Gifts		$ 20.00
Subtotal	$190.00	$ 20.00
Total Income		$210.00
Expenses		
Bus	$ 30.00	
School Supplies		$ 10.00
Food		$ 45.00
Phone	$ 9.99	
Entertainment		$ 25.00
Clothes		$ 25.00
Dues	$ 5.00	
Miscellaneous		$ 50.00
Charity	$ 5.00	
Savings	$ 5.00	
Subtotal	$ 54.99	$155.00
Total Expenses		$209.99

Is it a surplus or a deficit?

What could you do to start spending less and saving more?

Ways to Increase Income

It can be hard for a teen to find income, let alone find a way to earn more. Most states have laws regulating how many hours a teen can work. Many employers don't want to hire anyone younger than 16. What if you want to increase your income?

★ Could you work more hours?

★ Could you ask for a raise?

★ Could you find a different job?

★ Could you work more than one job?

★ Could you ask your parents for a larger allowance?

One way teens can earn more money is to start their own business. You might be able to mow lawns, babysit, or shovel snow. You might be able to fix computers for your neighbors, or sell jewelry you make yourself. What skills or talents do you have that you could use to start your own business?

Expecting the Unexpected

Remember that miscellaneous category you included in your budget? It's very important. It's the money you use to pay for unexpected expenses, such as when you need to pay your brother back for losing his basketball, or when you go over your allotted cell phone minutes. Unexpected expenses are not covered by the money in one of the regular expense categories. If you leave them out of your budget, you might not be able to pay for them when you need to.

When adults forget to budget for unexpected expenses, they can find themselves in trouble. The car might need a new transmission. A toothache might require oral surgery. If they don't have the money in the budget somewhere, they might be forced to use savings they wanted for something else, or cut back spending in another category, or charge the expense on a credit card to pay for later (with 20% added interest). Sometimes this can lead to even more financial trouble.

Of course, you might decide to use your miscellaneous expense money for something other than an emergency. What could you do with money you budgeted for miscellaneous expenses if you don't use it?

Money Madne$$

Each month you budget $25.00 for your entertainment expenses. This month, you already went to the movies once and rented two video games. Your friends want you to go with them to an arcade where every game is a dollar apiece. If a movie ticket costs $8.50, and the video games cost $2.50 each to rent, how many games can you afford to play at the arcade?

Tech Connect

If you'd like to keep track of your budget on a computer, there are quite a few software options available, such as Quicken®. Such programs not only allow you to track income and expenses. They also help you track investments, make online bank transactions, and create financial reports. In addition, there are several free personal budgeting programs available online.

Of course, if the only income you have is an allowance and you manage less than a hundred dollars a month in expenses, then professional money management software may be more than you need. You might do better to start with a spreadsheet or a blank sheet of paper for now and consider upgrading to a more complex system when your spending habits change.

Personal Money Management ■ Chapter 17

Staying on Budget

Staying on budget means using your money according to your plan to achieve your financial goals. It doesn't mean you have to earn exactly the amount you entered as income in your budget. It doesn't mean you have to spend exactly the amount you entered as expenses. There's room for flexibility.

For example, you might notice halfway through the month that you already spent the entire amount you budgeted for entertainment. Now, you have choices.

- Do you stop spending on entertainment?
- Do you use money you have budgeted for something else to pay for entertainment?
- If so, which money do you use? Savings? Charity? Food? Miscellaneous?

Of course, if you find that your budget is very different from your actual income and expenses every month, you might want to make some changes. You can evaluate your budget to decide if you need to adjust the amounts. Maybe you were being unrealistic when you set the budget amount for entertainment at $25.00, or maybe you really only need $30.00 a month for food.

What does a company do when it has a deficit? What kinds of cuts does it make?

Budgeting for the Business Owner

One of the best ways to learn how to use a budget is to start your own business. This can be as simple as setting up a lemonade stand, selling cookies at a sibling's baseball game, or offering to rake leaves around the neighborhood.

Take leaf raking as an example. Suppose you charge $20.00 per yard. That's your income. Now think about your expenses. You will need two rakes ($12.00 each) and bags ($6.00 per box). You will need to make copies of flyers to stuff in neighbors' mailboxes (50 copies at 5 cents each). Also, you hope to enlist the help of your sibling or friend and plan to split your earnings 60/40.

Make a monthly budget that shows your income and expenses. Assuming you can manage to clear two average-size yards per week, how much will *you* make by the end of the first month? What could you do to increase your sales or decrease your expenses?

Running in the Red

If you have a deficit, it means you are spending more money than you are earning every month. In the business world, that's called "running in the red." You might get away with a deficit for a month or two, but eventually you will run out of money. To have enough money to achieve your financial goals, you'll have to eliminate the deficit. That means making tough choices.

Are there expenses you can reduce or eliminate? How much can you save if you did the following:

- Watch free movies instead of going to the theater
- Bring snacks to school instead of buying them
- Shop for clothing, shoes, and accessories at a thrift store instead of buying new
- Drink tap water from a refillable bottle instead of buying bottled water

You might also be able to increase your income. If you have more money to spend, you might be able to cover the expenses the way you have them now, without a deficit.

Career Profile: Financial Planner — Finance Career Cluster

Job Summary

If you think you are good at planning budgets, and you enjoy giving people advice on how to save and spend, you might want to look for a career as a certified financial planner. A financial planner, or personal financial planner, helps people manage their finances by advising them on how to budget, spend, and save.

Certified financial planners must pass a series of certification exams such as those offered by the Certified Financial Planner Board of Standards. Most study financial planning at the college level, as well.

Use the Internet, library, or your school's guidance resources to learn more about a career in financial planning. Write a job description, including the educational requirements, career pathway, and potential salary range.

Personal Money Management ■ Chapter 17

Analyzing Your Paycheck

Up to now, most of the jobs you have had probably paid you in cash. When you work for a business, however, you receive a **paycheck**. Your paycheck is a written document that tells your employer's bank how much money to give to you. It is your payment for working.

Most paychecks have two parts: the paycheck itself and the **pay stub**. The paycheck is the part you take to the bank to exchange for cash or to transfer funds into your account. The pay stub provides information about your pay, including how much you have earned and how much your employer has deducted from your pay. A **deduction**—which is sometimes called withholding—is an amount that your employer withholds from your earnings to pay for things such as taxes or insurance.

Many employers directly deposit your paycheck in your account instead of giving you a paycheck. Even if you have a direct deposit arrangement, you will receive a pay stub so that you know how much you earned and how much was deducted. It's important to understand the parts of a paycheck so that you know where some of your money goes even before it gets to you.

How You Will Be Paid

You receive a paycheck for every pay period. A pay period is the number of days for which you are being paid. Pay periods are different depending on the employer.

Career Tips

You generally have six months to cash any check you are given. However, you should never wait that long. Whoever paid you, whether it is a paycheck from your employer or a birthday check from a family member, expects you to take that check to the bank. If you hang on to it, you might cause problems with the payer's record keeping. Always cash or deposit your checks in a timely manner.

When Will I Get Paid?
When you are considering a job, one of the things to think about is how often you will get paid. Your employer may pay you:

✔ Once a week
✔ Once every two weeks
✔ Once a month
✔ Immediately after you complete your work
✔ Never—you are a volunteer or intern

Understanding Your Paycheck

A paycheck looks like a personal check you might write. Most paychecks include the following:

- Your employer's name and address in the upper-left corner
- A check number
- The date of the check
- Your name as the recipient, or payee
- The amount of the check written in numbers and spelled out
- The signature of a person authorized by your employer in the lower-right corner

Your paycheck is just a piece of paper until you deposit it or cash it. You must **endorse**—sign—the back of the check to make it valid. It's a good idea to sign it at the bank, because once you sign it, anyone can cash it or deposit it.

There are two types of endorsements.

- A blank endorsement is when you just sign your name. The check can then be cashed or deposited by anyone, for any purpose.
- A restrictive endorsement is when you sign your name and below it write how the check should be used. For example, you can write "For Deposit Only," if you want the check to be deposited, or you can write "Pay to the Order of Jack Smith" if you want the check to be used by Jack Smith.

If you lose your paycheck, you will have to follow your company's procedure for canceling it and issuing a new one.

Understanding Your Pay Stub

A pay stub is attached to the paycheck and provides a lot of information about the money you earned. It has no cash value, and you cannot exchange it for money at the bank. Most pay stubs have four sections.

- *Personal information about the employee.* This usually includes the check number and date, the starting and ending dates of the pay period, your name, and your employee number.
- *Information about earnings.* This usually includes the number of hours you worked, your hourly wage, and the amount you are being paid this pay period.
- *Information about deductions.* Employers withhold state and federal taxes, as well as **FICA**, which is Social Security. They may also withhold local taxes. The amount of these deductions is based on a percentage of how much you earn. You might have other deductions such as contributions to a health care plan or to a savings account.

What did you do or what will you do with the money from your first paycheck?

Myth I'm too young to pay taxes.

Truth Even teenagers have to pay taxes. Age doesn't matter when it comes to paying taxes. If you receive a paycheck from a company, that company will withhold taxes and Social Security. If you make over a certain amount, you will even have to file an income tax form.

Even money earned babysitting or mowing lawns is potentially taxable. This is considered self-employment income. If you earn more than $400.00 in self employment income in a year, you are required to report it to the Internal Revenue Service (IRS).

Personal Money Management ■ **Chapter 17**

- *Information summarizing earnings and withholdings.* You usually get this information for the current pay period and the year-to-date (YTD), which is since January 1 of the current year. In this section, you can see your **gross income**, which is your pay before withholdings, and your **net income**, which is your pay after withholdings. Some people call net pay take-home pay, because it is the amount you actually take home.

It is important to look closely at your pay stub to make sure the information is correct, particularly the number of hours worked. If there are any errors, you will have to follow your company's procedure for correcting them.

What can you learn by looking at your pay stub?

21st Century Learning

CAREER COUNSEL

Anthony's school counselor suggested he take accelerated classes next year with the hopes of getting advanced placement credit for college. Though Anthony did well in his advanced math classes last year, he also had to study hard. Often when his friends would be out playing soccer, he would be huddled over a mountain of homework.

Taking accelerated classes next year might put him on the college track, but he is afraid of how much more work would be involved. If the classes are too difficult, he might spend all of his time studying and still do poorly. He knows that bad grades, even in tough classes, won't help him get into college. Anthony's parents want him to take the harder classes.

What should Anthony do? Should he take the advanced courses even if it means more work? Or should he take easier classes that he knows he can do well in and have a little more free time? Use your 21st Century Skills, such as decision making, goal setting, and being flexible, to pick a path for Anthony's education. Make up a list of pros and cons for each decision. Form small groups and debate Anthony's options, then come back together and discuss as a class.

Choosing a Method of Payment

You use money to pay for expenses. At the beginning of this chapter, you read that there are many types of money including cash, checks, credit cards, debit cards, and electronic funds. The type of money you use to make a payment depends on many things, including the type of purchase.

- For a purchase you make in person, you might use cash, a debit card, or a credit card.
- For a payment you send in the mail, you might use a check.
- For a payment you make over the phone, you might use a debit card or credit card.
- For a payment you make over the Internet, you might use a debit card, credit card, or electronic funds.

Other factors include the amount of the purchase and convenience. For example, if something costs a lot of money, you might choose to use a credit card because you can pay for it later. A debit card might be convenient because you do not have to keep going to the bank for more cash.

Each method has benefits and drawbacks. You can use your decision-making skills to compare and contrast each option before choosing a method of payment.

Payment Method	Benefits	Drawbacks
Cash	You know you can afford what you are buying.	If you lose it, there is no way to get it back.
Check	Generally safer than cash, especially when mailed.	Often requires identification. Some businesses don't take personal or out-of-town checks.
Debit card	Convenient. Withdraws directly from checking account, so no interest is charged.	Can overdraw checking account.
Credit card	Convenient. Allows you to make purchases even when you don't have enough money available.	A form of credit that can end up costing a lot more in the long run due to interest charges.

When to Use Cash

Between debit cards, check cards, electronic funds, and credit cards, it almost seems as if no one uses cash anymore. But cash is a good option for most purchases. When you use cash, you know you have enough money to pay for the purchase. You don't have to worry about whether there is enough money in your bank account to cover a check or a debit, or if you will have enough money to pay a credit card bill in the future.

Cash is also universal. Some businesses take only certain credit cards or no cards at all. Some won't take a personal check. Just about everyone takes cash. Cash is useful if you need to buy a drink from a vending machine, pay for a parking meter, or pay a toll. You may also need it to tip for service.

The biggest drawback to using cash is that it can be lost or stolen. You have very little chance of recovering cash once someone else has it. You may also find it inconvenient to keep going to the bank or ATM to withdraw cash so you have it when you need it. And, of course, you cannot make purchases online with cash.

When to Write a Check

When you write a check, you are **authorizing**—giving permission to—your bank to transfer money from your account to the account of the **payee**—the person whose name you write on the check. In order to pay by check, you must have a checking account with a bank, and you must have enough money in the account to cover the amount of the check.

Paying by check can be convenient.

- You do not have to carry cash.
- You can send a check through the mail.
- If a check is lost or stolen, your bank can stop payment, or cancel the check so that it cannot be exchanged for money.

Why is it important to always keep track of your bank balance?

If you use a checking account, it is important to keep the account **balanced**. That means you must keep track of how much money is in your account, as well as the amounts of all the payments you have made. If you make more payments than you have money, your account will be **overdrawn**. Most banks charge a large fee if you overdraw your account.

Checks have other disadvantages that you should be aware of.

- Banks may charge a fee for every check you write.
- Not every business accepts checks.
- Some businesses require a valid photo I.D. with each check.
- Some businesses will not accept a check from a bank in a different city or state.
- Checks take time to clear—time for the check to be verified and the bank to make the transfer.

How Will You Be Paying For This?

Look over the following purchases. Which method of payment would you use for each?

- ✔ Downloading music from the Internet
- ✔ Buying a new sweater from a catalog
- ✔ Paying a phone bill by mail
- ✔ Buying a snack at a ball game
- ✔ Purchasing a drink from a vending machine
- ✔ Buying a new computer at a store

Credit and debit cards may look the same, but they don't work the same way. What are the main differences?

When to Use a Debit Card

A **debit card** is just like cash except it is more convenient. You must have a bank account in order to use a debit card, and you must have enough money in the account to cover your purchases. Most cards will be declined if there is not enough money in the account to cover the expense.

When you use a debit card, the money is automatically and immediately transferred out of your bank account, just as if you withdrew it as cash. Most debit cards require you to enter a **personal identification number** (PIN) with each use. By entering the PIN, you prove that you are the person authorized to use the card.

The biggest risk of a debit card is that if someone else gets hold of your debit card and knows your PIN, he or she can use it to access your money. Because they are just like cash, debit cards are accepted for payment almost everywhere you can buy something, including stores, restaurants, and online.

When to Use a Credit Card

A credit card lets you use **credit**—a loan—to buy now and pay later. Every time you use a credit card, you are borrowing money from the business that issued the card, such as the bank, store, or credit card company. The business pays for the purchase, and then you repay the business by paying your credit card bill.

Credit cards may be convenient, because you do not have to have enough money to pay at the time you make the purchase. You can also make a lot of purchases, and then pay them all at once with a single check or electronic fund transfer by paying your credit card bill. However, if you don't pay your credit card bill in full each month, you will be charged interest, usually between 20 and 30 percent on the balance. You can end up spending a lot more than the original purchase price.

Using an Installment Plan

Another payment option—especially for items that you can't afford to pay for upfront—is *installment buying*. With this method of payment, you buy something, such as a car or a piece of furniture, on credit, with a series of future payments to be made at specified intervals. Of course with such a payment plan, you don't actually own the thing you purchased until your last payment is made, and if you fail to make your payments, the seller can take that car or couch back. While most often used with automobiles and major appliances, many stores offer layaway programs that also allow you to pay for items over time.

Personal Money Management ■ **Chapter 17**

Using Electronic Funds

You use **electronic fund transfers** (EFT) to transfer money from one bank account to another. To use an EFT, you must have electronic access to your bank account and enough money in your bank account to cover your purchases. There are three common methods of using EFT.

- *A debit card.* When you use your debit card, the money is transferred from your account to the seller.
- *Online banking.* If your bank account is set up for online banking, you can use the bank's online banking software to transfer money from your account to other accounts. You can make the transfer from your home computer or from a bank machine. You can transfer the money to almost any other account, so you can pay bills such as utilities, credit cards, and even taxes.
- *Online payment services.* Many online retailers such as eBay have their own online banking system that you can use to transfer funds electronically. You must set up an account and make sure you have money available to cover your purchases.

Tech Connect

Even now, many people are nervous about using the Internet to transfer funds, make purchases, or pay bills. However, advances in technology have made most online transactions quite safe. The key as a consumer is to know if the Web site you are using is secure.

A secure Web site uses **encryption**—codes—and authentication standards to protect online transaction information. That means your personal information including debit card numbers and PINs are safe when you shop online.

You can tell that you are viewing a secure Web site because the letters https display to the left of the Web site name in the Address bar of your browser. On an unsecured site, there is no "s." Also, a small lock icon displays in your Web browser's status bar. You can double-click the lock icon to display details about the site's security system.

Why do you think it is important to use a secure Web site when you are shopping or banking online?

Is shopping online just as safe as going to the store or the mall?

Dos and Don'ts of Credit Card Use

While credit cards can be convenient, they can also be dangerous. Think about how you will use a credit card once you get one.

👎 Paying only the minimum amount each month

👎 Charging much more than you can reasonably expect to pay off

👎 Paying your credit card bill late

👎 Applying for a new credit card when the first has reached its spending limit

👍 Choosing credit cards with the lowest interest rates

👍 Choosing credit cards with no annual or hidden fees

👍 Paying the balance off in full each month

👍 Only making purchases that you know you will be able to pay for in the future

Though you may not be old enough to have one yet, it won't be too long before you get your first credit card offer. Do you think you are old enough to use a credit card responsibly?

Case Study

Alisha received a preapproved credit card in the mail. She started using it right away to buy clothes and jewelry and to take her friends out for lunch. She bought things she didn't need and things she would never use. She bought gifts for her family and friends. She felt rich. When the first bill arrived, Alisha owed $650.00!

At first she thought the bill was wrong, but when she added up everything she had bought, she realized it was correct. Alisha started to panic. She knew she didn't have that much money in her bank account. She did not know what to do.

- What problems and choices are Alisha facing?
- Why do you think she spent so much?
- What do you recommend Alisha do now??

Answer It!

1. What is money?
2. What currency do we use in the United States?
3. What is purchasing power?
4. List four factors to consider when you make a financial decision.
5. What is a budget?
6. List three types of income.
7. List three types of fixed expenses.
8. What is the difference between a deficit and a surplus?
9. List three things that are printed on a pay stub.
10. List three drawbacks to paying by check.

Web Extra

When most teens get their first paycheck, they are shocked at how little they actually get to take home after all of the deductions are taken out. But they might also be surprised to see how much other people have taken from their paychecks.

Use the Internet to learn more about federal income and Social Security taxes. Who pays the most? What does Social Security cover? Why do you think people who earn more pay a higher percent in taxes? Do you think this is fair? Write a paragraph explaining why or why not and share it with your class.

Personal Money Management ■ **Chapter 17**

Write Now

Can you imagine how you will be spending money twenty years from now? Will you have an income? What type of lifestyle will you have? What types of expenses will you have? Will you have achieved any of the financial goals you have now? Will you have new goals? Will you have the same needs and wants that you have today? Will you have financial security? How about financial freedom?

Use the questions in the previous paragraph as a guide for writing a personal essay about your future self's personal finances. Include an explanation of how your personal budget reflects your desired lifestyle. Share your essay with the class.

COLLEGE-READY PRACTICES

The salary you earn in the future depends dictates your personal budget. College-ready individuals recognize that where they attend college and what they study will affect their income throughout their careers.

Conduct research using available resources to compare and contrast salaries of at least three careers in your career interest area. Write a paragraph explaining how salary might affect your personal budget and lifestyle choices.

Career-Ready Practices

Career-ready individuals take regular action to contribute to their financial well-being. They understand that personal financial security provides the peace of mind they need to pursue their own career success.

Divide into teams of four or five. Spend 10 to 15 minutes thinking of and writing down all the things on which you spend money in a typical week. Be sure to include things you need, things you want, and things you buy on impulse. Write down the amount you spend on each item. Then, for each item, think of an alternative that would either eliminate the expense or reduce it. There may be some items for which there is no alternative, like medications or bus fare. How much could you save? Share your findings with the class.

Career Portfolio

A career portfolio should showcase your successes. That includes any successes you've had running your own business. Think about all the ways you can earn extra income as a teenager, from babysitting to cleaning gutters to taking care of animals while their owners are away. This is the only kind of employment history many people your age have to draw from. Earning money through self-employment shows that you are motivated, organized, and responsible.

But how do you represent these in your portfolio? First, don't underestimate the kindness of satisfied customers. See if one of your regular customers or clients will write you a letter of recommendation. Or, you could create and print out your own comment cards and have people fill them out. Finally, if you do create a budget for your "business," you should include that as well.

What else could you include in your portfolio to show off your skills and successes if you don't have much work experience?

18 Personal Financial Planning

Skills in This Chapter . . .

- **Choosing a Bank and a Bank Account**
- **Managing Your Bank Account**
- **Saving and Investing**
- **Comparing and Contrasting Forms of Credit**
- **Paying Taxes**
- **Keeping Your Personal and Financial Information Safe**
- **Analyzing Banking and Credit Regulations**

GETTING STARTED

When asked, most people will say that they wouldn't mind being rich but that mostly they just want to "live comfortably." What does it mean to live comfortably? For most people, it means knowing they will have enough money to pay their bills and afford most of the things they want in life. It means having financial security. People who are financially secure know how to earn money, save for the future, and avoid debt. They don't worry about money, because they have enough to live a lifestyle that makes them happy.

▶ On a sheet of paper, write down all of the things you would like to own and activities you would like to do when you become an adult. This could include a house, a car, nice clothes, electronics, vacations, and leisure activities. Feel free to dream a little, but try to be realistic (no solid gold mansions or trips to the moon). Next to each item on your list, estimate how much you think it would cost. Ask your teacher for help. When you have finished, add up your costs. Then compare your totals with those of your classmates. Who had the highest total? Who had the lowest? How do people's goals affect their financial planning needs? Discuss your findings as a class.

343

Choosing a Bank and a Bank Account

A **bank** is a financial institution, which is a fancy way of saying a business that stores and manages money for individuals and other businesses. The bank invests that money or loans it to people and other businesses so that the bank can earn even more money.

There are basically three reasons to put your money in a bank.

1. *To keep it safe.* When your cash is in a bank account, it cannot be lost or stolen.
2. *To earn interest.* For some account types, the bank pays you for using your money. That means you earn money just by having a bank account.
3. *To make financial transactions easier.* Having a bank account makes it easier to cash paychecks, pay bills, and manage your money.

There are many types of banks and bank accounts. Understanding the differences helps you choose the one that is right for you.

Myth The bank will pay a check I wrote even if I don't have enough money in the account.

Truth While every bank is different, most banks will send the check back to the payee without paying. Then, the bank charges you an *overdraft fee*—a fee for spending more than you have in your account. It may also charge you a fee for returning the check. The payee may *then* charge you a fee for the returned check and still require you to pay the original amount. Be sure you have the money in your account before writing a check, or you'll end up owing a lot more in the long run.

Currency Exchange

Recall that most countries have their own currency. While some countries will accept U.S. dollars, if you are traveling internationally it is a good idea to have some local currency.

Banks can help you with this. Many banks will exchange your U.S. dollars for other currencies, such as euros, pesos, rupees, and yen, based on current exchange rates. For less commonly asked-for currencies, such as Icelandic kronur or Malaysian ringgits, you may have to go to your bank's central branch. Using the Internet, locate a site that lists exchange rates. See how much the dollar is worth in another currency.

Types of Banks

Do you have a savings account or a debit card? If you do, it's probably at a retail bank. Most of the banks that offer basic banking services to individuals are called **retail banks**. They provide services such as checking and savings accounts, credit cards, education loans, car loans, and home loans—also called **mortgages**. Most retail banks also provide resources for online banking. Commercial banks are like retail banks, except they provide their services to businesses.

Personal Financial Planning ■ Chapter 18

Credit unions also offer basic banking services. They are nonprofit banks that are owned by the customers, or members. **Nonprofit** means their goal is not to earn money for the business, but to earn money for the customers. Credit unions are formed by a group of people with something in common, such as employees of the same company or residents of the same town.

If you use the Internet to do all of your banking, you might have an account at an online bank. **Online banks** are retail banks that operate only on the Internet. They do not have buildings where you can go to talk to a banker or make a deposit. You do all of your banking electronically.

Are all banks the same? Why is it important to compare banks and their services?

Investment banks help businesses and other organizations raise money by issuing stocks and bonds. They also provide consulting services when one business wants to buy another business.

A **central bank** is an organization responsible for managing the banking activity of a particular country. In the United States, the Federal Reserve is the central bank.

Which Bank Is Right for You?

Most communities have several banks to choose from, not to mention all of the options available online. So, how do you decide which bank is best for you? If the bank has a **branch**—local office—nearby, you can go in and talk to the manager to learn more about the bank. You can also look online at a bank's Web site. Consider the following when choosing a bank.

- *Convenience.* You want a bank that is easy to get to and open when you need it. Does the bank have a branch nearby? Is it open in the evenings or on Saturdays? Does it provide free and secure ATM (Automated Teller Machine) service?
- *Services.* Different banks offer different services. Does the bank offer the type of account you need? Are there age limits for getting a debit card or ATM card? Does it offer online banking?
- *Interest rates.* Interest rates vary from bank to bank, and so does the way the bank credits the interest into your account. Does the bank offer the highest interest rate? How does the bank credit the interest? If monthly, the interest is added to your balance at the end of each month. If quarterly, interest is added to your balance at the end of every three months.

Finally, there are fees to consider. All banks charge fees—that's one way they make money. Before you choose a bank, you can compare the fees each bank charges for common services and transactions, including monthly service fees, check fees, and ATM transaction fees.

Some banks waive the monthly service fees if you maintain a minimum balance or sign up for direct deposit of your paycheck. Some allow you to use a certain number of transactions for free each month. Some banks also have low-fee or no-fee accounts for students.

NUMBERS GAME

How much should you save? Many financial experts recommend that you save at least 10% of your income.

To find the percentage of a number:

1. Change the percentage to a decimal by replacing the percent sign with a decimal point and moving the decimal point two spaces to the left.
2. Multiply the decimal by the number. For example, to find 20% of 150:

20% = .20

150 × .20 = 30

So imagine you receive $75.00 for your birthday—how much is 10%?

10% = .10

75.00 × .10 = $7.50

What if you wanted to save 15%?

Types of Bank Accounts

Banks keep track of the money you deposit and withdraw by using a bank account. A **bank account** is a record of the transactions between you and the bank. When you *deposit* money, the amount is *added* to your current bank account balance. This is known as a **credit**. When you *withdraw* money, the amount is *subtracted* from the balance. This is known as a **debit**.

The most common types of personal bank accounts are checking accounts and savings accounts.

Most banks have several different kinds of accounts. How do you know which one is right for you?

- **Checking accounts** are set up so that you can access your money by writing checks. You can also sometimes use a debit card or automated teller machine (ATM) card. Most people use checking accounts so they can pay for things without carrying cash. For example, you can write a check to pay for groceries or use your debit card to buy movie tickets. You must have enough money in the account to cover all of the checks and debit transactions.

- **Savings accounts** are not linked to checks or debit cards, although you may be able to deposit or withdraw money using an ATM card. Savings accounts earn interest, which means the bank pays you a percentage of your **balance**—the amount of money in the account. That interest is an incentive for you to keep your money at that bank. People use savings accounts to save money for future use.

Of course there are many different kinds of savings and checking accounts, including certificates of deposit (CD), money market accounts, and Individual Retirement Accounts (IRAs). These will be discussed later in this chapter.

Career Trend

More and more people are having their paychecks automatically deposited into their bank accounts. People prefer direct deposit because it saves them the time of going to the bank. Businesses prefer it because they don't have to print checks or worry about them getting lost. Banks prefer it because they know that money will come into the account, at least for a short time. In fact, many banks offer incentives or higher interest rates if you agree to have your paycheck automatically deposited.

Opening a Banking Account

As mentioned, there are many types of banking accounts. For most of them, the process of opening the account is similar. You open an account by going to a branch, filling out some forms, and making a deposit. Each bank has different rules, but usually you meet with a manager or assistant manager to show a picture I.D., such as a student I.D., a driver's license, or a passport. You may also need a birth certificate or Social Security card. Then you will do the following.

- Fill out a signature card that the bank will keep on file. A signature card helps keep your account safe, because it identifies the person authorized to use the account.
- Make the first deposit into the account when you open it. Some banks have a minimum deposit requirement, so it's a good idea to ask ahead of time how much money you will need.
- Fill out forms for ordering personalized checks. If you are opening a checking account, you will need to pick out your checks. The cost of the checks is automatically subtracted from your account balance, though some banks offer checks for free.
- Select personal identification numbers (PIN) for any debit or ATM cards, or for your online banking access, if available.

Why is it important that you understand all the rules and limitations of the type of account you choose?

Using Basic Transactions

The three basic types of transactions are deposits, withdrawals, and transfers. You deposit money into an account, withdraw money out of it, and transfer funds between accounts. With most accounts, you can deposit or withdraw money using cash or a check. You transfer funds electronically.

- *Depositing cash.* You can deposit cash in person at a branch of your bank, by mail, or at an automated teller machine (ATM). Usually, you fill out and submit a deposit slip that includes the amount of the deposit and the bank account number.
- *Depositing checks.* You can also deposit a check in person, by mail, or at an ATM. In addition, if you have a mobile banking app, you can use your mobile phone or tablet to take and upload a picture of the check. Usually, you must endorse, or sign, the back of the check and write the bank account number on it. You may also have to include a deposit slip.
- *Withdrawing money.* You can withdraw cash in person at a branch of your bank or at an ATM. Usually, you submit your ATM or debit card and enter your PIN to gain access to the account funds. You may be asked to show a photo ID, or provide a signature. Some banks let you use a mobile app on your phone to withdraw money from an ATM. You sign in on your phone and enter the withdrawal amount. The app generates a code which you scan at the ATM instead of inserting your card and entering a PIN.

What If I'm Not Old Enough?

Some banks have a minimum age requirement for opening certain accounts. For example, you might have to be 16 to open a checking account. What if you aren't old enough to open an account?

★ An adult can co-sign the account, which means he or she will sign the signature card, too.

★ You may be able to open a savings account instead of a checking account.

★ If your parents or guardians have an account at the bank, you may be able to open an account that is linked to theirs.

★ Consider a different bank with different age requirements.

If an adult co-signs on the account, the account will have your name and the adult's name on it—a joint account. You can both make transactions. What are benefits and drawbacks to having a joint account?

Most banks allow you to withdraw money using a bank check, which may be called a cashier's check or bank draft. Instead of cash, the bank issues a check payable to you or to the person or institution you specify. Bank checks are often used for large transactions, such as withdrawing money for a down payment on a car. Some landlords ask to be paid with a bank check. The bank may charge a fee for a bank check.

- *Using electronic funds transfer (EFT).* EFT lets you move funds from one account to another. You can use it to move money between your own accounts and from your account to someone else's account. EFT is often used to move money from your savings account to a checking account, from a parent's account to a child's account, and to pay bills.

Often, you set up EFT in advance by providing the name, address, and telephone number of the bank or other business to which you want to transfer the funds, along with the account number. If you are transferring the money to a bank account, you must also provide the routing number, which is a nine-digit code that identifies the U.S. bank location where the account was opened. Once the EFT is set up, you can usually make a transfer online or at an ATM simply by selecting the account from which to take the funds, the account to which the funds should go, and the amount. You can also make a transfer in person at a branch of your bank.

Using a Checking Account

A check is a piece of paper that tells the bank how much money to take from your account and put in someone else's—the payee's—account. A debit card—sometimes called a check card—authorizes an electronic transfer from your account to someone else's. It is important to keep enough money in the account to cover the amounts of every check or debit transaction you make.

When you open a checking account, the bank sends you personalized checks. Each check has printed information on it, including your personal information such as name and address in the upper-left corner, the check number in the upper-right corner, and your bank information such as the bank name and your account number across the bottom.

Each check also has blanks for you to fill in the following:

A Date
B Amount in words
C Payee
D Your signature
E Amount in numbers
F A memo line for noting what the check is for

Each time you write a check, you should record the transaction details in your **account register**, which is a book the bank provides along with your personalized checks. The register has spaces where you can enter the date, check number, payee, and amount. Then, you subtract the amount from your current balance, so you can see how much money is left in the account.

Write the Right Check!

It is important to fill out a check properly to avoid fraud and so that the bank can process it.

✔ Write neatly so the bank can read it.

✔ Use a pen, not a pencil, so the information cannot be erased and changed.

✔ Do not cross out mistakes.

✔ Remember to fill in all blanks, including the date.

✔ Make sure the numerical dollar amount matches the written dollar amount.

✔ Remember to sign the check in cursive—no printing!

If you make a mistake, tear up the check and write a new one. In your account register, enter the torn-up check as VOID, which means it is not available for use. What are some benefits of paying by check? What are some drawbacks?

Personal Financial Planning ■ **Chapter 18**

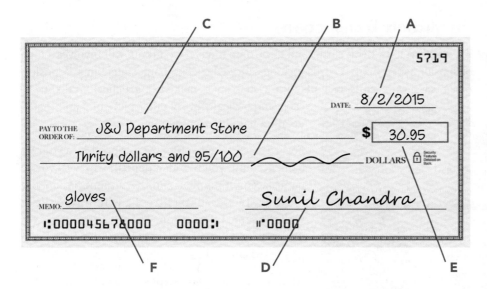

Can I Use a Debit Card?

If you have a debit card linked to your checking account, you can use it instead of writing a check. The card comes with your name and account number stamped on it. Some include your picture, too. There is a place on the back for your signature to help protect your identity.

To use the card, you run it through a machine at the store. You then punch in your PIN to authorize the transaction. The money is immediately transferred out of your account. Be sure to record debit transactions in your checking account register the same way you record checks.

Managing Your Bank Account

There are two important parts to managing your bank account:

- Keeping records of every transaction you make
- Making sure you and the bank enter each transaction correctly

Managing your bank accounts gives you control over your money. It allows you to make healthy financial decisions and to achieve your financial goals, because you know where your money is and how you are using it.

Why would you bother writing anything in the memo line?

Write the Right Check!

It is important to fill out a check properly to avoid fraud and so that the bank can process it.

✔ Write neatly so the bank can read it.

✔ Use a pen, not a pencil, so the information cannot be erased and changed.

✔ Do not cross out mistakes.

✔ Remember to fill in all blanks, including the date.

✔ Make sure the numerical dollar amount matches the written dollar amount.

✔ Remember to sign the check in cursive—no printing!

✔ If you make a mistake, tear up the check and write a new one. In your account register, enter the torn-up check as VOID, which means it is not available for use.

What are some benefits of paying by check? What are some drawbacks?

Recording Transactions

Even with ATMs, debit cards, and electronic transfers, banking still involves a lot of slips of paper. Bank records include canceled checks, bank statements, deposit receipts, ATM receipts, receipts for purchases you make using a check or debit card, and basically every other piece of paper that details a banking transaction.

For example, when you make a deposit into your bank account, you fill out a **deposit slip** with your name, account number, and the amount. When you make a withdrawal, you fill out a **withdrawal slip** with the same information. The teller uses the information to make sure the transaction is processed correctly.

After any transaction—deposit, withdrawal, debit card or check purchase—you receive a receipt that shows the transaction details. Keep that receipt! You can use it to make sure the transaction is correct, and then save the receipt to compare to your bank statement. If you ever find an error on your bank statement, you can show the bank your receipt, and the bank can correct the information.

Also, be sure to record every transaction in your register. Remember to record all fees and charges for services such as account maintenance or check ordering. Do this as soon as possible. If you forget, it will not only make it harder to balance your accounts, it could also cost you more money in overdraft fees.

How will keeping your accounts balanced help you reach your financial goals?

Balancing Your Account

Every month or so, the bank sends you a bank statement, which is a list of all the transactions for your bank account. The statement may come in the mail, the bank may send it via e-mail, or you may be able to access it online.

When you receive your statement, you can compare it to your account register and your transaction receipts to see if your information matches the bank's information. This is called *balancing, or reconciling, your account*. If you keep complete and accurate records, you will be able to balance your account correctly with each statement.

Ask your parents if you can look at their next bank statement. Are you surprised at how much they spend?

The bank statement shows the following information:

- Starting balance: the amount you had at the beginning of the month
- Ending balance: the amount you have at the end of the month
- Deposits and credits added to your balance
- Withdrawals and fees subtracted from your balance

Personal Financial Planning ■ **Chapter 18**

For checking accounts, the bank statement lists each transaction that has cleared, including the date, check number or debit location and amount, and sometimes the payee. Some banks also send a picture of every check that has been processed or allow you to view them online. Checking account statements usually come once a month.

For savings accounts, the statement also shows the interest that is credited to your account. Savings account statements may come once a month or four times a year (quarterly).

Career Tips

Want to be sure you have enough money in your account before making a purchase? Most banks have a number you can call to check your balance and even make transfers between accounts, or you may be able to look it up online. Sometimes, the bank charges a fee for these services.

Using an ATM

An Automated Teller Machine (ATM) lets you perform bank transactions without going to a branch office and waiting for a teller. You can use an ATM to:

- Deposit cash or checks
- Withdraw cash
- Transfer money from one account to another
- Check your account balance

What should you consider before withdrawing money from an ATM?

My Account Doesn't Balance!

There are a number of reasons your account might not balance with your bank statement.

- 👎 You forgot to record a transaction you made.
- 👎 You forgot to record an automatic deposit or withdrawal.
- 👎 You forgot to record bank fees or credits.
- 👎 You recorded the incorrect amount.
- 👎 You made a math error adding a deposit or subtracting a withdrawal.
- 👎 The bank made a mistake.
- 👍 You carefully reviewed the bank statement and your records and found that the account does balance.

If the account still doesn't balance after you check it again, take your receipts and the bank statement to the bank and ask a manager or assistant manager to review it with you. If the bank is responsible for the error, they will correct it. What steps can you take to make sure your account will balance with your bank statement?

To use an ATM, you need an ATM card that is linked to your bank account. The card has your name and account information stamped on it. When you apply for the ATM card at the bank, you choose a PIN. Every time you use the card, you must enter your PIN. Then you follow the on-screen instructions to choose a transaction type and your account. Most ATMs have instructions for hearing- and visually-impaired customers.

Remember: You must have enough money in your account to make a withdrawal at an ATM. Be sure to record the transaction in your account register and to adjust the balance by subtracting a withdrawal or adding a deposit. Some banks charge a fee each time you use an ATM, or if you use an ATM owned by a different bank, so subtract any fees from your account balance, too.

Money Madne$$

Greg started the month with $200.00 in his checking account. He wrote three checks this month, one for $25.00, one for $18.00 and the other for $15.00. He also withdrew $30.00 from an ATM (no fee was charged) and made a deposit of $120.00. When he got his statement, the bank listed his ending balance at $232.00. But he had calculated his ending balance at $262.00.

Can you explain the difference? Is the bank correct, or is Greg?

Saving and Investing

Saving money is one of the most important steps toward financial security. You put the money someplace safe—such as a bank or a credit union—so that it is available when you need it for an unexpected expense (such as medical bills), a planned expense (such as education), or for retirement. As a bonus, the bank will pay you just for leaving the money in your account. The longer it stays in your account, the more money you will have to spend later.

You might decide to invest some of your savings. Investing is riskier than saving, but it also provides an opportunity to earn more money or to increase your savings at a faster rate.

Saving and investing provide opportunities to increase your wealth, leading to financial security and freedom.

Personal Financial Planning ■ Chapter 18

The Difference Between Saving and Investing

When you save, you put money aside on a regular basis. You might deposit it in a savings account, a checking account, or a certificate of deposit (CD). When you invest, you take a portion of your savings and purchase a product such as stocks, bonds, or mutual funds.

Investing is riskier than saving, because the deposit is usually not insured, and your **return on investment (ROI)**—the amount of money you earn compared to the amount of money you invest—is not guaranteed. It is possible that you will lose all of the money you invest. But you have the opportunity to earn more from an investment than from savings.

Most financial experts advise that you use a combination of saving and investing. Saving keeps your money safe and available, and investing gives you the chance to grow your wealth. The amount you save and the amount you invest depend on your financial goals, your financial resources, how old you are, and how much risk you are willing to take.

Career Tips

Ninety percent of the best-paying and fastest-growing jobs require some kind of postsecondary education, usually a two- or four-year degree. There are several ways to start saving for a college education. Most states offer college savings plans that are tax-friendly and may even offer rebates. Ask your parents to help you explore your options.

Using Savings Accounts

The safest, most reliable, and most convenient way to save is to deposit your money in a savings account at a retail bank or credit union. The money will earn a small amount of interest and be available when you need it. It is insured by the Federal Deposit Insurance Corporation (FDIC) or the National Credit Union Association (NCUA), so you won't lose the money.

There are three basic types of savings accounts.

- *Passbook accounts* are the standard type of savings account. They earn a small amount of interest but are flexible, so you can withdraw your money at any time. A passbook account is a good way to get started saving.

- *Time accounts* are savings accounts that require you to leave the money untouched for a set amount of time—or **term**. Usually, the longer the term, the higher the interest rate. If you withdraw the money before the end of the term, you may have to pay a penalty, or fee. **Certificates of deposit (CDs)** are an example of a time account. Many time accounts have a minimum deposit, often between $500 and $10,000 dollars.

NUMBERS GAME

Have you ever heard of "The Rule of 72"? This theory is that to figure out the amount of time required to double an investment at a given compounded interest rate, you just divide the interest rate into 72. For example, if you want to know how long it will take to double your money at 6% interest, divide 6 into 72 and you get 12 years. You can also do the rule backwards to figure out what interest rate you need. So, if you know that you want to double your money in 12 years, you would divide 12 into 72 and then see that you need to invest your money at a 6% interest rate. The rule is very reliable with interest rates less than 20%.

1. How long would it take to double your money at an 8% interest rate?

2. If you want to double your money in 10 years, at what interest rate would you have to invest your money?

Try creating a spreadsheet to make projections using these formulas.

Why is it important to manage the resources you have available?

- *Money market accounts* are accounts that offer a higher interest rate than a passbook savings account. They usually have a minimum balance requirement and may limit the number of times per month that you can withdraw money.

Savings accounts, and any other investment for that matter, earn interest one of two ways. *Simple interest* is calculated based on the principal balance only. *Compounded interest* is calculated based on the principal plus any interest that has already been earned. Of the two, compounded interest is preferable as it results in greater savings over the long run.

Opening a Savings Account

To start saving, simply open a savings account at a retail bank or credit union. There are a few questions you want to ask about any savings account you are considering.

- What is the interest rate?
- Is the interest simple or compounded?
- How often is interest calculated?
- How often is the interest credited (added) to your account?
- Are there service fees?
- Can you withdraw your money at any time?

Understanding Investments

Once you achieve your savings goals, you may decide to invest a portion of your savings to see if you can earn a higher return on investment. You may not be able to invest on your own until you are 18 years old, but you can ask a parent or other trusted adult to help you set up a joint investment.

Learning as much as possible about an investment before you commit money will go a long way toward minimizing your risk. Still, all investments involve risk, so it is a good idea to invest only as much as you are willing to lose. Here are two basic types of investments.

- **Equity investments**, in which you purchase **stock**—or ownership—in a company. Your ROI depends on the stock price. If the stock goes up—increases in value—you gain. If it goes down—decreases in value—you lose. Stock prices change by the minute, and you do not actually earn any money until you sell your stock.

Personal Financial Planning ■ Chapter 18

Does the idea of investing in stocks excite you? What stock would you invest in if you had the money?

- **Fixed income investments**, in which you lend money to a business or government agency in exchange for a **bond**. A bond provides fixed interest payments for a set period of time. The business or government agency uses your money for the term of the bond.

Bonds are generally safer than stocks. You are promised a specific rate of return, and most bonds are insured. However, stocks usually provide greater opportunity to earn money.

One way to minimize your risk when investing is to diversify—to purchase a variety of stocks and bonds. One of the easiest ways to diversify is to invest in a mutual fund. A **mutual fund** is a pool of money collected from many—maybe thousands—of investors, and then used to buy stocks, bonds, and other securities.

The success of a mutual fund is not dependent on a single company, the way a stock is. Some of the assets in the fund might decrease in value, but if overall the assets increase in value, then the fund increases in value. This diversity helps minimize your risk over time.

Planning for Retirement

Retirement is the stage of life after you stop working, when you can relax and have time to do things you enjoy. Of course, you still need money to pay for your needs and wants. That's why it is important to start saving for retirement as soon as possible.

While you can use regular savings and investment accounts to save for retirement, there are also some types of accounts designed specifically for retirement savings. Called **tax-deferred savings plans**, they have tax benefits that encourage people to save for the long term. Following are the two most common tax-deferred savings plans.

- An **individual retirement account (IRA)** is a personal savings plan that allows you to set aside money for retirement, usually up to $2,000.00 per year. You do not pay taxes on the money you deposit into the IRA. You can start withdrawing from your IRA when you are 59½ years old.

Investment Terms

If you plan to invest, it is important to speak the language of investors.

- ✔ **Bond**: A debt security issued by corporations, governments, or their agencies, in return for cash investors
- ✔ **Commodity**: An item such as gold, wheat, or coal that can be traded or sold
- ✔ **Portfolio**: All of the investments and accounts an investor owns
- ✔ **Return**: The money received annually from an investment, usually expressed as a percentage
- ✔ **Stock**: A share of ownership in a corporation
- ✔ **Risk**: The measurable likelihood of loss or less-than-expected returns
- ✔ **Security**: An asset that has financial value and that can be traded
- ✔ **Securities and Exchange Commission (SEC)**: The government agency that regulates the securities industry

- A **401(k)** or **403(b)** plan is a savings plan offered by an employer to an employee. The employee contributes a percentage of his or her earnings to the 401(k)/403(b) account each pay period. Some employers then match the contribution. Such plans are considered to be *tax sheltered annuities* because employees are allowed to invest a portion of their paycheck *before* taxes are taken out. Keep in mind that not all companies offer retirement packages. If they do not, then it is up to the employee to start one.

In addition to the saving and investments you make independently, you also contribute to **Social Security**, a government program that pays monthly benefits to retired workers in the United States who paid their Social Security tax. It also pays benefits if you become disabled and cannot work, and to the spouse and children of workers who die before they retire. The benefits are based on the amount of Social Security tax you pay during your working life and the age when you retire.

Social Security provides a welcome income to many retirees, but it is usually not enough to cover living expenses. Experts estimate that if you have average earnings throughout your life, Social Security benefits add up to about 40% of the yearly income you earned while you were working. In addition, many people think the Social Security system is in trouble and may run out of money before all of the people who have been contributing are able to collect benefits. That's why many people save for their own retirement.

Do you think it's too early to start planning for retirement?

Career Fact — Most large companies used to offer *pensions*—a retirement plan wherein employers set aside money to be granted to employees upon retirement. The number of companies offering pensions has declined in recent decades, forcing more workers to contribute to their own retirement. There are still some careers that offer traditional pensions, namely those in public service. Police officers, firefighters, teachers, and judges generally receive pensions, provided they put in the required years of service.

Personal Financial Planning ■ Chapter 18

Using a Money Management Service

Money management is the way you handle your money in order to get the most value from it. It includes investing, budgeting, spending, banking, and planning for and paying taxes. You can do all of these things on your own, or you can hire someone to do it for you. A money management service is a company that performs these tasks, for a fee. The company may make the actual transactions, or provide advice and counseling services. Some types of money management services include financial planners, brokers, banks, and online services.

An advantage to using a money management service is that it saves you time. You let someone else do the research and analysis for you. The company has experience and knowledge that it might take you years to gain. A disadvantage is that you have less direct control over your money. Another is that you must pay a fee.

When using a money management service it is important to understand how they charge. Some are fee-only, which means they charge an hourly rate or a flat fee depending on the service they provide. Some charge a percentage of the assets they manage. For example, a firm might charge 2% a year for an account with $75,000 in assets. That means you would pay $1,500 each year.

Comparing and Contrasting Forms of Credit

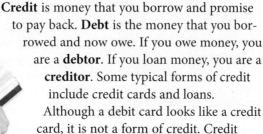

Credit is money that you borrow and promise to pay back. **Debt** is the money that you borrowed and now owe. If you owe money, you are a **debtor**. If you loan money, you are a **creditor**. Some typical forms of credit include credit cards and loans.

Although a debit card looks like a credit card, it is not a form of credit. Credit cards use money that you borrow. Debit cards use money that is already in your checking account.

When you use credit responsibly, it can be very convenient, because you can use it to buy things now and pay for them later. You might use a credit card to pay for a shirt, because you know your grandmother is going to give you money for your birthday next week. When you receive the money, you can pay the credit card bill. When you use credit irresponsibly, you might wind up with so much debt that you can never pay it back. What happens if you buy a shirt that costs more than your grandmother gives you? How will you pay the credit card bill?

Managing credit and debt responsibly shows that you are trustworthy, you understand the difference between needs and wants, and you can make healthy financial decisions.

Can I Borrow 20 Bucks?

Borrowing from friends and family can help you achieve certain financial goals. You might be able to get a loan from a relative even if you don't qualify at the bank. But, it's still a loan, and you still have to pay it back. Taking advantage of a friend or relative by ignoring your responsibilities is likely to hurt your relationship.

- 👎 You skip payments.
- 👎 You use the money for frivolous wants.
- 👎 You borrow for something you can afford on your own.
- 👎 You forget to put the terms in writing.
- 👍 You make all payments on time and in full until the loan is repaid.

Borrowing from or loaning to a friend or relative can cause conflict in your relationship. How would you feel if you lent your friend money and she never paid you back?

Chapter 18 ■ Personal Financial Planning

Career Profile: Broker

Finance Career Cluster

Job Summary

A broker, also called a securities and commodities sales agent, buys and sells stocks, bonds, and other products for clients who want to invest in the stock market. They explain the advantages and disadvantages of investments and keep their clients up to date with price quotes and economic changes. Some agents help clients with insurance, tax planning, and estate planning.

Being a broker is a stressful and demanding job. Brokers generally work long hours and are always under pressure to make important financial decisions. The job can also be very rewarding: It's not unusual for brokers to earn more than six figures per year. Most brokers have a bachelor's degree in business or finance.

Use the Internet, library, or your school's guidance resources to learn more about a career in financial planning. Write a job description, including the educational requirements, career pathway, and potential salary range.

What skills and interests do you think would be helpful for someone pursuing a career as a broker?

Getting a Loan

A **loan** is a transaction in which the lender agrees to give the borrower money and expects to be repaid in full. You can get a loan from a bank, business, credit company, insurance company, relative, or friend.

Most loans have **terms** that are written in a contract that both the lender and the borrower sign. For example, you might agree to repay the loan by a certain date, and you might agree to pay a specific rate of interest. A **contract** protects both the lender and the borrower. It spells out all the conditions so everyone involved knows what to expect. Such conditions may include the need for a *cosigner* (someone who is responsible for the loan if you can't pay) or for *collateral* (something you own, such as a house or car, that the lender can take possession of if the loan is not repaid). Generally speaking, borrowing more money than you can repay is not good for your well-being. Your credit history will suffer, you will be unable to get new loans, and personal relationships with the lender will suffer.

At some point, you will want to buy something you can't afford. Where will you get the money?

Using a Credit Card

A credit card is useful and convenient. It lets you make a bunch of purchases and then pay for them all at the same time by writing a single check. A credit card can come in handy in an emergency, such as paying for repairs or towing if your car breaks down. If you use it responsibly, it helps you establish credit so you can get loans in the future.

Responsible credit card use means:

- Paying the entire balance by the due date each month
- Only using it to buy items that you can afford
- Using your credit card sometimes, but also using cash, checks, or your debit card
- Having only one or two credit cards in your name
- Saving all receipts and checking your statement carefully for errors each month

While it may be tempting to apply for more than one credit card, there are drawbacks to having too much plastic in your purse or wallet. If lenders believe you have too much credit, they may reject your loan application. You might not be able to borrow money for education or to buy a car.

Keep in mind that credit card companies are in business to make money. Each card comes with a set of terms and conditions that states the fees, penalties, and other charges that the company can impose. Common terms and conditions include:

Having too many cards can cause problems in the future. Why do you think some people have multiple credit cards?

- *Credit limit:* the maximum amount of purchases you can charge to the card
- *Annual percentage rate:* the interest you pay on your outstanding balance
- *Annual fee:* the yearly charge for using credit
- *Grace period:* the length of time from when you make a purchase to when you start accumulating interest
- *Late fee:* the fee charged when you make a payment past the due date
- *Over-limit fee:* the fee you pay when you charge an amount over your credit limit
- *Minimum payment:* a percentage of the outstanding balance that must be paid each month

The only way to avoid paying additional fees and charges is to pay your balance in full and on time every month. If you don't, the fees and charges are added to your balance, which continues to grow, even if you don't make any new purchases. If you only pay the minimum amount required, you will be paying that same bill for many years. In fact, you may never pay it off. This also becomes part of your credit history, and banks, landlords, and even employers will doubt your ability to make responsible decisions and repay loans.

Money Madne$$

Credit card companies charge an annual percentage rate on any outstanding balance. This rate varies tremendously, from as low as 5% to as high as 35%.

Imagine you made $150.00 in charges this month and you are only able to make the minimum payment of $20.00. Your credit card has an interest rate of 22%. How much will your finance charge be? Use a spreadsheet program to calculate what your financing costs will be.

How Do I Establish Credit?

At some point in your life, you will need a loan. You might need an education loan to go to college, an auto loan to buy a car, or a mortgage to buy a house. You might apply for a credit card to use to pay for purchases. You might apply for a small business loan to help start up a new business.

Before a bank or retailer will give you a loan or issue you a credit card, it checks your credit history to find out if you know how to manage money and make healthy financial decisions. Landlords check your credit history before they rent you an apartment. Some employers check your credit report before offering you a job. When it comes to small business loans, lenders scrutinize your credit history. No matter how good your business plan is or how brilliant your idea for a new

Personal Financial Planning ■ Chapter 18

product or service, a bank won't give you the money to get off the ground if you haven't managed your own personal finances well. After all, if you can't pay your own bills, how can they expect you to pay them back?

It takes time to establish credit, but it is worth the effort. There are positive steps for establishing credit.

- Maintaining bank account balances
- Earning a paycheck
- Paying bills on time
- Paying rent
- Using a credit card responsibly

What's a Credit Report?

A **credit report** is a summary of your credit history—usually for the past seven to ten years. If you have a good credit history, lenders will loan you money. If you have a poor credit history, they won't. For example, if you have a few loans and credit cards, and you pay all of your bills on time every month, banks, landlords, and other lenders will believe you are responsible. If you have many credit cards with high outstanding balances,
and you regularly miss payments or pay late, the lenders will believe you are irresponsible.

A credit report lists the following.

- All of the loans you have received, including all credit cards
- Your outstanding balances—the money you owe
- Whether you have paid your bills on time
- The names of agencies or companies who have recently requested copies of your credit report
- Your credit score, which is a three-digit number that ranks your likelihood of repaying your loans

Credit reports also include **derogatory marks**, which are negative reports that can take from seven to ten years to clear from your credit history. Examples of derogatory marks include bankruptcy, unpaid taxes, and civil judgements, such as the result of a law suit for which you must pay damages. Derogatory marks are red flags to lenders, as they are indicators of someone who is unreliable and untrustworthy, and possibly criminal.

The information is collected by a credit bureau, or consumer credit reporting company, and sold to banks and other lenders to help them determine whether they should lend you money.

Ways to Protect Your Credit

You can keep your credit safe by making sure your credit information does not get into the hands of an unauthorized person. To keep your credit safe:

✔ Keep a list of your account numbers.

✔ Keep a list of the contact information for your banks and lenders.

✔ Keep your credit, debit, and ATM cards in a safe place.

✔ Report lost or stolen cards immediately.

✔ Do not give your account number to anyone who calls, texts, or e-mails you.

✔ Do not give anyone your PIN.

✔ Shred documents that have your account number on them.

Why is it important to check your credit report every year?

Some potential employers may request to see your credit report before offering you a job. Access to your credit report is governed by the Fair Credit Reporting Act, a U.S. federal law which sets limitations on when and who your credit information is accessed. Some states have additional restrictions. For potential employers, the following federal restrictions are in place:

- Employers must ask the applicant for permission, and the applicant must consent.
- The applicant's credit score is not included in the report sent to the employer.
- If the employer denies employment based on the information in the credit report, the employer must first provide the applicant with a copy of the credit report to review.

Potential employers cannot access your credit score, but a credit report can provide information about potential illegal activity, high debt, and a general lack of responsibility. These factors might influence the potential employer's decision on whether or not to offer you the job. Of course, a clean credit report with no late fees, manageable debt, or derogatory marks will show a potential employer that you are responsible, financially sound, and trustworthy.

There are three nationwide consumer credit reporting companies in the United States: Equifax, Experian, and TransUnion. By law, you can obtain one copy of your credit report for free, once a year, from each of the three credit bureaus. Looking over your credit report gives you the opportunity to see if there are any mistakes that might hurt your chances of getting a loan. You can also see if any unauthorized people have been trying to get credit in your name. If you find an error, notify the reporting agency immediately.

Myth All debt is bad.

Truth Some debt is inevitable. Very few people can afford to pay cash for a house, for example. A mortgage is a form of debt that allows you to work towards owning a home. Thus it is also a form of investment. The key is to manage your debt responsibly—not let it get out of hand.

Managing Debt

Sometimes borrowing money is the easy part; paying it back is more difficult. As long as you make your payments in full and on time, you are managing your debt responsibly. The problems come when debt grows to the point where you do not have enough money to cover your payments as well as your other expenses.

Excessive debt may be caused by irresponsible use of credit cards and loans, but it can also be caused by unexpected expenses such as medical bills. The following tips can help you keep your debt in control.

- Do not purchase more than you can afford.
- Set financial goals and stick to them.
- Pay all of your bills in full and on time every month.
- Keep accurate records, so you always know how much you have and how much you owe.
- Use a budget to track and analyze your expenses.
- Save at least 10 percent of your income every pay period.

Personal Financial Planning ■ Chapter 18

If debt grows to the point that you cannot pay more than the monthly minimum, or maybe not even that, it will affect your well-being. It is not easy, but you can get out of debt.

- Figure out how much you owe.
- Stop using your credit cards.
- Analyze your budget and cut out all nonessential expenses.
- Allocate as much as possible to paying your bills.

If you have spent all of your money and cannot pay any of your bills, you may be faced with bankruptcy. **Bankruptcy** is a process in which you declare yourself legally unable to pay your outstanding debts. A judge approves your request, and most of your debts are cleared so you can start over. Bankruptcy negatively affects your credit history, making it very difficult to get credit again for a long time.

The World Bank

The World Bank is not really a bank at all. It is a part of the United Nations called a specialized agency. It includes two organizations: the International Bank for Reconstruction and Development and the International Development Association. Together, they support efforts to improve the lives of people in developing countries by providing low-interest loans, interest-free credit, and grants (monetary aid given to support a cause). The funds are used in a wide range of areas such as agriculture, water supply, science, and education.

Using resources in your library or on the Internet, research the World Bank and the areas that it supports. Select a project for which it provided a loan or credit and report on it to your class.

CAREER COUNSEL

Keyshan loves to fix things. He spends many Saturday afternoons working on upgrades to his home computer or helping his dad make repairs around the house. His favorite subjects in school are technology education, art, and computer science, though he is also pretty good at math. He is starting his junior year, which means he needs to think seriously about his future after high school. His grades might be good enough to get into one of the community colleges in town. He's also thinking about forgoing college and going directly to work after graduation, maybe as an apprentice.

Use your 21st Century Skills, such as goal setting, decision making, and critical thinking, to help Keyshan choose a career path that not only meets his interests, but also makes the most of his available resources. Make a list of five or ten options, then rank them from most to least desirable. Compare your rankings to other members of your class. Which option seems to be the best overall? What would you do if you were Keyshan?

Too Much Debt?

In the United States, the number of credit cards in circulation ranks in the hundreds of millions. In fact, most Americans have between 3 and 12 credit cards in their pockets or purses. Unfortunately, the convenience of buying on credit comes with a cost.

★ As a nation, we owe over $850 billion dollars in credit card debt.

★ The average credit card balance per consumer is almost $3,800.

★ The average U.S. college graduate begins his or her career with more than $2,000 in credit card debt.

While most Americans do use credit cards responsibly, it's clear that many struggle to keep their plastic paid off. Do credit cards do more harm than good? Do you think it would be possible to live the lifestyle you want without them?

21st Century Learning

Paying Taxes

A **tax** is money we pay the government. The government uses the money to pay for public resources. People complain about taxes all the time, but if we didn't pay them we wouldn't have things such as streets, sidewalks, parks, schools, and libraries. The government wouldn't be able to run programs to help citizens in need, pay for the military, or send astronauts into space.

There are three basic categories of taxes.

- *Income taxes* are based on wages and other earnings.
- *Consumption taxes* are based on things we buy, such as computers or gasoline.
- *Asset taxes* are based on things we already own, such as houses or cars.

You have to keep taxes in mind when making purchases, planning a budget, or setting financial goals. While a 5% sales tax on a candy bar won't have a big impact on your budget, that same sales tax on a $16,000 car would. There are all sorts of rules and measures in place to make sure that we pay our taxes when they are due.

One of the key issues prompting the Revolutionary War and the birth of our nation was unfair taxation. Do you think it's fair for a government to tax its people?

Not everything you buy is taxed. For example, most food that you purchase in a grocery store is tax-free. Also, many states have "tax-free holidays"—usually weekends when you can make purchases without paying sales tax. Research the sales tax laws for various states online. Is there one state where it's cheaper to go shopping than the others?

Personal Financial Planning ■ Chapter 18

Where Do Your Tax Dollars Go?

Governments use our tax dollars to pay for public resources. We pay different taxes at different rates according to who is collecting and using the money.

- Some taxes we pay to the federal—national—government. Federal taxes are the same for everyone, no matter where you live. We all pay income tax, Social Security tax, and federal fuel tax for gasoline.

- Some taxes we pay to the state government. State taxes are different depending on where you live. Each state sets its own rates, so one state might have a 5% sales tax, while another state might have an 8% sales tax.

- Some taxes we pay to the local government—the community where we live. Some communities have property taxes. Each community sets its own tax rates.

Often we don't even notice when we are paying taxes. Gas tax is already included in the price per gallon. Property taxes are usually a part of a homeowner's monthly mortgage payment. But it's always important to be aware of how taxes will affect your income and expenses.

Paying Income Tax

Income tax is a percentage of your income that you pay to the government. The federal government collects income tax, and most—but not all—state governments do, too. Income tax is automatically withheld from your paycheck by your employer.

The United States has a *progressive tax system*, which means the more you earn, the more you pay. Income is categorized into levels, called **tax brackets**. If your income falls into the lowest tax bracket, you pay 0%. If it falls in the next tax bracket, you pay 10%. Currently, the highest tax bracket pays 39.6%. The government changes the levels and percentages as necessary to keep up with economic changes, such as salary levels and the cost of living.

Your tax bracket is based on your **taxable income**, which is not the same as your annual salary. To calculate your taxable income, you add up all of your income and then subtract **tax deductions**, which are expenses that you are allowed to deduct from your income, such as contributions to charities and interest you pay on a mortgage. The remainder is your taxable income.

Once you know your tax bracket, you can calculate the amount of actual tax you owe. From that, you may be able to subtract tax credits, which are expenses you are allowed to subtract from your actual tax payment. Some typical tax credits include expenses for college and childcare.

Do you ever think about the sales tax when you buy something? Is everything you buy taxed?

Filing Income Tax Returns

Each year by April 15, you must file income tax returns for the income you earned the previous year. **Income tax returns** are forms on which you calculate the amount of income tax you owe. You file federal tax returns to the **Internal Revenue Service (IRS)**, the agency responsible for collecting federal taxes, and you file state tax returns with your state's revenue department.

Once you calculate the amount you owe based on your tax bracket, you subtract the amount that your employer withheld throughout the year. If it is more than the amount you owe, the government sends you a refund check. If it is less than the amount you owe, you must send the government a check for the difference.

Do you think it would be better if everyone paid the same percent on income tax?

According to recent data, the top 50% of U.S. taxpayers paid 98% of all income taxes collected, and the bottom half paid only 2%.

What If We Just Stopped Paying?

Some believe that without taxes we wouldn't have any government-funded programs. For example, what if people decided not to pay their local taxes? There would be no:

★ Teachers and schools
★ Firefighters
★ Police officers
★ Trash collection and recycling
★ Street plowing and cleaning
★ Municipal parks

Now think about what your community could be like if people were able to pay ten times the amount in local taxes that they do now.

Tech Connect

The days of taxpayers poring over reams of paper with W-2 forms flying and paper cuts galore are fading. More and more people are using software to file their taxes. Programs such as TurboTax® not only do the calculations for you, they also walk you through the process, helping you to find important deductions and exemptions that can save you money. In addition, tax returns computed and filed electronically usually have fewer errors, which is better for the taxpayer and the IRS alike.

In fact, the number of people filing their taxes online is up to 99 million and growing. To learn more about how to file your tax return online, go to irs.gov/Filing. Read about all of the options available to individual taxpayers. Then make a list of the pros and cons of filing your taxes using each method. If possible, prepare a sample EZ income tax form and 1040A income tax form both manually and electronically. Present your findings to the class.

Keeping Your Personal and Financial Information Safe

When was the last time you gave out personal information? Personal information includes your name, address, phone number, and e-mail address. It also includes your birth date, Social Security number (SSN)—the number assigned to you by the U.S. government to track your income and employment history—bank and credit card account numbers, and medical and insurance records. Protecting yourself from identity theft and fraud means protecting all of that information.

Identity theft occurs when someone steals your personal information and uses it to commit fraud. The thief might open new charge accounts, rent an apartment, make purchases, and take out loans. He might get a job using your Social Security number, or apply for a driver's license or passport. She might obtain medical care using your insurance information. These actions will negatively impact your credit history making it nearly impossible for you to obtain credit. It can take years and cost a significant amount of money to clear your name and correct your credit and your credit report. The **Federal Trade Commission (FTC)**—the government agency responsible for protecting consumers—estimates that as many as 9 million Americans have their identities stolen each year.

Keeping your personal and financial information safe is an important part of financial responsibility.

Protect Your Records

Identity thieves will do almost anything to get their hands on your personal, financial, and medical records. They'll go through your trash, bribe clerks, and peek over your shoulder when you use your computer. It is important to keep track of these documents and make sure they are stored safely.

What types of documents should you secure?

- Personal records such as a birth certificate and Social Security card
- Medical records showing when you had immunizations
- School records showing what school you attend, report cards and standardized tests results, and your student identification.
- Financial records such as bank statements, credit and debit card numbers, and loan information
- Copies of tax returns, pay stubs, and insurance records
- Driver's license, marriage certificate, and passport

You can keep these items in a safe, or in a **safe deposit box**, which is a box in a fireproof vault that you can rent from your bank. The information will be safe from thieves, and you will know where to find it when you need it.

Do you have anything that you value highly, such as a diary or a journal or important papers? Where do you keep important items?

Who Is Responsible?

You cannot assume the banks, lenders, and regulatory agencies are always going to know what is going on with your accounts. You have to take responsibility as a consumer. What if you notice an unauthorized charge on your credit card statement?

★ Notify the credit card company by phone and in writing immediately.

★ Check your other statements to see if there are any other unauthorized charges.

★ Check your credit report to see if there are any signs of fraud.

★ Pay the undisputed charges in full and on time.

What would you do if you found an error on your credit report? Who would you contact if you felt a credit card company was treating you unfairly?

Protect Your Wallet

Think about the items you carry in your purse or wallet: cash, credit and debit cards, insurance card, driver's license or student I.D., cell phone, receipts, and keys. What could a thief do with all of that?

If you lose your wallet or purse, you should immediately take steps to protect your identity and your finances. If your wallet or purse is missing:

- File a report with the police immediately and keep a copy.
- Report the loss to the fraud department at your bank and other financial institutions.
- Cancel your credit, debit, and ATM cards immediately.
- Apply for new cards with new account numbers.
- Call the fraud departments of the three major credit reporting agencies and ask each agency to put out a fraud alert on your accounts. A fraud alert is like a red flag on your accounts that shows your information may be in the hands of an unauthorized person.
- Review your credit reports regularly and have them corrected when necessary.
- Report a missing driver's license to your state's department of motor vehicles.
- Change the locks on your home and car if your keys are missing.
- Notify your cell phone service provider that your phone is missing.

Protect Yourself Online

It is certainly true that the Internet has made information more accessible. Unfortunately, it has also made your personal information easier to get as well. Experts believe that the wired world has led to the dramatic increase in identity theft over the last decade. **Phishing** is a scam designed to steal your personal and financial information over the Internet. Criminals use e-mail messages and fake Web sites to trick you into entering the information, which they can sell or use themselves.

What steps can you take to make sure your identity is safe, even online?

You can protect yourself by being wary.

- Never reply to an e-mail message, text message, or pop-up browser window that asks for your personal information, even if the message looks as if it comes from a bank or credit card company you use.
- Never call a phone number included in an e-mail message or text to contact your bank or credit card company. Use the phone number on the back of your account statements or credit card.
- Keep your antivirus and antispyware software up to date, and use a firewall.
- Keep your e-mail and online account passwords in a safe place and change them often. Don't use the same password for every account.

Analyzing Banking and Credit Regulations

The federal and state governments have passed many laws and regulations designed to protect you and your money. They have set up agencies such as the **FDIC—Federal Deposit Insurance Corporation**—to provide insurance for your deposits, and the **FTC—Federal Trade Commission**—to make sure businesses treat you fairly. They allow you to report problems, such as lenders that discriminate based on your race, and to ask questions, such as "What's my credit score?"

Regulating Banks

Banks must be chartered to operate in the United States. A **bank charter** is an agreement that controls how the bank operates. The agency that charters the bank is responsible for making sure the bank follows all rules and regulations. Each state has a regulatory agency responsible for regulating state-chartered banks. In addition, three federal agencies regulate banks:

- The Office of the Comptroller of the Currency (OCC) (occ.treas.gov)
- The Federal Reserve Board (federalreserve.gov)
- The Federal Deposit Insurance Corporation (FDIC) (fdic.gov)

All of these agencies have consumer protection departments. You can contact them with questions or problems at any time.

Protecting Credit

Consumer credit laws are designed to help consumers and creditors meet their legal responsibilities. Some federal acts regulating credit include:

- *The Fair Credit Reporting Act.* This act is enforced by the Federal Trade Commission (FTC). It promotes accuracy and ensures privacy of the information used in consumer reports such as your credit report.
- *The Equal Credit Opportunity Act.* This act protects you from discrimination by creditors for any reason other than your creditworthiness. Creditors cannot ask your race, sex, country of origin, religion, or age.
- *The Fair Credit Billing Act.* This act sets rules for when credit card bills are mailed and payments are credited, as well as procedures for settling billing errors.
- *Truth in Lending Act.* This law is designed to make sure you have all of the information you need to make a healthy decision about borrowing money. It lists the information that a creditor must disclose before lending you money. This is also called the *Consumer Credit Protection Act.*
- *The Credit Card Act of 2009.* This act restricts credit card issuers from raising interest rates without warning and charging excessive fees for late payments. It also set the minimum age for obtaining a credit card to 21.

Myth Deposits in all savings and investments account are insured by the federal government.

Truth Investment accounts are not insured by the government the way savings accounts are.

However, the **Securities and Exchange Commission (SEC)** is a government agency responsible for supervising and regulating companies to protect investors. The SEC monitors companies to make sure they provide investors with accurate and timely information. The SEC gives consumers rights to sue if the information is incomplete or inaccurate.

Case Study

Maggie has been putting money in a savings account for five years to help pay for college. She has $2,200.00. The savings account earns about 1.2% interest. She thinks she has enough saved now that she might be able to earn more interest if she puts the money in a different type of account, or invests it.

The bank manager recommends putting $1,500.00 in a CD for 12 months at 2.5% interest. Maggie is considering the CD, when her older brother offers to set up a stock trading account online for her. He recommends buying shares of the company where he works. The current stock price is $10.00 per share, so she has enough money to buy at least 200 shares, as well as pay the fees. He tells her the stock will go up; then she'll be able to sell and make a nice profit.

- What options does Maggie have?
- What are the pros and cons of each choice?
- What information does she need to make the best choice?
- What should Maggie do?

Answer It!

1. What are the three basic reasons for putting your money in a bank?
2. What is the main difference between a retail bank and a credit union?
3. Explain which is riskier, saving or investing.
4. What are the advantages of investing in a mutual fund?
5. What is the difference between credit and debt?
6. List three positive steps for establishing credit.
7. Explain why a credit score is important, and how a poor credit score affects both personal finance and career opportunities.
8. What is the effect of identify theft on credit?
9. What government agency is responsible for protecting investors?
10. What is the purpose of consumer credit laws?

Web Extra

Many government Web sites provide consumers with information about finances and other money issues. Some, such as www.USA.gov, has a section specifically for teens.

Use the Internet to locate useful government Web sites, and make a directory of the sites. Include a brief description of the information available on the site. Post the directory on your school Web site, or make it available in your school library.

Personal Financial Planning ■ Chapter 18

Write Now

According to the Credit Card Act of 2009, you must be 21 to obtain a credit card. Do you think that's fair? Is there another age that makes more sense? Is there an age limit for obtaining a debit card?

Conduct research to learn more about the different types of credit available to students and adults, including credit and debit cards. Then write an essay explaining what you have learned. Illustrate the essay with a chart that compares and contrasts the different types of credit.

COLLEGE-READY PRACTICES

College is expensive. College-ready individuals understand that it is never too early to start planning for college expenses.

Conduct research into saving and investing options specifically for college. For example, many states offer educational savings accounts. Also research money management services to see if there are some that specialize in planning for the cost of college. Identify the advantages and disadvantages of different money management services. When your research is complete, write a report or create a presentation explaining the options for college savings, including the information on money management services.

Career-Ready Practices

Career-ready individuals have experience managing their money using different types of bank accounts. They know how to select a bank, open an account, conduct and record transactions and fees, and reconcile their statements.

Working in teams of four or five, prepare training materials that teach students how to use a bank to manage their money. Develop a written handbook that discusses banking options, includes definitions of key banking terms, and explains how to perform banking tasks. Illustrate the handbook by creating one or more videos that show how to open and maintain different types of accounts, make deposits and withdrawals, and reconcile a statement, including fees and services. If you cannot make videos, develop and perform skits, instead.

As a class, watch each team's videos or skits. Discuss the importance of understanding financial management and how it can impact your college and career options.

Career Portfolio

A career portfolio is a treasure-trove of personal information. It often includes names, addresses, phone numbers, work history, transcripts, grade reports, and identification numbers. While you are encouraged to make a digital version of your portfolio, you also need to be aware of how you will make it available and to whom. For example, if you choose to post your portfolio online, is there a chance that someone could use the information in the files for fraudulent purposes? Are there ways you can limit access to your online portfolio or at least restrict the kinds of personal information you include with it?

A number of Web sites allow you to create an online portfolio or e-portfolio and store it on a secure server (try visualcv.com for starters—or check with your school to see if they have their own service). You can then restrict who can access the portfolio by sending them an e-mail or requiring a password. In addition, your school may have a secure place on their Web site for creating, storing, and posting your digital portfolio.

Explore your options for posting your portfolio online. Report your findings to the class.

A Language Arts Review

Being able to use basic grammar and punctuation in your writing is essential. How will you get the job without being able to fill out the application, write the cover letter, and create a resume? Written communication is an indispensable skill.

Grammar

Subjects and Verbs

The basic unit of writing is the simple sentence. A simple sentence expresses a complete thought. Every simple sentence must contain two basic building blocks: a **subject** and a **verb**. The subject is the main topic of the sentence—who or what the sentence is about. The verb says what the subject does.

Example of a simple sentence:
Donna bakes cupcakes for the New York Cupcake Café.

Subject: Who? <u>Donna</u>

Verb: What does Donna do? She <u>bakes</u>.

All sentences are built from subjects and verbs, so understanding how to find them is an important first step in mastering other sentence skills.

> **NOTE**
> Since not all verbs express action, it is helpful to keep in mind that verbs are also words that change their forms to indicate the tense or time of a sentence.
>
> For example:
>   *Reading is my favorite activity.*
>
> What is the verb?
>
> Change to past tense:
>   *Reading was my favorite activity.*
>
> Which word has changed?
> So what is the verb?

Find the Subject

When you write a sentence, you write about someone or something: the subject. To find the subject of a simple sentence, ask who or what the sentence is about. Subjects can be people, places, things, or ideas.

- Maddie eats cheese every day.
  Who? <u>Maddie</u> (person)
- Expensive restaurants intimidate me.
  What? Expensive <u>restaurants</u> (place)

Find the Verb

When writing about someone or something (the subject), to complete the thought you must write what the subject does: the **verb**. A verb often conveys the idea of action. To find the verb, ask what the subject does.

Action Verbs

Verbs often show action. These verbs are called **action verbs**. They tell us what somebody or something *does*.

- Maddie eats cheese every day.
  What does Maddie do? She <u>eats</u>.
- Expensive restaurants intimidate me.
  What do expensive restaurants do? They <u>intimidate</u>.

Language Arts Review ■ Appendix A

Linking Verbs

Some verbs, called **linking verbs**, do not show action. Instead, a linking verb connects a noun in front of the verb with a word or group of words that comes after it. In doing so, the linking verb tells something about the subject: what the subject is or was.

Some Common Linking Verbs

am	be	feel
is	become	seem
are		look
was		appear
were		

> **NOTE**
> Linking verbs sometimes relate to the five senses: sight, sound, taste, touch, and smell.

- Pizza <u>is</u> my favorite food. (connects *pizza* with *favorite food*)
- Your outfit <u>looks</u> good. (connects *outfit* with *good*)

Prepositional Phrases

A **prepositional phrase** is a group of words beginning with a part of speech called a preposition and ending with a noun. Writers often use prepositional phrases to show time or location, as in before the game, during the party, below the table, or inside the box.

Common Prepositions

about	behind	despite	in	onto	until
above	before	down	inside	over	up
across	beneath	during	into	through	upon
after	beside	except	of	to	with
around	between	for	off	toward	within
at	by	from	on	under	without

> **NOTE**
> Prepositional phrases often answer the question *when* or *where*.
>
> For example:
> When?
>   *before breakfast*
> Where?
>   *over the fence*

The subject never appears within a prepositional phrase, so you should ignore prepositional phrases when looking for the subject of a sentence. In the examples below, the prepositional phrases are crossed out.

- ~~Through the night~~, we heard a strange tapping sound.
  Subject: Who? <u>We</u>
  Verb: What did we do? We <u>heard</u>.

- The music ~~at the party~~ was boring.
  Subject: What? The <u>music</u>
  Verb: What about the music? It <u>was</u>.

Helping Verbs and Verb Phrases

Both action and linking verbs often are accompanied by other special verbs called **helping verbs**. Helping verbs frequently show time. Listed below are some frequently used helping verbs.

Helping Verbs

can	may	shall	will
could	might	should	
have	must	used to	
is	need	was	

Main verbs accompanied by one or more helping verbs are called **verb phrases**. For example, following are some verb phrases formed by adding helping verbs to the main verb learn:

Helping Verbs and Verb Phrases

is learning	has learned	should have learned
was learning	will learn	should have been learned
	had been learning	had learned
	has been learned	should have been learning

Below are sentences that contain verb phrases:

- Eliza will be moving to Washington next week.
  Subject: Who? <u>Eliza</u>
  Verb phrase: What about her? She <u>will be moving</u>.
- We should have left hours ago.
  Subject: Who? <u>We</u>
  Verb phrase: What about us? We <u>should have left</u>.

NOTE
Helping verbs must always accompany (help) another verb.

NOTE
Words like *not*, *never*, *always*, and *just* are not considered part of the verb even though they may be in the middle of the verb.
  I *will* not *be going with you today*.
  Kate *had* never *flown on an airplane before*

Fragments

A word group that lacks a subject or a verb and that does not show a complete thought is called a **fragment**. Because fragments are incomplete thoughts punctuated as complete ones, they can confuse readers and must be avoided. One key to eliminating fragments from your writing is knowing the difference between two types of word-groups: **phrases** and **clauses**.

> **NOTE**
> **Fragment** literally means "a part broken off."

Phrases and Clauses

A group of words without a subject/verb unit is called a **phrase**. A group of words with a subject/verb unit is called a **clause**.

- **Phrase (Fragment):**
  My relatives in Chicago. (no verb)

  Clause:
  My relatives live in Chicago. (contains a subject, *relatives*, and a verb, *live*)

- **Phrase (Fragment):**
  Rounding the corner. (no subject)

  Clause:
  A red convertible was rounding the corner.
  (contains a subject, *convertible*, and a complete verb phrase, *was rounding*)

Independent and Dependent Clauses

Though clauses contain both a subject and a verb, that does not mean that all clauses are complete sentences. There are two types of clauses: **independent clauses** and **dependent clauses**.

A clause that can stand alone as a complete sentence is called an **independent clause**.

A **dependent clause** cannot stand alone. Dependent clauses always start with a word called a **subordinator**. Words like *although* and *since* are subordinators. Because they are incomplete thoughts, dependent clauses must be attached to independent clauses.

> **NOTE**
> Use a comma after a dependent clause if it comes at the beginning of a sentence:
> *After I started exercising, I had a lot more energy.*
> Do not use a comma if the dependent clause comes at the end of a sentence:
> *I had a lot more energy after I started exercising.*

Common Subordinators

after	if	until	wherever
although	in order that	what	whether
as	since	whatever	which
because	that	when	while
before	though	whenever	who
even though	unless	where	whose

For example:

- **Independent Clause:**
  Dwane likes professional basketball.

  Dependent Clause (Fragment):
  Although Dwane likes professional basketball.

  Correction:
  Although Dwane likes professional basketball, he enjoys watching football even more. (attached dependent clause)

- **Independent Clause:**
  Richard came home for summer vacation.

  Dependent Clause (Fragment):
  Since Richard came home for summer vacation.

  Correction:
  Since Richard came home for summer vacation, he has not done a single thing. (attached dependent clause)

Correcting Fragments

Sentence fragments are phrases or dependent clauses punctuated as if they were complete sentences. Fragment literally means a "part broken off." In keeping with this definition, a sentence fragment can usually be fixed by attaching the fragment to a sentence. Thus, the "broken part" is glued back to its original position.

For example:

- **Phrase (Fragment):**
  Stanley has no patience for people. Especially his sister-in-law Blanche.

  Correction:
  Stanley has no patience for people, especially his sister-in-law Blanche.

- **Dependent Clause (Fragment):**
  Because Donna took so long to get ready. We all missed the first act of the play.

  Correction:
  Because Donna took so long to get ready, we all missed the first act of the play.

Hints on Proofreading for Fragments

Focus on each sentence separately, reading slowly. Do not be tempted to skim over your work.

Try reading out loud when you are trying to decide whether a group of words is complete or incomplete. You can often "hear it" when something is incomplete.

Learn to identify phrases and dependent clauses in writing. If you see a word group that is unattached, you know it must be a fragment.

Language Arts Review ■ Appendix A

What Are Run-on Sentences?

While some writers make the mistake of not putting enough information in a sentence, resulting in sentence fragments, others try to cram too much into their sentences. **Run-on sentences** result when two complete sentences (independent clauses) are joined with either no punctuation or only a comma. This construction makes it unclear where one thought ends and the next one begins.

Like fragments, run-ons cause readers to become confused. People often write run-ons when they sense that two thoughts belong together *logically* but do not realize that the two thoughts are separate sentences *grammatically*, as in the following examples.

- **Run-on:**
  Dave and Rhonda are crazy about figure skating they watch it on television constantly.

- **Run-on:**
  Dave and Rhonda are crazy about figure skating, they watch it on television constantly.

- **Correction:**
  Dave and Rhonda are crazy about figure skating. They watch it on television constantly.

Generally, good writers will often join sentences which are logically related to make a more complex point. However, complete sentences *cannot* be joined with just a comma or no punctuation at all.

Correcting Run-on Sentences

Following are three useful ways to correct run-ons:

- Make two separate sentences of the run-on thoughts by inserting a **period and a capital letter**.
  Dave and Rhonda are crazy about figure skating. They watch it on television constantly.

- Use a **comma plus a coordinating conjunction** (*and, but, for, or, nor, so, yet*) to connect the two complete thoughts. Coordinating conjunctions are joining words that, when used with a comma, show the logical connection between two closely related thoughts.
  Dave and Rhonda are crazy about figure skating, *so* they watch it on television constantly.

- Use a **subordinator**. You can also show the relationship between two sentences by using a subordinator (words like *after, because,* and *although*) to change one of them into a dependent clause.
  Because Dave and Rhonda are crazy about figure skating, they watch it on television constantly.

> **NOTE**
> Technically, run-ons that have no punctuation to mark the break between complete thoughts are called *fused sentences*.
>
> Run-ons that use only a comma to attempt to join complete thoughts are called *comma splices*.

> **NOTE**
> If you have trouble recognizing the seven coordinating conjunctions, remember them by the handy acronym "FANBOYS": For, And, Nor, But, Or, Yet, So.

Common Subordinators

after	even though	unless	whenever
although	if	until	where
as	in order that	what	wherever
because	since	whatever	whether
before	though	when	while

Words That Often Lead to Run-on Sentences

Frequently the second sentence in a run-on begins with one of the words in the following list. These words often refer to something in the first sentence or seem like joining words. Beware of run-ons whenever you use one of these words in your writing.

Words That Often Lead to Run-ons

I	we	there	now
you	they	this	then
he		that	next
she			
it			

Making Subjects and Verbs Agree

Being able to identify subjects and verbs is important. But you must also make sure that the subjects and verbs agree in number. This grammatical rule is called **subject-verb agreement**.

- A singular subject (one person or one thing) is used with a singular verb.
  For example:
  Her habit annoys me. (singular)
  The plane was late. (singular)

- A plural subject (more than one person or thing) is used with a plural verb.
  For example:
  Her habits annoy me. (plural)
  The planes were late. (plural)

Writers sometimes make mistakes in subject-verb agreement in sentences with more than one subject—a compound subject—or with verbs separated from subjects. It's also common for writers to confuse subject-verb agreement when using pronouns (I, you, he, she, it, we, you, and then), either/or, neither/nor connectors, or "there" sentences. Examples of these follow.

NOTE
The rules of subject-verb agreement apply mostly to *present tense* verbs since the form of a verb changes in the present tense singular.

Present	Past
I write	I wrote
you write	you wrote
he/she *writes*	he/she wrote
we write	we wrote
they write	they wrote

NOTE
The verb *to be* causes agreement problems in the past tense as well as the present because it has many different forms.

Present	Past
I *am*	I *was*
you *are*	you *were*
he/she *is*	he/she *was*
we *are*	we *were*
they *are*	they *were*

You must learn these forms to avoid any agreement problems in your writing.

Language Arts Review — Appendix A

Compound Subjects

Subjects joined by **and** are typically paired with a plural verb. These are called **compound subjects.** The only exception to this rule would be subjects considered singular because they are taken as a single unit, such as *Rock 'n' Roll*.

- For example:
    John and Tina are very close. (plural)
    Corned beef and cabbage is my favorite meal. (singular)
    Hot cocoa and a good book make Sandra happy. (plural)
    Hide-and-seek is played by almost all children. (singular)

Verbs Separated from Subjects

When words, such as prepositional phrases, come between the subject and verb, the interrupting words do not change subject-verb agreement. The verb still must agree with the subject of the sentence.

- For example:
    The coins on the table are mine. (plural)
    The price of the dining room chairs is ridiculous. (singular)
    That woman with plaid bell bottoms seems strange. (singular)
    Those shirts, as well as that coat, need a thorough cleaning. (plural)

Punctuation

Commas

Writers use commas to mark slight pauses or breaks in sentences. When used properly, commas clarify meaning in a sentence. When overused, however, commas can interrupt the smooth flow of sentences and cause confusion. Whenever you add a comma to a sentence, you should be conscious of the specific comma usage rule you are applying. All good writers should know the six primary comma rules covered in this topic.

1. Use a comma after an introductory word or word group that leads into the main sentence.
    - Strolling down the nature trail, Zac saw a brown bear.
    - When you have finished eating your broccoli, you may leave the table.

A dependent clause that comes at the beginning of a sentence always needs to be followed by a comma. The second example above illustrates this concept. However, if the dependent clause comes at the end of the sentence, no comma is necessary:

- You may leave the table when you have finished eating your broccoli.

NOTE
Inexperienced writers tend to overuse commas rather than omit them. If you cannot think of a specific reason to use a comma, *leave it out.*

2. Use commas to enclose a word or word group that interrupts the flow of a sentence.
   - Jane, however, will not be coming tonight.
   - Richard, knowing that it was going to rain, bought a new umbrella.

If you are unsure whether Rule 2 applies to a sentence, try reading the sentence without the interrupting word or words. If the sentence still makes sense without the missing material, set off the interrupting expression with commas. Note how **nonessential information** is set off with commas in the following example:

   - Marty Lasorda, who sat next to me in high school, is now a trader on Wall Street.

The words *who sat next to me in high school* are added information and not needed to identify the subject of the sentence, *Marty Lasorda*. However, in the next sentence the added information is necessary:

   - The guy who sat next to me in high school is now a trader on Wall Street.

The words *who sat next to me in high school* are **essential** to the sentence. Without them, we would have no idea to which *guy* the writer is referring.

3. Use commas to separate items in a series.
   - Steve ordered a large coke, large fries, and a double cheeseburger.
   - Tanya did her laundry, cleaned the bedroom, washed the dishes, and painted the kitchen on Sunday.

Use a comma between descriptive words in a series if *and* sounds natural between them, as in the following:

   - We immediately left the crowded, noisy restaurant.
     (We immediately left the crowded *and* noisy restaurant.)
   - Pablo wore an expensive, well-tailored suit to the party.
     (Pablo wore an expensive *and* well-tailored suit to the party.)

Notice, however, how commas are not necessary in the following sentences:

   - Brenda bit into a juicy red apple.

     Awkward:

     Brenda bit into a juicy *and* red apple.

In the above example *and* does not sound natural between descriptive words, so no comma is used.

4. Use a comma before the conjunctions and, but, for, or, nor, yet, or so when they connect two independent clauses.
   - Dwane thought he had enough money for the movie, but he was fifty cents short.
   - The running back broke through the line for a thirty yard gain, and the home crowd began cheering wildly.

> **NOTE**
> Nonessential information: information that can be removed from a sentence without changing its meaning.

Language Arts Review ■ **Appendix A**

5. Use commas around direct addresses.

When addressing a person, set off the person's name or title with commas. If the direct address comes at the beginning or end of a sentence, only one comma is necessary.

- Ernest, your pants are on backwards.
- Ladies and gentlemen, you are cordially invited to a reception after the show.

6. Use a comma to set off a direct quotation.

A comma separates what is said from who said it.

- "Never tell me the odds," said Han Solo.
- "Seeing the movie version," continued Samantha, "is never as good as reading the book."

NOTE
Commas and end punctuation marks go on the *inside* of quotation marks.

Possessive Apostrophes: Singular Nouns

The possessive form of a noun shows ownership—or possession. There are several ways to show ownership without changing the noun itself, such as:

- the sweater belonging to the *girl*
  OR
- the sweater of the *girl*

However, a simpler, more efficient way to show ownership is to change the possessive noun using a punctuation mark called an **apostrophe** ('):

- the *girl's* sweater

Rule 1: To make a singular noun possessive, add an apostrophe and an s ('s).

- the test of the *student* = the *student's* test
- the ending of the *movie* = the *movie's* ending

Be careful: Do not use 's when you are simply forming a plural.

- Incorrect:
  Barbecue short rib's are the specialty here.
- Correct:
  Barbecue short ribs are the specialty here.

NOTE
When using an apostrophe to show possession, the owner is always followed by the thing possessed. To determine the owner simply ask, *"To whom does it belong?"* The answer to this question takes the 's or the '.

Possessive Apostrophes: Plural Nouns

A plural noun names two or more persons, places, things, or ideas. Most commonly a noun is made plural by adding an *s*: one *girl* becomes several *girls*; one *book* becomes several *books*. Making a plural noun ending in *s* possessive is simple:

Rule 2: To make a plural noun ending in s possessive, place an apostrophe after the s (s').

- the tests of the *students* = the *students'* tests
- the endings of the *movies* = the *movies'* endings

Some nouns change their spellings to form the plural: *child* becomes *children*; *woman* becomes *women*, for example. To make this kind of plural noun possessive simply add an apostrophe and an *s* ('s).

- the children's toys
- the women's self defense class

Contractions

Sometimes writers combine two words to form a single shorter word. Such a construction is called a **contraction**. An apostrophe is added to show where letters have been omitted. For example:

- I + am = I'm (the apostrophe replaces the missing *a*)
- you + will = you'll (the apostrophe replaces the missing *w* and *i*)

Here are some other common contractions:

cannot = can't	is not = isn't	they have = they've
did not = didn't	it is = it's	was not = wasn't
do not = don't	let us = let's	we are = we're
he is = he's	she is = she's	we have = we've
I will = I'll	there is = there's	will not = won't

Be careful: The possessive form of the word *it is its*. Do **not** add an apostrophe to show possession in this case; *it's* always means *it is* or *it has*.

- Vlada's car blew out *its* right front tire. (possessive)
- The plant outgrew *its* pot. (possessive)
- *It's* been a pleasure to meet you. (contraction: *it has*)
- I think *it's* time to go home now. (contraction: *it is*)

NOTE
When a proper noun (a name) ends in s, you may choose to add either 's or ', depending on your preference in pronunciation.

For example:
  Charles's room
OR
  Charles' room

NOTE
Although contractions are very common in everyday speech, try to avoid them in formal writing.

NOTE
To test if "its" is correct in a sentence, substitute the word "its" with "his."

For example:
  Give the dog <u>its</u> bone.
  Give the dog <u>his</u> bone.

Other Punctuation Marks

Punctuation is necessary to help make sentence meanings clearer. Commas and apostrophes are the most commonly misused punctuation marks. However, they are not the only marks that give writers trouble. Listed below are the rules for other punctuation marks that are used in writing.

Period

Use a **period** (.) at the end of all sentences except for direct questions and exclamations.

Use a period at the end of any indirect question.

- **Example:**
  John asked Beth why there were no cookies left.

Use a period after most abbreviations.

- **Example:**
  Dr.
  Ms.
  Jr.

NOTE
Before the advent of computers, two spaces were always inserted after a period. However, the current trend is to use only one space after a period.

NOTE
Do not use periods in acronyms (abbreviations made up of the first letter from a series of words).

Example:
NATO

Question Mark

Use a **question mark** (?) at the end of a direct question. But, as illustrated above, do not use a question mark to end an indirect question.

- **Examples:**
  How cold is it outside?
  When was the Civil War fought?
  John asked Beth, "Why are there no cookies left?"
  "Why are there no cookies left?" asked John.

Exclamation Point

An **exclamation point** (!) is used to at the end of a statement of strong feeling or after an interjection.

- **Example:**
  Look out for that truck!
  Hey! Somebody stole my wallet!

NOTE
Never use exclamation points in formal writing. Save them for casual e-mail correspondence and written dialog.

Colon

Use a **colon** (:) to introduce a list. The words that come before the colon must be a complete sentence.

- **Incorrect:**

    Two things that I hate are: rainy days and Sundays.

- **Correct:**

    There are two things that I hate: rainy days and Sundays.

A colon is used to help explain the statement that precedes the colon. It is also used to set off an explanation or final word.

- **Examples:**

    There are only two things I like to do on Sundays: go to the movies and have pizza for dinner.

    We all had the same goal: success.

Use a colon after salutations in business correspondence, even if you address the person by their first name.

- **Examples:**

    Dear Ms. Smith:

    Dear Verna:

Semicolon

NOTE
Do not use semicolons to set off a list; use a colon.

A **semicolon** (;) is used to separate closely related independent clauses. Often the semicolon is used in place of the word *because*.

- **Example:**

    Sarah was excited about the party; she knew that Greg was going to be there.

Use a semicolon to separate items in a series when the items themselves contain commas.

- **Example:**

    There are four pizza toppings that I enjoy: pepperoni, sausage, and mushrooms; green peppers, onions, and olives; eggplant, garlic, and anchovies; and spinach, goat cheese, and sun-dried tomatoes.

Hyphens

NOTE
If you are not sure if you should hyphenate a compound word, try looking the word up in the dictionary. As a rule of thumb, words ending in *ly* are not hyphenated.

Incorrect: *freshly-cut flowers*
Correct: *fresh-cut flowers*

Hyphenating documents has become much easier on the computer since you can automatically hyphenate the document, move the word to the next line, or compress the word to keep it on one line. However, hyphens have other functions, as discussed below.

Use a **hyphen** (-) to combine two nouns when they are acting as a singular, descriptive word. To see if the two words should be hyphenated into one descriptive word, ask yourself "what kind" of noun is being described.

- **Examples:**
  a three-legged dog (What kind of dog? *three-legged*)
  a four-day convention (What kind of convention? *four-day*)
  Incorrect:
  I went to a convention that lasted four-days.
  (Four days does not answer the question "what kind.")

In writing, hyphenate the numbers twenty-one through ninety-nine and all fractions:

- **Examples:**
  thirty-three
  one-half

Do not hyphenate three-word numbers.

- **Example:**
  four hundred five

Use a hyphen with the prefix *mid* when referring to time.

- **Example:**
  the mid-sixties

Do not hyphenate the prefix *mid* when referring to other things:

- **Example:**
  midlife crisis

Hyphenate the prefix *re* only for ease of reading.

- **Examples:**
  re-edited
  re-evaluated
  restated (no hyphen needed)

Dashes

Use a **dash** (—) to show a sudden break in thought or to set off parenthetical information. The dash is also called an **em dash** because it is about the width of a capital M.

Use an **em dash** (—) to interrupt a sentence or to add additional information. Em dashes can be used in place of commas to add additional drama to the sentence. Em dashes can also be used to set off contrasting remarks.

- **Examples:**
  The bank robbers—with guns in hand—fired the first shot.
  We plan to revise the book in two—not three—months.

An **en dash** (–) is longer than a hyphen and shorter than an em dash. Its length is about the width of a capital N. Use an en dash to show continuation in time, dates, or other numbers. Think of using an en dash instead of the words *to* or *through*.

> **NOTE**
> In Microsoft Word you can create an em dash (—) by typing two hyphens (--) with no spaces between them or before or after them. Word will automatically create an em dash for you.

- **Examples:**
  9:00 AM–5:00 PM   1990–94
  March–May   pages 220–284

Parentheses

Like dashes or commas, **parentheses ( )** are used to set off information that is extra or inessential to the meaning of the sentence.

- **Example:**
  The chapter on medieval art (pages 172–184) is very interesting.

> **NOTE**
> Do not overuse parentheses in your writing.

Quotation Marks

Quotation Marks ("") are used to set off someone's exact words. A comma always separates what is said from who said it. Periods and commas go inside of quotation marks.

- **Examples:**
  "There are too many rules to punctuation," he stated.
  The clerk told me, "There are no more bananas today."

Use **single quotes** (' ') to enclose titles of poems, stories, movie titles, or other quoted material within quoted material.

- **Examples:**
  "'Survivor' is my favorite TV show," the teenage girl proclaimed.
  "If you call Johnny 'stinky toes' one more time," Mom told Suzy, "you're going to your room."

> **NOTE**
> Quotation marks always come in pairs.

B Math Review

Knowing basic math concepts and knowing when to apply them are essential skills. You should know how to add, subtract, multiply, divide, calculate percentages, and manipulate fractions. This section will help you accurately apply basic math concepts and skills necessary to your success at school, work, and to manage your personal finances at home.

Numbers

Numbers are expressed in different forms:

- Whole numbers, which are the counting numbers and zero. Whole numbers do not contain decimals or fractions. *Examples:* 1, 2, 3, 10, 15, 18, 0. Ignore zeros before whole numbers. For example, 025 is the same as 25.
- Nonwhole numbers, which are numbers that have decimals, such as 6.25 or 9.85.
- Mixed numbers, which are numbers that combine whole numbers and a fraction, such as $6^{1}/_{4}$ or $7^{2}/_{3}$.
- Percentages, which are portions in relation to a whole, such as 65% or 22%.

Place Value

Numbers that have more than one digit are defined by their place value. Place value is the value of a digit based on where it is in a number. For example, the number 7777 is given the following values:

This number is described by saying "seven thousand, seven hundred seventy-seven." Numbers are written with a comma placed to the left of every third digit. The number 7777 is properly written 7,777.

Rounding Whole Numbers

1. Find the place to be rounded.
2. If the digit to the right is 5 or more, add 1 to the place to be rounded. If the digit to the right is 4 or less, leave the place to be rounded unchanged.
3. Change all digits to the right of the rounded place to zeroes.

For example, to round 687 to the nearest ten:

1. The place to be rounded is the number in the tens column.
2. The digit to the right of the tens column is greater than 5, so you add 1 to the digit in the place to be rounded, making it 9.
3. Change the 7 to a zero. The result is 690.

Math Review ■ Appendix B

Working with Decimals and Percentages

Recall that a percentage is a portion of a whole. 100 percent = the whole or all of something. The symbol for percent is %. In the figure to the right, four turtles = 100% or all of the turtles. One turtle is ¼ of the total, or 25%. Two turtles = ½ of the total, or 50%. Three turtles = ¾ or 75%. Numbers to the left of a decimal are whole numbers. Numbers to the right of a decimal are less than one.

Four turtles = 100%.

To divide decimals by whole numbers:

$24.5 \div 4 = 4\overline{)24.5}$ Place the decimal point in the answer directly above the decimal point in the dividend.

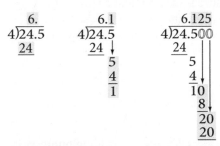

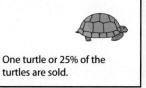
One turtle or 25% of the turtles are sold.

Three turtles or 75% of the turtles are left.

Numbers to the right of a decimal point indicate less than one whole. For example, 1.5 is the same as 1½.

If there were five turtles in the figure, what percentage would one turtle represent?

Examples:

Decimal		Percentage		Fraction(s)
0.10	=	10%	=	$^{10}/_{100}$ or $^{1}/_{10}$
0.25	=	25%	=	$^{25}/_{100}$ or ¼
6.25	=	625%	=	$6\,^{25}/_{100}$ or 6¼

To change a decimal number into a percentage: Move the decimal two places to the right.

Examples:

0.5 = 0.5 0. = 50%

30.0 = 30.0 0. = 3000%

0.04 = .0 4. = 4%

To change a percentage into a decimal number: Replace the percent sign with a decimal and move the decimal two spaces to the left.

To find the percentage of a number:

1. Change the percentage to a decimal.
2. Multiply that decimal by the number.

Examples:

15% of 63   change 15% to 0.15 then multiply 0.15 × 63 = 9.45

9.45 is 15% of 63

20% of 100   change 20% to 0.20 then multiply 0.20 × 100 = 20

20 is 20% of 100

Addition

Addition is the totaling of two or more numbers. Each number you are adding is called an *addend*. The result is called the *sum* or *total*. For example, 2 computers in the reception area + 3 computers in accounting = a total of 5 computers in the office.

To add:

1. Write the addends in columns, making sure each digit lines up correctly according to its place value. That means putting ones in the ones column, tens in the tens column, hundreds in the hundreds column, and so on.
2. Draw a line under the last addend in the column.
3. Add all the addends above the line and write the total or sum below the line.

Examples:

```
   2      34     <— addends
 + 3    + 4     <— addends
   5      38     <— totals
```

When adding more than one column of numbers, always start adding numbers in the right column—place value ones—first.

Adding by Carrying Numbers

When the numbers in a column add up to more than 10, you must *carry* all digits to the left of the place value ones column. To carry means to move the digits to the top of the next place value column. You then add the carried digits with all the numbers in that column.

For example, to add 81 + 384 + 10 + 9:

1. Write the numbers in columns, lining them up according to place value.
2. Add the numbers in the ones column. The total is 14.
3. Write the 4 in the ones column below the total line, and carry the 1 to the top of the tens column.
4. Add the numbers in the tens column, including the 1 you carried. The total is 18.
5. Write the 8 in the tens column below the total line, and carry the 1 to the top of the hundreds column.
6. Add the numbers in the hundreds column, including the 1 you carried. The total is 4.
7. Write the 4 in the hundreds column below the total line. The sum of the numbers is 484.

Example 1:

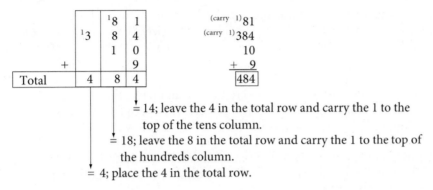

= 14; leave the 4 in the total row and carry the 1 to the top of the tens column.
= 18; leave the 8 in the total row and carry the 1 to the top of the hundreds column.
= 4; place the 4 in the total row.

Example 2:

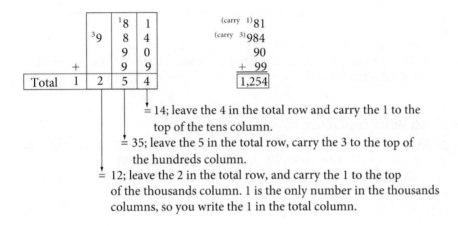

= 14; leave the 4 in the total row and carry the 1 to the top of the tens column.
= 35; leave the 5 in the total row, carry the 3 to the top of the hundreds column.
= 12; leave the 2 in the total row, and carry the 1 to the top of the thousands column. 1 is the only number in the thousands columns, so you write the 1 in the total column.

Subtraction

Subtraction is the opposite of addition. Subtracting numbers means taking one number—called the *subtrahend*—away from another number—called the *minuend*. The result is called the *difference*. Write a subtraction problem in columns, just like you write an addition problem. Put the minuend on top, and the subtrahend below it. Make sure you line up the digits in the proper place value column.

Subtract the number in the ones column first, and write the difference below the line. Then, subtract the number in the tens column, then the hundreds column, and so on.

```
    84        <— minuend —>        136
  - 23        <— subtrahend —>    - 12
   [61]       <— difference —>    [124]
```

Check your answer by adding the difference to the subtrahend. If your answer is correct, your total will equal the minuend

For example, you can check your subtraction for the first equation above by adding 61 (the answer) to 23 (the number you subtracted). 61 + 23 = 84, which is the number you started with at the top of the equation. The answer is correct.

Subtracting by Borrowing Numbers

When the number you are subtracting is greater than the number above it in the column, you can borrow a one from the column to the left. Borrowing makes the number at the top of the equation greater, so you can complete the subtraction.

When you borrow the one, you must do two things:

1. Cross out the number in the column to the left, and replace it with a number that is one less than the number that was there. For example, if there is a 3 in the column to the left, you cross out the 3 and replace it with a 2. If there is a 7, you cross out the 7 and replace it with a 6.
2. Write the borrowed 1 to the left of the number that is currently in the column you are subtracting. So, if there is a 5 in the column you are subtracting, you write the borrowed 1 to the left of the 5; the 5 becomes 15.

Examples:

$$\begin{array}{r} {}^2\cancel{3}\,{}^15 \\ -\ 1\ 8 \\ \hline \boxed{1\ 7} \end{array}$$

8 is greater than 5, so you can borrow a one from the tens column.
Cross out the 3 in the tens column and replace it with a 2.
Write the borrowed 1 to the left of the 5, making it 15.
Subtract 8 from 15 and write the result (7) below the line in the ones column.
Subtract 1 from 2 and write the result (1) below the line in the tens column.

Adding and Subtracting Decimals

For decimals, the process of adding and subtracting numbers is similar to the process for adding and subtracting whole numbers. The only difference is how the numbers are aligned in the column. Numbers should be aligned by the decimal point as in the following examples:

$$\begin{array}{r} 12.136 \\ +\ 10.246 \\ \hline 22.382 \end{array}$$

$$\begin{array}{r} 23.453 \\ -\ 10.37 \\ \hline 13.083 \end{array}$$

Adding and Subtracting Fractions

A fraction is made up of two parts: a *numerator*, which is the number on top, and a *denominator*, or the number on the bottom. To add or subtract a fraction, the denominator must be the same. Consider the following examples:

Example 1:

$$\frac{1}{4} + \frac{3}{4} = \frac{1+3}{4} = \frac{4}{4} = \frac{1}{1} = 1$$

Example 2:

$$\frac{1}{5} + \frac{2}{5} + 12\frac{4}{5} = \frac{1}{5} + \frac{2}{5} + \frac{64}{5} = \frac{1+2+64}{5} = \frac{67}{5} = 13\frac{2}{5}$$

Math Review ■ Appendix B

Example 3:

$$\frac{2}{3} - \frac{1}{3} = \frac{2-1}{3} = \frac{1}{3}$$

Note that Example 1 shows a simplified fraction, or a fraction that does not have any common factors (other than 1) for the numerator and denominator. Example 2 shows a mixed number, which is converted to a fraction to be added or subtracted.

In some cases, fractions will need to be converted so each fraction being added has a common denominator. Consider the following example:

$$\frac{1}{3} + \frac{1}{2} = ?$$

Step 1: Convert each fraction so they have a common denominator.

The denominators, 2 and 3, are factors of 6, so multiply both the nominator and the denominator of each fraction by the number that makes the denominator equal to 6.

$$\frac{1}{3} \times \frac{2}{2} = \frac{2}{6}$$

$$\frac{1}{2} \times \frac{3}{3} = \frac{3}{6}$$

Step 2: Add fractions.

$$\frac{2}{6} + \frac{3}{6} = \frac{5}{6}$$

For subtraction, follow the same process to make fractions that have the same denominator. Simplify fractions after adding and subtracting as needed.

$$\frac{3}{4} - \frac{1}{3} = ?$$

Step 1: Convert each fraction so they have a common denominator.

$$\frac{3}{4} \times \frac{3}{3} = \frac{9}{12}$$

$$\frac{1}{3} \times \frac{4}{4} = \frac{4}{12}$$

Step 2: Subtract fractions.

$$\frac{9}{12} - \frac{4}{12} = \frac{5}{12}$$

Add or subtract the following:

3.45 + 2.34 + 5.6

$3\frac{3}{4} + \frac{2}{3} + \frac{4}{5}$

3.4 − 1.28

$2\frac{3}{4} - 1\frac{1}{3}$

What will happen if you don't convert fractions to a common denominator before adding or subtracting them?

Multiplication

Multiplication, the process of finding the product of two factors, is a quick, easy way to add. For example, 9 + 9 + 9 = 27, but an easier process is 3 × 9 = 27.

To multiply numbers easily, memorize the multiplication table.

Multiplication Table

	1	2	3	4	5	6	7	8	9	10
1	1	2	3	4	5	6	7	8	9	10
2	2	4	6	8	10	12	14	16	18	20
3	3	6	9	12	15	18	21	24	27	30
4	4	8	12	16	20	24	28	32	36	40
5	5	10	15	20	25	30	35	40	45	50
6	6	12	18	24	30	36	42	48	54	60
7	7	14	21	28	35	42	49	56	63	70
8	8	16	24	32	40	48	56	64	72	80
9	9	18	27	36	45	54	63	72	81	90
10	10	20	30	40	50	60	70	80	90	100

Math Review ■ **Appendix B**

Multiplying Fractions

To multiply fractions:

1. Multiply the first **numerator**—the number above the line in the fraction—by the second numerator. Write the answer as the numerator in the result.
2. Multiply the first **denominator**—the number below the line in the fraction—by the second denominator. Write the answer as the denominator in the result.

The numerators, or top numbers of the fractions, are multiplied; and the denominators, or bottom numbers of the fractions, are multiplied. For example:

$$\frac{3}{4} \times \frac{12}{15} = \frac{36}{60}$$

If possible, reduce the fraction, which means dividing both the numerator and the denominator by the largest number than goes into both. In the following example, 12 is the largest number that goes into both. 36 divided by 12 equals 3. 60 divided by 12 equals 5. You can reduce the fraction to $3/5$.

$$\frac{36}{60} \text{ to } \frac{3}{5}$$

Glossary

401(k) plan	a savings plan offered by an employer to an employee
403(b) plan	a savings plan offered by an employer to an employee
abstinence	avoiding or refraining from certain behavior
accommodation	a change that makes it possible for a disabled employee to perform his or her job responsibilities in a safe and accessible work environment
accomplishments	things you have achieved at home, in school, or in the community
account register	a book the bank provides to customers to keep track of their transactions
addend	the numbers totaled in an addition equation
addiction	a need to continue using a substance even though it has negative consequences
addition	the totaling of two or more numbers
aggressive	forcing your opinions on others
Americans with Disabilities Act (ADA)	a law that makes it illegal to discriminate based on age or physical disability
apprentice	someone who works with a professional to learn a skill or trade
apprenticeship	a paid training program in a given trade
assertive	stand up for yourself in a strong and positive way
associate's degree	a degree that usually requires two years of study after high school to earn
attitude	the way you think, feel, or behave, particularly when you are with other people
auditory learners	people who take in information best when they hear it
authorizing	giving permission or allowing
automation software	programs that enable computers to perform repetitive, time-consuming tasks
average	the sum of two or more quantities divided by the number of quantities
bachelor's degree	a degree that usually requires four years of study after high school to earn
balance	the amount of money in an account
balanced	in banking, keeping track of money coming in and going out of an account
bank account	a record of the transactions between a bank and its client
bank charter	an agreement that controls how a bank operates
bank	a business that stores and manages money for individuals and other businesses
bankruptcy	a legal process in which you declare yourself legally unable to pay your outstanding debts
bar graph	a graph that displays comparisons among individual items

barter	negotiate the value of traded goods
base pay	hourly wage or annual salary that you earn
biometric time clock	a time-clock that uses a fingerprint or handprint to record when an employee comes and goes
bond	a debt security issued by corporations, governments, or their agencies, in return for cash from lenders and investors
bonus	a lump sum paid in addition to base pay
Boolean operators	words, terms, or symbols that set criteria for an Internet search
branch	local bank office
budget	a plan for spending and saving money
business opportunity	a consumer need or want that can be met by a new business
business	an organization that provides goods or services, usually to earn money
candidate	possible employee
capitol resources	nonhuman resources, such as money or cars
career cluster	a category used to identify occupations in a similar area or industry
career fair	an event organized so that people searching for jobs may meet and collect information from a large number of employers in one place; also called a job fair
career plan	a map that shows you the way to achieve your career goals
career requirements	the responsibilities that you must perform in order to succeed in the career
career self-assessment	a process that helps you identify career-related interests, values, and abilities
career	a chosen field of work in which you try to advance over time by gaining responsibility and earning more money; another word for career is occupation
central bank	an organization responsible for managing banking activity
certificates of deposit (CDs)	time account with minimum balance and set term
character	the personal qualities or traits that make you unique
chart	a picture that displays data; also called a graph
check	written order to a bank to transfer funds from one account to another
checking account	a bank account that allows you to withdraw funds using checks and/or a debit card
civic responsibility	service that you owe to a community of which you are a part
Civil Rights Act of 1964	a law which states that employers may not use race, skin color, religion, sex, or national origin as a reason to promote, not promote, hire, or fire an employee
cloud computing	a type of computing in which all files and programs are stored online and accessed through the Internet
co-curricular	activities such as athletic teams, clubs, and career student organizations, offered to students in addition to class subjects

Glossary

collaborative software	programs that allow more than one person to work on a document at a time
colleagues	co-workers
commission	a payment based on a percentage of sales
communication barriers	differences that can be obstacles to effective communication
community resources	services that the government provides, such as public parks, public schools, libraries, and police and fire departments
community	a group of people who have a common goal
compassionate	caring about the suffering of others
compensation	wages and benefits provided in exchange for work
computer application	a software program that enables you to perform specific tasks using a computer
computer program	a software application that enables you to perform specific tasks using a computer
conflict	a disagreement between two or more people who have different ideas
consequences	results that happen in response to a decision or action
constructive criticism	helpful suggestions from others on how to improve
contract	a legal document that defines the conditions of an agreement between two parties; an agreement between competent parties in which each promises to take or avoid a specified action
coping skills	actions and thoughts that help you deal with or overcome problems and difficulties that you might not be able to solve
cost/benefit analysis	the process of adding up all the expected benefits of an opportunity and subtracting all the expected costs
cover letter	a short, one-page summary of who you are, why you are applying for the job, and why you are the best person for it, that you send to a potential employer with your resume
credit card	a card you use in place of cash to buy something and pay for it later
credit limit	the maximum amount you can charge to a credit card or borrow from a bank
credit report	a summary of your credit history
credit union	nonprofit banks that are owned by the customers or members
credit	money that you borrow and promise to pay back
creditor	someone who loans money
critical reading	thinking about what is written and why, while you read
critical thinking	the ability to be honest, rational, and open-minded
criticism	advice about how to make positive changes in your actions or behavior
culture	language, beliefs, attitudes, customs, manners, and habits that define a group of people or community
currency	the unit of money used in a country

data analysis	a type of math used to collect, organize, and analyze data; another term for statistics
data	information
debit	an amount withdrawn or subtracted, particularly money from an account
debit card	card you use in place of cash to make purchases or withdrawals
debt	the money that you borrowed and owe
debtor	someone who owes money
decimal point	the period used to separate whole numbers and numbers less than one in a written decimal
decimal	a fraction written using a decimal point to separate whole numbers and numbers less than one, representing multiples of ten
decision	the process of choosing one option over one or more alternative options
deductible	a set amount an insurance beneficiary must pay toward a claim before the insurance company pays any money
deduction	an amount that your employer withholds from your earnings to pay for things such as taxes or insurance
deference	courteous regard or respect
deficit	when more money is spent than earned
degree	a certificate awarded by an institution such as a college or university when you complete the requirements of an academic or vocational program
demographics	facts about a population
denomination	the face value of money
denominator	the number below the fraction bar in a written fraction
deposit slip	a form you fill out when you deposit money in a bank account
derogatory marks	negative reports such as bankruptcy on a credit report that can take from seven to ten years to clear from your credit history
devil's advocate	a label used to describe taking a position you do not necessarily agree with, but pretend you do for the sake of argument
diet	the food you eat
difference	the measurement between two values; in math, you find the difference by subtracting one value from another
diploma	the document which certifies that you successfully completed your high school study
direct deposit	paycheck that is automatically deposited into an account
discrimination	unfair treatment of a person or group based on age, gender, race, or religion
diversity	an environment that includes variety, such as people of different ages, cultural backgrounds, and race
dividend	the total or whole in a division equation; also, the amount of a corporation's after-tax earnings that it pays to its shareholders

Glossary

division	the process of separating a whole into parts; the opposite of multiplication
divisor	the number by which you divide a dividend in a division equation
docent	guide
dropout	someone who leaves school without receiving a diploma
dual-income family	a family in which both parents have paying jobs
dues	fees paid for membership
dysfunctional	not working correctly
e-commerce (electronic commerce)	buying and selling goods or services over the Internet
economics	the study of the way people produce, distribute, and use goods and services and other resources to achieve their needs and wants
economy	activity related to the production and distribution of goods and services
elapsed time	the amount of time that has passed from one time to another
electives	classes you choose because you are interested in the subject
electronic fund transfer (EFT)	transfer of money from one bank account to another electronically
emotional well-being	well-being that depends on your ability to deal with problems and stress
empathy	being sensitive to the thoughts and feelings of other people
employability	having and using your life skills and abilities to be hired and stay hired
employee handbook	a document that describes company policies and procedures
employee	a person who works in a business owned by someone else
emulate	copy in a respectful manner
encryption	code
endorse	approve or sign
entrepreneur	a person who organizes and runs his or her own business
entrepreneurial thinking	the process of starting a new business, including thinking of, creating, and developing a brand-new product or service
entrepreneurship	the process of starting a new business
environmental well-being	well-being that depends on your comfort and satisfaction with the environment in which you live and work
Equal Employment Opportunity Commission (EEOC)	the federal agency responsible for investigating charges of discrimination against employers
Equal Pay Act of 1963	a law that prohibits employers from paying workers less simply because of their gender
equity investments	stock—or ownership—in a company
estimate	guess, using past experience and knowledge as a guide
ethics	a set of beliefs about what is right and what is wrong

etiquette	a culture's social rules and customs
exchange rate	a calculation that determines how much foreign currency you can buy for one dollar
excise tax	a consumption tax imposed on a specific product or activity
exit interview	a meeting with your supervisor or human resources representative to discuss why you are leaving
expenses	all the ways you use money
face value	the number printed or stamped on money
factors	numbers multiplied in a multiplication equation
fair use	a legal provision that allows for the limited quotation of a copyrighted work without permission from or payment to the copyright holder
Family and Medical Leave Act (FMLA)	a law that provides certain employees with up to 12 weeks of unpaid, job-protected leave per year
FDIC (Federal Deposit Insurance Corporation)	the government agency that provides insurance for bank deposits
Federal Reserve	institution responsible for creating and tracking all of the country's money
FICA	Social Security tax
financial aid	money provided by others to help you pay for education
financial decision	a decision about how to manage your money
financial goals	plans you have for using your money
financial needs	the things you must buy in order to survive
financial wants	the things you need to maintain a certain standard of living
firewalls	software that restricts unauthorized access to a site
fixed expense	known expenses that stay the same every month
fixed income investments	money lent to a business or government agency in exchange for a bond
fixed income	income that does not change month to month
flexible (variable) expense	expenses that change from month to month
flexible (variable) income	income that changes month to month
flexible standards	standards that you can adapt to different situations
formula	rule or method of doing something
fraction bar	the line between a numerator and denominator in a written fraction
fraction	a ratio, or comparison of two whole numbers, or one whole number divided by another whole number
franchisee	someone who buys a franchise
franchisor	a company that sells franchises

Glossary

FTC (Federal Trade Commission)	the federal agency responsible for the fair treatment of consumers
fungible	the ability to trade one item or quality for another
geometry	math that deals with the size, position, and shape of 2- and 3-dimensional objects
give notice	tell your employer you are leaving your job
goal	something you are trying to achieve
governance	making decisions about how government will act
grade point average	the average of all grades that you receive
grant	financial aid you do not have to repay
graph	a picture that displays data; also called a chart
graphical representation	way of conveying information without words
gross income	pay before withholdings
harassment	unwanted, repeated behavior or communication that bothers, annoys, frightens, or stresses another person
help wanted ads	job listings for specific positions; also called classified ads
human resources	resources people provide that things cannot
hydraulic arm	a machine that creates power by pumping water through a confined space
identity theft	when someone steals your personal information and uses it to commit fraud
impulsive	inclined to act without thinking
income tax returns	forms you fill out and send in to the government stating how much income you earned and how much tax you must pay
income	all the money that you earn or receive
indecisive	unable to make a decision
individual retirement account (IRA)	a personal retirement savings plan
influence	something that affects the way you think and act
information technology (IT)	computer and telecommunication systems that transmit and store information
informational interview	the opportunity to sit down with someone who is employed to ask questions and learn about his or her career or industry
infrastructure	the underlying foundation or basic framework of a system or organization
infringement	violating the rights provided by a copyright or patent
instrumental value	value based on an item's usefulness for acquiring something else
insurance	an investment that protects you financially against everyday risks
integral	necessary or essential

intellectual property	artistic and industrial creations of the mind
intellectual well-being	well-being that depends on your ability to think and learn new things
interests	subjects or activities that attract your attention and that you enjoy doing or learning about
Internal Revenue Service (IRS)	the federal government agency responsible for collecting federal taxes
internship	a temporary job, usually for students, that may or may not pay a salary, but provides other useful benefits, such as the opportunity to gain experience
intrapreneurship	acting like an entrepreneur while working as an employee
intrinsic value	value based on an item's own merits
inventory	goods you have available to sell
investment bank	banks that help businesses and other organizations raise money by issuing stocks and bonds
irradiation	a process that exposes fresh food to a source of radiation in order to kill bacteria and insects
job application	a standard form you fill out when you apply for a job
job interview	a meeting between a job seeker and a potential employer—the interviewer
job lead	information that might lead you to find a job
job outlook	a prediction of future opportunities for a particular occupation
job search resources	tools designed to help you find job leads
job security	knowing you will have a job for a long time
job shadowing	to follow someone around at work for a day, or part of a day
job	any activity that you do in exchange for money or other payment
job-share	a position in which two employees share the responsibilities for one job
labor market	any situation in which workers compete for jobs and employers compete for workers
layoff	a job loss caused when a company has no work for certain employees for a period of time
leader	someone who unites people to work toward common goals
learning style	a way of taking in information so that you remember it and can use it in a practical way
least common denominator (LCD)	the smallest number that two different denominators can divide into evenly
leisure	fun activities outside of work
letter of resignation	a brief, formal, and positive letter that states that you are leaving the job, the date your employment will end, and thanks your employer for the opportunity to have worked there
lifestyle factors	things about your job that affect the way you live your life
lifestyle	the way you think and behave every day

Glossary

loan	a transaction in which the lender agrees to give the borrower money and expects to be repaid in full
local time	current time in the time zone where a person is located
long-term goals	something you want to achieve in the more distant future
majority	a greater number or part
manager	someone who makes decisions, solves problems, and uses resources to achieve specific goals
manufacturing	using machines to design and build products
marriage	an institution that grants a couple a unique set of legal, social, and religious rights and responsibilities
mentor	someone knowledgeable and experienced in your field who is willing to teach you, advise you, and help you reach your goals
milestones	a series of short-term goals achieved in the process of working toward a long-term goal
minority	a lesser number or part
monetary compensation	compensation in the form of money
money	anything you exchange for goods or services
moral values	personal values based on what one thinks is good compared to what one thinks is wrong
mortgage	a loan to purchase a home
motivator	something that encourages you to set goals and make decisions
motive	a reason for doing something
multiplication	a quick method of adding numbers by repeating an addition problem in one step
mutual fund	a pool of money collected from multiple investors and then used to buy stocks, bonds, and other securities
natural resources	things that exist in nature and are available for everyone
need	something you can't live without
negative attitude	behavior that shows someone is unhappy, has little confidence in himself or herself or in others, and believes that life is unfair
negotiate	discuss options that will lead to an agreement or compromise that will satisfy everyone
net income	pay after withholdings
networking	sharing information about yourself and your career goals with personal contacts
nonhuman resources	things, such as money and cars
nonprofit organizations	companies that are in business to serve the good of society, not to make a profit
nonrenewable resources	natural resources that are available in limited quantities and may one day be used up
nontraditional occupation	any job that a man or woman does that is usually done by someone of the other gender
numerator	the number above the fraction bar in a written fraction

nutrients	the parts of food that your body requires
nutrition	a science that studies the way the food you eat nourishes your body
objective	fairly, without emotion or prejudice; also, another word for goal
occupation	a job or career
Occupational Outlook Handbook (OOH)	a book published by the U.S. Bureau of Labor Statistics that describes more than 200 occupations, including responsibilities, working conditions, education requirements, salary ranges, and job outlook
Occupational Safety and Health Act	a law requiring all employers to provide a safe and healthful workplace
Occupational Safety and Health Administration (OSHA)	the federal agency responsible for inspecting companies and enforcing safety laws
on-call	available at any time to cope with emergencies
online bank	retail bank that operates only on the Internet
opportunity cost	the value of what you are willing to give up to achieve something else
overdrawn	the condition when there are not enough funds in an account to cover expenses
passive reading	reading without thinking critically about the written words; reading to take in information or be entertained
pathway	a group of jobs within a career cluster
pay stub	a printed record of information about your pay, including how much you have earned and how much your employer has deducted from your pay
paycheck	a written document that tells your employer's bank how much money to give to you
payee	the person a check is written to
peer pressure	when peers such as friends, classmates, or co-workers influence you to do something
peers	people your own age
pension	a regular payment given to a retired person by a former employer
percentage	a part of a whole
performance review	a report that rates how well you do your job
perks	compensation in addition to wages or salary
personal academic plan	a document that you use to set goals for the things you want to accomplish while you are in school; some schools call it a personal education plan, a personal career plan, or even a personal life plan
personal digital assistant	a hand-held computing device usually used for storing and retrieving information, such as calendars and address books
personal identification number (PIN)	identification number used with a debit card
personal information card	an index card—or other piece of paper—on which you write or print the information you might need to fill out a job application accurately

Glossary

personal qualities	the characteristics and traits that make you unique
personal well-being	well-being that depends on how satisfied and confident you are with yourself
phishing	a scam designed to steal your personal and financial information over the Internet
physical well-being	well-being that depends on your health
population	the number of people in a country or area
portfolio	a collection of information and documents; for investors, all the investment and accounts they own
positive attitude	behavior that shows someone is happy, has confidence in himself or herself or in others, and believes that life is good
postsecondary education	school after high school
prejudice	negative opinions that are not based on fact
prestige	positive recognition and admiration
prioritize	organize in order of importance
probability	the chance that something will happen
problem	a difficulty or challenge that you must resolve before you can make progress
process	a series of steps that leads to a conclusion
product	the result of a multiplication equation; also, goods manufactured and sold
productivity	the amount of work an employee accomplishes
professional development	training in your chosen career
professional organization	an association of people who are all employed in the same field or industry
professionalism	the ability to show respect to everyone around you while you perform your responsibilities as best you can
promotion	an advance in your career that includes a new job title and additional responsibilities; also, actions that raise awareness of a product
proportion	two ratios that equal each other
punctual	on time
purchasing power	the number of goods or services that can be purchased with a unit of currency
quotient	the result of a division equation
raise	an increase in pay
ratio	a comparison between two values
reciting	speaking out loud
recreation	fun activities outside of work
reference	someone who knows you well and is willing to speak to employers about your qualifications
relating	interacting with someone else

relationship	a connection or association between two or more people
Renaissance	a historical time period of artistic and intellectual revival in Europe
renewable resources	natural resources that can be recreated in unlimited quantities, such as air and sunlight
requirements	things you need or must have
resent	to blame someone else for your unhappiness
resources	things you can use to get something else
responsibility	something people expect you to do, or something you must accomplish
resume	a document that summarizes you, your skills, and your abilities and provides a snapshot image of your qualifications for potential employers
retail bank	a for-profit financial institution providing services such as checking and savings accounts, credit cards, education loans, car loans, and home loans
retirement	when someone leaves a job or career after many years and does not return to a paid job
return on investment	the amount of money you earn compared to the amount of money you invest
rewards	things you receive in return for something you do
risk	the chance of losing something; also, the measurable likelihood of loss or less-than-expected returns
risky behavior	activity that puts your health and wellness in danger
roadblock	something that gets in the way and interferes with your progress
role	the way a person behaves in a specific situation
room and board	cost of meals and a place to live
rule	a written or unwritten statement about how something is supposed to be done
safe deposit box	a box in a fireproof vault that you can rent from your bank
salary	the money that you receive in exchange for work performed
savings account	interest-bearing bank account
scale	to change the size of something
scholarship	a type of grant, usually based on specific criteria
Securities and Exchange Commission (SEC)	the government agency that regulates the securities industry
severance package	compensation a company pays an employee when the employee is dismissed
short-term goals	goals you can accomplish in the near future—maybe even today
skill	an ability or talent—something you do well
smartphone	a cell phone that has the ability to run computer programs
Social Security	a government program that pays monthly benefits to workers who pay the Social Security tax (FICA)
social well-being	well-being that depends on how you get along with other people

Glossary

stage of life	how old you are
standard of living	a measure of how comfortable you are based on the things you own
standards	guidelines for whether or not something meets expectations
statistics	a type of math used to collect, organize, and analyze data; also called data analysis
stock	a share of ownership in a corporation
strategies	careful plans and methods
strengths	positive qualities and skills
stress	the way your body reacts to a difficult or demanding situation
stressor	a cause of stress
subjective	affected by existing opinions, feelings, and beliefs
subordinates	employees supervised by a manager
subsidy	a cash payment toward a specific work-related need
substance abuse	using substances—such as food, drugs, or alcohol—that are bad for your health and wellness
subtle	unmentioned or implied
subtraction	taking one number away from another number; the opposite of addition
success	setting and achieving a goal
sum	the result of an addition equation
surplus	when more money is earned than spent
SWOT analysis	a business method used to evaluate four areas: Strengths, Weaknesses, Opportunities, and Threats
tactile learners	people who learn best by touching things
tax brackets	levels dictating how much tax is due based on income
tax deductions	expenses that you are allowed to deduct from your income when calculating taxes
tax	money citizens pay the government
taxable income	income minus deductions
tax-deferred savings plans	retirement savings plans with tax benefits
team	a group of two or more people who work together to achieve a common goal
technological resources	resources provided by technology such as computers, automated teller machines (ATMs), or medical equipment
technology infrastructure	the computer systems, networking, and other devices that form the core or backbone of an organization's technological capabilities
technology	the use of resources such as scientific knowledge and methods to make lives easier or better
telecommuter	an employee who works from home

templates	sample documents
term	the set amount of time you have to earn the return on an investment or pay back a loan
terms	the conditions of an agreement, such as a loan, included in the contract
time clock	a machine that automatically records the time an employee arrives and leaves
time sheet	a table used for entering the time in, time out, and total hours spent at work
time zones	geographic regions that use the same standard time
tolerant	willing to consider the opinions of others
total	the result of an addition equation; also, the whole of something
trade barrier	a governmental restriction on international trade
trade union	an organization of people who do similar work, that acts on behalf of members regarding employment issues such as compensation and working conditions
trade-off	a compromise, or giving up one thing in order to get something
transcript	a record of the courses you took in high school and the grades you earned
transferable skills	skills that can be used on almost any job; they are called transferable skills because you can transfer them from one situation or career to another
trend	a general move in a certain direction
tuition	cost of education
unemployment rate	the percentage that measures the number of unemployed people looking for jobs
unique	one of a kind
value	a measure of the quality or importance of something
values	the thoughts, ideas, and actions that are important to you
videoconference	a meeting conducted over distances using video cameras and an Internet connection so that people do not have to travel to the same location
virtues	positive character traits, such as honesty, compassion, and loyalty
visual aids	pictures, graphs, charts, or anything without words that conveys a message or idea
visual learners	people who take in information best when they see it
volunteering	doing work without pay
volunteers	unpaid workers
W-4 (Employee's Withholding Allowance Certificate)	a form employees fill out to provide information the employer needs in order to calculate how much money to withhold from wages to pay taxes
wages	the money that you receive in exchange for work performed
want	something you desire
well-being	the feeling and understanding that things are going right
wellness	an overall feeling of well-being

Glossary

Wi-Fi	a technology that enables connection to the Internet without wires or cords
withdrawal slip	a form you fill out when you withdraw money from a bank account
work ethics	beliefs and behaviors about what is right and wrong in a work environment
work history	a list of jobs you have held from the past through the present, showing your experience as an employee
work week	the number of hours a person works per week
work-related values	values that refer to how you like to work and the results that you produce
work-study program	the opportunity to work while you are in school so you can earn money to help pay your costs

Index

A

Abstinence, 313
Academic planning, 115–138
    career-related opportunities, 132–135
    developing skills in school, 124–125
    high school, 120–121
    learning beyond high school, 126
    learning styles, 118–119
    personal academic plans, 122–124
    postsecondary education, 127–130
    value of school, 116–117
Accommodations, 245
Accomplishments, 10
Accountants, 58, 328
Accounting, 328
Account registers, 348
ACT (American College Test), 129
Action verbs, 374
Action words, 234–235
Active listening, 148–149
Actuaries, 186
ADA (Americans with Disabilities Act), 285
Addends, 176, 392
Addictions, 313
Addition, 176, 212, 392–395
    by carrying numbers, 392–393
    of decimals, 394
    of fractions, 394–395
Advanced courses, 125
Advanced Placement (AP) courses, 125
Advancement, 263–265
    earning raises or promotions, 264–265
    importance of, 263
Advertisements, 222–223
Age
    of business owners, 79
    requirements for bank accounts, 347
Aggressiveness, 170
Aggressive speech, 145
Agriculture
    career and technical student organizations, 135
    technology in, 199
Agriculture, Food & Natural Resources career cluster, 55
Alcohol abuse, 313
Algorithms, 195
Alternative energy, 195
American College Test (ACT), 129
American Red Cross, 307
Americans with Disabilities Act (ADA), 285
Anger management, 169
Annual fees, 360
Annual percentage rate, 360
Annual raises, 264
Antivirus and anti-spyware programs, 368
AP (Advanced Placement) courses, 125
Apologies, 169
Apostrophes
    for contractions, 384
    for possessives, 383–384
Appearance
    for job interviews, 238
    judging attitude based on, 14
Applying for jobs, 235–236
    filling out applications, 236
    personal information cards, 236
    tips for, 235
Apprentices and apprenticeships, 126, 133, 303
Apps, 203
Architecture & Construction career cluster, 56
    pathways, 56
    sample occupations, 56, 86
Area charts, 187
Armed Services Vocational Aptitude Battery (ASVAB), 129
Articulation agreements, 125
Arts, Audio/Video Technology & Communications career cluster, 57
    pathways, 57
    sample occupations, 57, 165
Assertiveness, 170
Assertive speech, 145
Asset taxes, 364
Associate's degrees, 126
ASVAB (Armed Services Vocational Aptitude Battery), 129
ATMs (Automated Teller Machines), 351–352
Attitude
    conflict and, 168
    defined, 14
    judging based on appearance, 14
    negative, 14
    positive, 14
Auditors, 58
Auditory learners, 118
Authorizing, 337
Automated Teller Machines (ATMs), 351–352
Automation software, 194
Average (mean), 124, 184, 299

B

Bachelor's degrees, 126
Balance (bank account), 337
    checking, 351
    earning interest on, 346
    reconciling accounts, 350–351
Balance (in life), 294–300
    balancing relationships, 295–296
    balancing roles, 294–295
    balancing work and leisure, 298
    career choices and, 296–297
Bank accounts, 346–352
    age requirements for, 347
    balancing, 350–351
    checking accounts, 346, 348–349, 351
    debit cards, 349
    defined, 346
    opening, 347
    overdrawn accounts, 337, 344
    recording transactions, 350
    savings accounts, 346, 351, 353–354
    transactions, 347–348
    types of, 346
Bankruptcy, 363
Banks, 344–345. *See also* Bank accounts
    ATMs, 351–352
    branches, 345
    choosing, 345
    defined, 344
    fees, 345
    online banking, 339, 345
    regulation of, 369
    types of, 344–345
Bar charts, 187, 194

415

Bartering, 320
Base pay, 242
Beneficiaries, 244
Benefits, 243
Biometric time clocks, 43
Birth certificates, 251, 347, 367
Blogs, 88, 142, 309
BLS (U. S. Bureau of Labor Statistics), 65, 72–73, 242. *See also* Career trends
Body language, 150–152
   negative, 151–152
   overcoming miscommunication, 155
   positive, 151–152
Bonds, 355
Bonuses, 242
Boolean operators, 200
BPA (Business Professionals of America), 134
Branches, 345
Brokers, 358
Budgets, 326–332
   for business owners, 331
   deficits, 332
   defined, 326
   fixed versus flexible, 328
   parts of, 327
   setting up, 328–329
   skills related to, 327
   sticking to, 331
   unexpected expenses, 330
Businesses. *See also* Business opportunities; Entrepreneurs and entrepreneurship
   age of business owners, 79
   defined, 78
   education of business owners, 81
   turning opportunities into, 90
Business knowledge, 85
Business Management & Administration career cluster, 58
   pathways, 58
   sample occupations, 17, 58, 221
Business opportunities
   defined, 88
   evaluating, 89
   finding ideas for, 88
   turning ideas into, 88
   turning into businesses, 90
Business Professionals of America (BPA), 134
Business skills, 84

C

Cafeteria-style (flexible) benefits, 243
Calculators, 181
Calories, 312
Cancer, 313
Capital (nonhuman) resources, 30
Career and job fairs, 222
Career and Technical Student Organizations (CTSOs), 134–135
CareerBuilder.com, 220
Career centers and agencies, 221–222
Career changes, 265–269
   finding new jobs, 268–269
   layoffs, 267–268
   leaving jobs, 266
   putting career on hold, 269
Career clusters, 53–76
   Agriculture, Food & Natural Resources, 55
   Architecture & Construction, 56, 86
   Arts, Audio/Video Technology & Communications, 57, 165
   Business Management & Administration, 17, 58, 221
   Education & Training, 59, 131
   Energy, 68
   Finance, 60, 186, 332, 358
   Government & Public Administration, 61, 147
   Health Science, 62–63, 264, 281
   Hospitality & Tourism, 47, 64
   Human Services, 65, 241
   Information Technology, 66, 104, 198, 300
   Law, Public Safety, Corrections & Security, 67
   Manufacturing, 68
   Marketing, 69
   Science, Technology, Engineering & Mathematics, 33, 70
   Transportation, Distribution & Logistics, 71
Career counselors, 17, 100
Career planning, 209–226
   career plans, 210–211
   career-related goals, 210
   career self-assessments, 213–217
   events that change, 296
   resources for, 218–223
   steps for, 210
Career plans, 210–211
   defined, 211
   flexibility of, 211

   what to include in, 211
Career portfolios, 16, 230
Career profiles
   actuaries, 186
   brokers, 358
   college admissions officers, 131
   computer software engineers, 300
   computer support specialists, 104
   electricians, 86
   employment and placement managers, 17–18
   environmental scientists, 33
   financial planners, 332
   flight attendants, 47
   human resources managers, 221
   image consultants, 241
   information technology managers, 198
   nurses, 63
   occupational therapists, 281
   politicians, 147
   veterinarians, 264
   video directors, 165
Career requirements, 4–5
Career rewards, 5
Careers. *See also* Career planning; Career profiles; Career trends
   changing, 15
   defined, 4
   employer expectations, 275
   exploring, 85
   identifying through job-related activities, 72
   most dangerous, 287
   origin of word, 5
   planning for, 5
   qualifications for, 65
   reasons for choosing, 63
Career self-assessments, 213–217
   critical thinking, 213
   developing worksheets for, 213–214
   identifying abilities, 215
   identifying interests, 216
Career trends, 44–49
   accountants and auditors, 58
   accounting, 328
   changing lifestyles, 46
   direct deposit, 346
   diversity in the workforce, 49
   economic trends, 45
   entrepreneurship, 48
   environmental science industry, 33
   flexible benefits, 243
   flexible lifestyles, 47–48

Index

growing industries, 45
healthcare-related occupations, 14
impact of technology, 46
informational interviews, 73
information technology, 198
job jumpers, 254
math skills, 177
nontraditional occupations, 48
nursing, 63
peer reviews, 256
population shifts and, 48
self-reviews, 256
stress management, 310
teams, 163
writing, 142
Cash, 336–337
Cell phones, 202
Central banks, 345
Certificates of deposit (CDs), 353
Certification, 127
Challenge, as work-related value, 6
Change, counting, 180
Character, 11–14
    benefits of, 13
    defined, 11, 83
    developing, 12
    entrepreneurs and, 13
    personal qualities, 11–12
    positive attitude, 14
    recognizing, 12
Charts (graphs), 187, 194
    defined, 187
    jobs that use, 187
    types of, 187
Chat rooms, 295
Checking accounts, 346, 348–349, 351
Checks, 333–337
    cashing promptly, 333
    defined, 320
    filling out properly, 348–349
    paychecks, 333–335
Children, choosing to have, 299
Cigarette smoking, 313
Circle (pie) charts, 187
Citizenship
    good, 304–305
    rights and, 302
Civic responsibility, 305
Civil Rights Act of 1964, 285
Claims on insurance, 244
Clarity, in verbal communication, 143
Clauses, 377–378
    dependent, 377–378
    independent, 377–378

Clocking-in/clocking-out, 43
Cloud computing, 203
COBRA, 268
Co-curricular activities, 122
Collaborative software, 193
Collateral, 359
Colleagues, communication with, 144
College admissions officers, 131
College entrance exams, 212
Colleges and universities, 127–129, 301
    applying to, 128
    choosing, 128
    college-ready assessments, 129
    paying for, 130
    planning for, 5
    researching, 127
    types of, 128
Colons, 386
Column charts, 187
Commas, 381–383
Comma splices, 379
Commissions, 242, 245
Commodities, 355
Communication, 139–158
    characteristics of good communicators, 147
    conflict management, 282
    entrepreneurs and, 84
    listening skills, 148–149
    nonverbal, 150–152
    obstacles to, 153–155
    relationships and, 162
    skills in, 84, 125
    verbal, 143–147
    written, 140–142
Communication barriers, 154
Community, 304–306
    civic responsibility, 305
    defined, 304
    entrepreneurs and, 85
    environment, 306
    volunteering, 305
Community colleges, 128
Community resources, 30, 73
Compassion, 12
Compensation. *See also* Salaries and wages
    defined, 242
    negotiating, 244
Compliments, 161
Compound subjects, 381
Computer applications (programs), 203
Computers. *See also* Internet; Technology

entrepreneurs and, 84
skilled workers and, 196
transferable skills, 201
Computer software engineers, 300
Computer support specialists, 104
Conflict, 168–171, 282–283
    anger management, 169
    asserting yourself, 170
    attitude and, 168
    causes of, 169, 171, 283
    communication and, 282
    defined, 168
    peer pressure, 283
    resolving, 169–170
    team, 170
    types of, 168
    unresolved, 171
    valuing differences, 282
Consequences, 98–101
    defined, 98
    predicting, 100
Constructive criticism, 141–142
Consumer Credit Protection Act (Truth in Lending Act), 369
Consumption taxes, 364
Contractions, 384
Contracts, 359
Coping skills, 309
Cosigners, 359
Cost/benefit analysis, 89
Courage, 83
Cover letters, 142, 186, 228–229
Co-workers. *See* Relationships
Creativity, as work-related value, 6
Credit, 357–363
    credit cards, 359–360
    credit reports, 361–362
    defined, 338, 346, 357
    establishing, 360–361
    loans, 359
    protecting, 361
    regulation of, 369
Credit Card Act of 2009, 369
Credit cards, 338–339, 359–360
    average debt associated with, 363
    benefits and drawbacks of, 336
    unauthorized charges, 368
Credit limits, 360
Creditors, 357
Credit reports, 361–362
Credit unions, 345
Critical reading, 140–141
Critical thinking, 107–108
    defined, 107

emotions and, 108
objective versus subjective thinking, 107
Criticism
constructive, 141–142
defined, 15
CTSOs (Career and Technical Student Organizations), 134–135
Culture. *See* Diversity; Multi-cultural perspective
Curiosity, 83
Currency, 185, 321, 344

D

Dashes, 387–388
Data, defined, 184
Data analysis (statistics), 184
Debates, 146
Debit cards, 336, 338–339, 349
Debits, 346
Debt
average amount of, 363
defined, 82, 357
inevitable, 362
managing, 362–363
Debtors, 357
DECA, 134
Decimals, 178–179, 391
addition of, 394
subtraction of, 394
Decision-making, 98–101
for college and career readiness, 99
consequences of, 98, 100–101
entrepreneurs and, 84
financial, 323–324
hurtful, 99
overwhelmed by options, 100
process for, 98–99
uniqueness of decisions, 98
Deductibles, 244
Deductions (withholding), 333–335
Deference, 144
Deficits, 332
Demographics, 49
Denomination (face value), 320
Denominators, 179, 394, 397
Dependent clauses, 377–378
Deposits, 347
Deposit slips, 350
Derogatory marks, 361
Devil's advocate, 146
Difference, in subtraction, 117, 176, 393
Differences, valuing, 282
Direct deposit, 346

Disability insurance, 288
Discipline, 83
Discrimination, 284–285
Diversity
communication barriers, 154
trends in, 49
Dividends, 177
Division, 177
Divisors, 177
Docents, 133
Dropping out, 116
Dual-income families, 46
Dysfunctional relationships, 161

E

E-commerce, 196
Economics, defined, 42
Economic trends, 45
Economic value, 42
Economy, defined, 42
Education & Training career cluster, 59
pathways, 59
sample occupations, 59, 131
Educational value, 43–44
Education and training, 259–262. *See also* Academic planning; Colleges and universities
beyond high school, 66
developing new skills, 260
entrepreneurs and, 85
high school, 120–121
lifelong learning, 261–262, 301–303
as lifestyle factor, 38
mentors, 261
postsecondary education, 127–130, 353
value of school, 116–117
EEOC (Equal Employment Opportunity Commission), 285
EFTs (electronic funds transfers), 339, 348
Elapsed time, 7
Electives, 125
Electricians, 86
Electronic communication, 142
Electronic funds transfers (EFTs), 339, 348
Em dashes, 387
Emotions
critical thinking and, 108
emotional health, 308
emotional value of work, 43
emotional well-being, 26

Empathy, 83
Employability, 15–16
building employability skills, 16
characteristics of, 15
defined, 15
transferable skills, 16
Employee handbooks, 251
Employees, defined, 78
Employment and placement managers, 17–18
Employment scams, 232
Emulating, 12
Encryption, 339
En dashes, 387–388
Endorsements, 334
Energy career cluster, 68
Entrepreneurial thinking, 77–80, 134
Entrepreneurs and entrepreneurship, 77–94
age of business owners, 79
in Architecture & Construction career cluster, 56
career self-assessments, 217
career trends and, 48
character qualities of entrepreneurs, 13
communication skills, 153
education of business owners, 81
entrepreneurial thinking, 77–80, 134
keys to successful, 85
opportunities for, 88–90
qualities of entrepreneurs, 83
relationships and, 160
rewards of, 80–81
risks of, 80, 82
skills of, 84
technology and, 200
value of work and, 43
Environment
as lifestyle factor, 38
positive work environment, 42
taking care of, 306
as work-related value, 42
Environmental scientists, 33
Environmental well-being, 26
Equal Credit Opportunity Act, 369
Equal Employment Opportunity Commission (EEOC), 285
Equal Pay Act of 1963, 285
Equifax, 362
Equity investments, 354
Estimating, 328
Ethics
defined, 9, 167

Index

ethical relationships, 167
   technology and, 167
   work ethics, 167, 276–277
Exchange rates, 80, 185
Exclamation points, 385
Exercise, 311–312
Exit interviews, 266
Expecting too much, 162
Expenses, 327
Experian, 362
Experience, entrepreneurs and, 84
Eye contact, 151

F

Facebook, 220
Face value (denomination), 320
Factors, 177
Fair Credit Billing Act, 369
Fair Credit Reporting Act, 362, 369
Family, Career and Community Leaders of America (FCCLA), 16, 72, 134
Family and Medical Leave Act (FMLA), 285
Farming
   career and technical student organizations, 135
   technology in, 199
FBLA-PBL (Future Business Leaders of America-Phi Beta Lambda), 134
FCCLA (Family, Career and Community Leaders of America), 16, 72, 134
Federal Deposit Insurance Corporation (FDIC), 353, 369
Federal Reserve, 320, 369
Federal Trade Commission (FTC), 367, 369
Fees, 345
FFA Organization, 135
FICA, 334
Finance career cluster, 60
   pathways, 60
   sample occupations, 60, 186, 332, 358
Financial aid, 130
Financial decisions, 323
Financial goals, 324–326, 329
   prioritizing, 325
   short-term and long-term, 325
   spending and saving, 324, 326
Financial management. *See* Personal financial planning; Personal money management

Financial needs and wants, 322–324
   changing, 323
   decision-making and, 323–324
   establishing, 322
Financial planners, 332
Financial planning. *See* Personal financial planning
Financial skills, 85
Fixed budgets, 328
Fixed income investments, 355
Flexible (cafeteria-style) benefits, 243
Flexible (variable) budgets, 328
Flexible standards, 8
Flight attendants, 47
FMLA (Family and Medical Leave Act), 285
Foreign countries. *See* Multi-cultural perspective
Formulas, 101
401(k) plans, 356
403(b) plans, 356
Four-year colleges, 128
Fraction bar, 179
Fractions, 178–179, 212
   addition of, 394–395
   multiplication of, 397
   subtraction of, 394–395
Fragments, 377–378
   correcting, 378
   proofreading for, 378
Franchisees, 90
Franchises, 90
Franchisors, 90
Frequency, 184
FTC (Federal Trade Commission), 367, 369
Fungibility, 197
Fused sentences, 379
Future Business Leaders of America-Phi Beta Lambda (FBLA-PBL), 134

G

Gas taxes, 365
Gender
   as communication barrier, 154–155
   discrimination and harassment, 284–285
   nontraditional occupations and, 48
General strategies, 119
Geometry, 178
Giving notice, 266
Global warming, 31
Goals and goal-setting, 104–106

   career-related goals, 210
   financial, 324–326, 329
   goals, defined, 104
   long-term goals, 106
   meeting career goals, 117
   milestones, 106
   plans and, 106
   process for, 105
   short-term goals, 106
Goods, defined, 42
Google Calendar, 279
Government. *See also names of specific government agencies*
   job search resources, 220–221
   use of taxes, 365
Government & Public Administration career cluster, 61
   pathways, 61
   sample occupations, 61, 147
Grace periods, 360
Grade point average (GPA), 124
Grammar, 374–381
   clauses, 377–378
   fragments, 377–378
   phrases, 375–377
   run-on sentences, 379–380
   subjects, 374, 380–381
   verbs, 374–376, 380–381
Graphical representation, 152
Graphs. *See* Charts
Greenpeace, 307
Gross income, 335
Gutenberg, Johannes, 140

H

Habitat for Humanity International, 307
Harassment, 284–285
Health, 293–316
   avoiding risky behaviors, 313
   balanced lifestyle, 294–300
   community, 304–306
   emotional, 308
   lifelong learning, 301–303
   mental, 308
   nutrition, 310–311
   physical, 307
   physical activity, 311–312
   stress management, 308–309
   well-being, 26–28
Healthcare-related occupations, 14
Health Occupations Students of America (HOSA), 16, 72, 134
Health Science career cluster, 62

pathways, 62
sample occupations, 62–63, 264, 281
Heifer International, 307
Helping verbs, 376
Help wanted ads, 222–223
High school, 120–121
choosing, 121
meeting graduation requirements, 123
opportunities, 121
preparing for, 120–121
unknowns, 121
High technology, 194–199
in agriculture, 199
careers in, 195
in e-commerce, 196
in information technology systems, 197
in manufacturing, 195–196
Honesty, 83
HOSA (Health Occupations Students of America), 16, 72, 134
Hospitality & Tourism career cluster, 64
pathways, 64
sample occupations, 47, 64
Humane Society, 307
Human resources, 30
Human resources managers, 221
Human Services career cluster, 65
pathways, 65
sample occupations, 65, 241
Hydraulic arms, 195
Hyphens, 386–387

I

IB (International Baccalaureate) courses, 125
Identity theft, 367
Illiteracy, 141
Image consultants, 241
Impulse purchases, 325
Impulsiveness, 100
Income, 327
earning more, 330
gross, 335
net, 335
taxable, 365
Income taxes
defined, 364
filing returns, 366
paying, 364
Income tax returns, 366
Indecisiveness, 100
Independence, 6

Independent clauses, 377–378
Individual retirement accounts (IRAs), 355
Influences, 28–29
on decisions, 98
defined, 28
good versus bad, 29
major, 28–29
Informational interviews, 73
Information security, 367–368
protecting records, 367
protecting self online, 368
protecting wallet, 368
Information Technology career cluster, 66
pathways, 66
sample occupations, 66, 104, 198, 300
Information technology managers, 198
Information technology systems, 197–198
Infrastructure, 197
Initiative, 255
Installment plans, 338
Instrumental value, 6
Insurance, 243–244, 288
Insurance agents, 244
Insurance policies, 244
Integral abilities, 140
Intellectual well-being, 26
Interest rates, 345
Interests, 10
Internal Revenue Service (IRS), 366
International Baccalaureate (IB) courses, 125
Internet. *See also* Computers; Technology
blogs, 88, 142, 309
Boolean operators, 200
career searches and, 193
job search resources, 220
online banking and payment services, 339
online education, 262
protecting self online, 368
Interns and internships, 132, 303
Interviews
exit, 266
informational, 73
job, 237–241
Intrapreneurship, 78
Intrinsic value, 6
Inventions, 90
Investment banks, 345

Investments, 354–357
money management services, 357
retirement planning, 355–356
versus saving, 353
IRAs (individual retirement accounts), 355
Irradiation, 199
IRS (Internal Revenue Service), 366

J

Job applications, 235–236
defined, 236
filling out, 236
personal information cards, 236
practicing filling out, 236
tips for, 235
Job interviews, 237–241
defined, 237
following up on, 239
interviewers, 240
making the most of, 238
practicing for, 238
preparing for, 237
results of, 240–241
telephone interviews, 239
thank-you notes, 239
Job jumpers, 254
Job leads, 218
Job offers, 242–245
accommodations, 245
benefits, 243
factors to consider, 243
insurance, 243–244
negotiating compensation, 244
pay, 242
Jobs, defined, 4
Job satisfaction, 211
Job searches, 227–248
applying for jobs, 235–236
evaluating job offers, 242–245
job interviews, 237–241
length of, 230–231
materials for, 228–231
resumes, 233–234
Job search resources, 218–223
career and job fairs, 222
career centers and agencies, 221–222
help wanted ads, 222–223
networking, 218–220
online, 220
Job security, 6, 47
Job shadowing, 133
Job-sharing, 48
Junior colleges, 128

Index

L

Labor market, 44
Language arts, 373–388
    grammar, 374–381
    punctuation, 381–388
Late fees, 360
Law, Public Safety, Corrections & Security career cluster, 67
Layoffs, 267–268
LCD (least common denominator), 212
Leaders and leadership, 110–111
    defined, 110–111
    effective, 111
    opportunities for, as work-related value, 6
    teamwork and, 165, 255
Learning styles, 118–119
    auditory learners, 118
    study strategies, 119
    tactile learners, 119
    visual learners, 118
Least common denominator (LCD), 212
Leisure, 297–298
Letters
    cover, 142, 186, 228–229
    of resignation, 266
Licenses, 127
Lifelong learning, 261–262
Lifestyle factors, 38
Lifestyle goals
    realism of, 40
    setting, 39
Lifestyles, 294–300
    balancing relationships, 295–296
    balancing roles, 294–295
    balancing work and leisure, 298
    career choices and, 296–297
    changing, 46
    defined, 26
    flexible, 47–48
    future lifestyle changes, 299–300
    needs versus wants, 40–41
    trade-offs, 39
Line charts, 187
LinkedIn, 220
Linking verbs, 375
Listening, 144, 148–149
    hearing versus, 148
    on the job, 149
Listening strategies, 119
Loans, 60, 359
    for college, 130
    involving friends and family, 357

Local time, 280
Location
    as lifestyle factor, 38
    urban versus rural population, 39
Long-term goals, 106, 123–124, 325
Loyalty, 275

M

Majority, 162
Management, 109–111. *See also* Personal money management
    anger, 169
    conflict, 168–171
    effective managers, 109
    leadership, 110–111
    multi-cultural perspective, 110
    time, 278–281
Managers
    defined, 109
    effective, 109
    meaning of title, 110
Manufacturing
    defined, 195
    high technology in, 195–196
Manufacturing career cluster, 68
Marketing career cluster, 69
Marriage, 299
Math skills, 124, 175–190, 389–397
    addition, 176, 392–395
    calculators, 181
    charts (graphs), 187
    data analysis, 184
    decimals, 178–179, 391
    developing in school, 183
    division, 177
    entrepreneurs and, 84
    fractions, 178–179
    geometry, 178
    multiple ways to solve problems, 180
    multiplication, 177, 396–397
    numbers, 390–391
    percentages, 178, 180, 391
    place value, 390
    probability, 184–185
    proportions, 181, 183
    ratios, 181–182
    rounding whole numbers, 390
    subtraction, 176, 392–395
Mean (average), 124, 184, 299
Measurement systems, 178
Median, 184
Memory strategies, 119
Mental health, 308. *See also* Emotions
Mentoring programs, 133

Mentors, 261
Merit raises, 264
Metric system, 178
Microsoft Office Excel, 181
Microsoft Office Outlook, 279
Microsoft Office Word, 232
Milestones, 106
Military service, 126
Minimum payments, 360
Minority, 162
Minuends, 393
Mode, 184
Monetary compensation, 242
Money. *See also* Payment methods; Personal financial planning; Personal money management; Salaries and wages; Saving
    currency, 185, 321, 344
    defined, 320
    purchasing power, 321
    as resource, 32
    value of, 320–321
    worrying about, 321
Money management services, 357
Money market accounts, 354
Monster.com, 220
Moral values, 6
Mortgages, 344
Motivators, 40
Multi-cultural perspective
    college entrance exams, 212
    communication barriers, 154
    currency, 185, 321, 344
    entrepreneurial thinking, 80
    management, 110
    measurement systems, 178
    military service, 126
    multilingual employees, 11
    nonverbal communications, 150
    One Laptop per Child program, 200
    relationships, 170
    resources, 31
    rights, 302
    sweatshops, 245
    time zones and local time, 280
    vacation, 298
    volunteering overseas, 260
    working abroad, 55
    work week, 40
    World Bank, 363
Multilingual employees, 11
Multiplication, 177, 396–397
Mutual funds, 355
MyPlate, 311

N

National Association of State Directors of Career Technical Education Consortium (NASDCTEc), 53–54. *See also* Career clusters
National Credit Union Association (NCUA), 353
National Young Farmer Educational Association (NYFAE), 135
Natural resources, 30
Nature Conservancy, 307
NCUA (National Credit Union Association), 353
Needs
   defined, 40
   financial, 322–324
   versus wants, 40–41
Negative attitude, 14
Negative body language, 151–152
Negotiating, 244
Net income, 335
Networking, 218–220
New jobs, 250–255
   avoiding feeling confused, 250
   building relationships, 253–254
   finding, 268–269
   forms, 250–251
   leadership, 255
   making good impressions, 252
   making less stressful, 250
   orientation, 251
   probation period, 252
   teamwork, 255
Nonessential information, 382
Nonhuman (capital) resources, 30
Nonprofit, 345
Nonrenewable resources, 30
Nontraditional occupations, 48
Nonverbal communication, 150–152
   multi-cultural perspective, 150
   negative body language, 151–152
   positive body language, 151
   visual aids, 152
Note-taking strategies, 119
Numerators, 179, 394, 397
Nurses, 63
Nutrients, 310
Nutrition, 310–311
NYFAE (National Young Farmer Educational Association), 135

O

Obesity, 312
Objectivity, 107
OCC (Office of the Comptroller of the Currency), 369
Occupational Outlook Handbook (OOH), 48, 72–73
Occupational Safety and Health Act, 289
Occupational Safety and Health Administration (OSHA), 289
Occupational therapists, 281
Occupations, defined, 4
Office of the Comptroller of the Currency (OCC), 369
One Laptop per Child program, 200
Online banking, 339, 345
OOH (*Occupational Outlook Handbook*), 48, 72–73
Opportunities
   career-related, 132–135
   entrepreneurship, 88–90
   leadership, 6
Opportunity-cost analysis, 89
Oral presentations, 146–147
Organizational skills, 84
Organization strategies, 119
Orientation, 250–251
OSHA (Occupational Safety and Health Administration), 289
Overdraft fees, 344
Overdrawn, 337
Over-limit fees, 360

P

Parentheses, 388
Passbook accounts, 353
Passive reading, 140–141
Passive speech, 145
Pathways, 54. *See also* Career clusters
Paychecks, 333–335
   endorsements, 334
   frequency of, 333
   information on, 334
   pay periods, 333
   pay stubs, 333–335
Payees, 337
Payment methods, 336–339
   cash, 336–337
   checks, 336–337
   credit cards, 336, 338
   debit cards, 336, 338
   electronic funds transfers, 339, 348
   installment plans, 338
PDAs (personal digital assistants), 202, 278
Peer pressure, 166, 283, 313

Peer reviews, 256
Peers, defined, 22
Pensions, 47, 356
Percentages, 178, 391
   converting fractions to, 322
   defined, 180
   finding, 80, 140, 345
   increases and decreases, 277
Perfect jobs, 44
Performance reviews, 256–259
   failing to meet standards, 259
   meaning of, 258
   requesting, 258
   what to expect from, 259
Periods, 385
Perks, 263
Personal academic plans, 122–124
   creating, 123
   long-term and short-term goals, 123–124
   purpose of, 122
   things to include in, 122
Personal appearance
   for job interviews, 238
   judging attitude based on, 14
Personal digital assistants (PDAs), 202, 278
Personal financial planning, 343–372
   bank accounts, 346–352
   banks, 344–345, 369
   credit, 357–362, 369
   debt, 357, 362–363
   information security, 367–368
   investments, 353–355
   saving, 352–354
   taxes, 364–366
Personal identification numbers (PINs), 338
Personal information cards, 236
Personal money management, 319–342
   budgets, 326–332
   goal-setting, 324–326
   money, defined, 320
   needs and wants, 322–324
   paychecks, 333–335
   payment methods, 336–339
   purchasing power, 321
   value of money, 320–321
Personal strengths, 3–20
   career requirements, 4–5
   career rewards, 5
   character, 11–14
   employability, 15–16
   interests, 10

Index

planning for college and career, 5
  strengths, 9–10
  values, 6–9
Personal well-being, 26
Phishing, 368
Phrases, 375–377
Physical activity, 6, 311–312
Physical health, 307
Physical needs, 41
Physical well-being, 26
Pie (circle) charts, 187
PINs (personal identification numbers), 338
Place value, 390
Planning. *See* Academic planning; Career planning; Personal financial planning
Planning strategies, 119
Politicians, 147
Population, 49
Portfolios
  career, 16, 230
  investment, 355
Positive attitude, 14
Positive body language, 151–152
Positivity, in verbal communication, 143
Postsecondary education, 127–130, 353
  applying to college, 128
  choosing colleges, 128
  college-ready assessments, 129
  defined, 127
  paying for college, 130
  researching colleges, 127
Posture, 151
Praise, 161
Premiums, 244
Prepositional phrases, 375–376
Prestige
  defined, 31
  as work-related value, 6
Prioritizing
  budgets, 329
  financial goals, 325
  time management, 280
Privacy, 285
Probability, 101, 184–185
Probation period, 252
Problem-solving, 102–103, 275–276
  asking for and giving help, 103
  entrepreneurs and, 84
  owning problems, 102
  problems, defined, 102
  process for, 103

Product, in multiplication, 177
Productivity
  defined, 42
  measuring, 43
Professional development, 260
Professionalism, 274–275
Professional organizations, 287
Programs (computer applications), 203
Progressive tax system, 365
Promotions, 263–265
Property taxes, 365
Proportions, 181, 183
Psychological needs, 41
Punctuality, 281
Punctuation, 381–388
  colons, 386
  commas, 381–383
  contractions, 384
  dashes, 387–388
  exclamation points, 385
  hyphens, 386–387
  parentheses, 388
  periods, 385
  possessive apostrophes, 383–384
  question marks, 385
  quotation marks, 388
  semicolons, 386
Purchasing power, 321

Q

Question marks, 385
Quicken, 330
Quotation marks, 388
Quotients, 177

R

Race and ethnicity
  average employment by, 49
  discrimination and harassment, 284–285
Raises, 263–264
Range, 184
Ratios, 181–182, 253
Reading, 140–141
Reciting, 118
Recreation, 298
References, 132, 229–230
Relating to others, 160
Relationships, 159–174. *See also* Teams and teamwork
  balancing, 295–296
  building, 253
  communication and, 162
  conflict management, 168–171

  defined, 160
  entrepreneurs and, 84–85
  ethical relationships, 167
  multi-cultural perspective, 170
  qualities of healthy, 161
  relating to others, 160
  team relationships, 163–166
  as two-way street, 161
  types of, 254
  unhealthy, 295
Reliability, 276
Renaissance, 140
Renewable resources, 30
Requirements, defined, 4
Resources, 29–32
  available, 30
  Bureau of Labor Statistics, 72–73
  community, 30, 73
  defined, 29
  job search, 218–223
  making most of, 32
  managing, 31
  money as, 32
  multi-cultural perspective, 31
  at school, 72
  sharing, 30
  types of, 30
Respect, 255
Responsibilities
  changing, 24
  defined, 4–5, 22
  different, 22
  fulfilling, 24
  rules and, 25
  showing you can/cannot meet, 4
Resumes, 233–235
  action words, 234–235
  format of, 233
  sections of, 234
  templates for, 232–233
Retail banks, 344
Retirement, defined, 47
Retirement plans, 253, 355–356
Retooling, 259
Return on investment (ROI), 353
Returns, 355
Rewards
  career rewards, 5
  defined, 4
  of entrepreneurship, 80–81
Rights, 284–289
  discrimination and harassment, 284–285
  multi-cultural perspective, 302

professional organizations, 287
safety, 288–289
unions and trade organizations, 286
Risks, 80, 82, 355
Risky behaviors, 313
Roadblocks, 27
ROI (return on investment), 353
Roles
assigned versus acquired, 23
balancing, 294–295
changing, 24
defined, 22
different, 22
main areas on life and, 23
time spent in each, 25
Room and board, 127
Rounding whole numbers, 390
Royalties, 140
RSS feeds, 295
Rule of 72, 353
Run-on sentences, 379–380
correcting, 379–380
words that often lead to, 380

S

Safe deposit boxes, 367
Safety
rights and responsibilities, 288–289
as work-related value, 6
Salaries and wages
comparing, 7
good, 6
as lifestyle factor, 38
defined, 5
wealth and, 326
Sales taxes, 364
SAT (Scholastic Aptitude Test), 129
Saving, 352–354
versus investing, 353
savings accounts, 353–354
Saving goals, 324, 326
Savings accounts, 346, 351
opening, 354
types of, 353–354
SBA (U. S. Small Business Administration), 85
Scatter charts, 187
Scholarships, 130
Scholastic Aptitude Test (SAT), 129
School. *See* Academic planning; Colleges and universities; Education and training
School clubs, 133

Science, Technology, Engineering & Mathematics career cluster, 70
pathways, 70
sample occupations, 33, 70
Screening, 239
Securities, 355
Securities and Exchange Commission (SEC), 355, 369
Self-employment, 266
Self-reviews, 256
Semicolons, 386
Services, defined, 42
Severance packages, 268
Short-term goals, 106, 123–124, 325
Sign language, 150
Skills, 97–114
critical thinking, 107–108
decision-making, 98–101
defined, 9, 83
developing in school, 124
employability, 15–16
entrepreneurial, 84
goal-setting, 105–106
listening, 148–149
management, 109–111
problem-solving, 102–103
transferable, 16
SkillsUSA, 135
Small businesses, defined, 85
Smartphones, 202, 278
Smoking, 313
Social networking sites, 220, 295
Social Security, 334, 356
Social Security cards, 347, 367
Social Security numbers (SSNs), 367
Social well-being, 26
Spelling errors, 231
Spending goals, 324, 326
SSNs (Social Security numbers), 367
Stages of life, 24
Standard of living, 322
Standards
defined, 8
failing to meet, 259
flexibility of, 8
Statistics (data analysis), 184
Stocks, 354–355
Strengths. *See also* Personal strengths
defined, 9
identifying, 9–10
Stress
defined, 82, 308
managing, 308–310
signs of, 309

Stressors, 308
Student organizations, 134–135
Study strategies, 119
Subjectivity, 107
Subjects, 374
agreement with verb, 380
compound, 381
verbs separated from, 381
Subject-verb agreement, 380
Subordinates, 144
Subordinators, 377, 379–380
Subsidies, 243
Substance abuse, 313
Subtle cues, 149
Subtraction, 176, 392–395
by borrowing numbers, 394
of decimals, 394
of fractions, 394–395
Subtrahends, 393
Sums (totals), 176, 392
Sweatshops, 245
SWOT analysis, 89

T

Tactile learners, 119
Tape measures, 183
Taxable income, 365
Tax brackets, 365
Tax deductions, 365
Tax-deferred savings plans, 355
Taxes, 364–366
age and, 334
defined, 364
filing income tax returns, 366
government use of, 365
paying income tax, 365
Tax-free holidays, 364
Tax preparation software, 366
Tax sheltered annuities, 356
Teaching certification, 59
Teams and teamwork, 163–166
behavior and success of, 163
challenges of teamwork, 163
conflict and, 170
leadership and, 255
majority rule, 162
peer pressure, 166
qualities of effective teams, 166
successful teams, 164
team leaders, 165
team members, 165
teams, defined, 163
Technical colleges, 128
Technological resources, 30

Index

Technology, 191–206. *See also* Internet
   agriculture, 199
   budgeting software, 330
   calendars and planners, 32
   career trends and, 46
   communication barriers, 153
   communications devices, 202
   community involvement, 307
   computer applications, 203
   entrepreneurship and, 81
   ethics and, 167
   high technology in industry, 194–199
   impact of, 192–194
   management and, 111
   measuring productivity, 43
   newspapers, 223
   online education, 262
   in school, 120
   secure Web sites, 339
   smartphones, 278
   socializing online, 295
   spreadsheet programs, 181
   tax preparation software, 366
   transferrable computer skills, 201
   use of in different jobs, 54
   working from home, 8
Technology Student Association (TSA), 135
Telecommuters, 193
Telephone interviews, 239
Templates, 232–233
Temp work, 46
Term, in time accounts, 353
Terms, of loans, 359
Test strategies, 119
Texas Success Initiative (TSI), 129
Texting, 202, 295
Thank-you notes, 239
Thinking
   critical, 107–108
   entrepreneurial, 77–80, 134
   relationship to verbal communication, 146
   as work-related value, 6
Time. *See also* Time management
   American work hours, 279
   amount worked by entrepreneurs, 81
   average spent in school-related activities, 23
   average spent in work-related activities, 23
   calculating hours, 7
   as lifestyle factor, 38
   spent in different roles, 25
Time accounts, 353
Time clocks, 43
Time management, 278–281
   tips for, 280–281
   tools for, 279
Time sheets, 43
Time zones, 280
Tolerance, 8
Totals (sums), 176, 392
Trade barriers, 80
Trade-offs, 39
Trade organizations, 286
Transcripts, 128
Transferable skills, 16
Transfers, 347
Transportation, Distribution & Logistics career cluster, 71
TransUnion, 362
Trends, 44. *See also* Career trends
Truth in Lending Act (Consumer Credit Protection Act), 369
TSA (Technology Student Association), 135
TSI (Texas Success Initiative), 129
Tuition, 127
Tuition reimbursement programs, 260
TurboTax, 366

U

Unemployment, 267–269
Unemployment rate, 45
Unions
   credit, 345
   labor, 286–287
Uniqueness
   of decisions, 98
   defined, 22
Universities, 128. *See also* Colleges and universities
U. S. Bureau of Labor Statistics (BLS), 65, 72–73, 242. *See also* Career trends
U. S. customary system, 178
U. S. Department of Agriculture (USDA), 311
U. S. Department of Education. *See* Career clusters
U. S. Small Business Administration (SBA), 85

V

Vacation, 298
Value of school, 116–117
   dropping out, 116
   effects on employment, 117
Value of work, 42–44
   defined, 42
   economic, 42
   educational, 43–44
   emotional, 43
   entrepreneurship and, 43
Values, 6–9
   changing, 7–8
   common, 6
   communication barriers, 154–155
   defined, 6
   ethics, 9
   flexibility in standards, 8
   showing, 9
   types of, 6–7
Variable (flexible) budgets, 328
Variety, as work-related value, 6
Verbal communication, 143–147
   delivering oral presentations, 146
   process for, 143–144
   relationship to thinking, 146
   steps for being great speakers, 146–147
   types of, 144–145
Verb phrases, 376
Verbs, 374–376
   action, 374
   helping, 376
   linking, 375
   separated from subjects, 381
   subject-verb agreement, 380
   verb phrases, 376
Veterinarians, 264
Videoconferencing, 193
Video directors, 165
Virtues, 9
Visual aids, 150, 152
Visual learners, 118
Vocational schools, 128
Volunteers and volunteering, 72
   community involvement and, 305
   volunteering overseas, 260
   volunteers, defined, 133

W

W-4 forms, 250
Wages. *See* Salaries and wages
Wants
   defined, 40
   financial, 322–324
   versus needs, 40–41

Well-being, 26–28
  contributors to, 26
  defined, 26
  promoting own, 27–28
  roadblocks to, 27
  whose to put first, 26
Wellness, 26
Wi-Fi, 194
Withdrawals, 347–348
Withdrawal slips, 350
Withholding (deductions), 333–335

Work environment
  as lifestyle factor, 38
  positive, 42
  as work-related value, 6
Work history, 132
Working abroad, 55
Working with people, as work-related value, 6
Work/leisure balance, 298
Work-related values, 6–7

Work safety
  rights and responsibilities, 288–289
  as work-related value, 6
Work-study programs, 130
Work week, 40
World Bank, 363
Written communication, 140–142
  electronic communication, 142
  reading, 140–141
  writing, 141–142

Photo Credits

Title page: olesiabilkei/Fotolia; Andres Rodriguez/Fotolia; michaeljung/Fotolia; Monkey Business/Fotolia

Part I: p 1 ryanking999/Fotolia

Chapter 1: p 3 Patrick Hermans /Fotolia; p 4 imagedb.com/Fotolia; p 5 wellphoto/Fotolia; Monkey Business/Fotolia; p 6 Monkey Business/Fotolia; p 7 Minerva Studio/Fotolia; Tsiumpa/Fotolia; p 8 Darren Baker/Fotolia; ra2 studio/Fotolia; p 9 Gabriel Blaj/Fotolia; p 10 Nick Freund/Fotolia; EastWest Imaging/Fotolia; p 12 Monkey Business/Fotolia; p 14 Jenner/Fotolia; p 15 Andres Rodriguez/Fotolia; p Rido/Fotolia; p 17 viappy/Fotolia; emde71/Fotolia; Monkey Business/Fotolia; Andres Rodriguez/Fotolia; vadymvdrobot/Fotolia; p 18 ryflip/Fotolia; p 19 Andres Rodriguez/Fotolia; Maya Kruchancova/Fotolia

Chapter 2: p 21 Darrin Henry/Fotolia; p 22 chunumunu/Fotolia; p 23 Africa Studio/Fotolia; p 24 ndoeljindoel/Fotolia; p 26 juiceteam2013/Fotolia; p 27 michaeljung/Fotolia; p 28 zoomyimages/Fotolia; micromonkey/Fotolia; p 29 Marco Govel/Fotolia; p 31 serranostock/Fotolia; p 32 Tsiumpa/Fotolia; ra2 studio/Fotolia; p 33 viappy/Fotolia; emde71/Fotolia; Monkey Business/Fotolia; Monkey Business/Fotolia; Andres Rodriguez/Fotolia; olesiabilkei/Fotolia; kerdazz/Fotolia; p 34 ryflip/Fotolia; p 35 Andres Rodriguez/Fotolia; Actionpics/Fotolia

Chapter 3: p 37 Ljupco Smokovski/Fotolia; p 38 vadymvdrobot/Fotolia; p 39 Ljupco Smokovski/Fotolia; p 40 Jelena Ivanovic/Fotolia; p 41 yanlev/Fotolia; pressmaster/Fotolia; p 42 Kurhan/Fotolia; p 43 ra2 studio/Fotolia; p 44 Klaus Eppele/Fotolia; Minerva Studio/Fotolia; p 45 Pixsooz/Fotolia; BlueOrange Studio/Fotolia; p 47 viappy/Fotolia; emde71/Fotolia; Monkey Business/Fotolia; Monkey Business/Fotolia; Andres Rodriguez/Fotolia; Andres Rodriguez/Fotolia; kerdazz/Fotolia; p 48 desaif/Fotolia; p 49 FotolEdhar/Fotolia; p 50 ryflip/Fotolia; p 51 Andres Rodriguez/Fotolia; Monkey Business/Fotolia

Chapter 4: p 53 Andres Rodriguez/Fotolia; p 54 Africa Studio/Fotolia; ra2 studio/Fotolia; p 55 José 16/Fotolia; p 56 ndoeljindoel/Fotolia; p 57 Dirima/Fotolia; p 58 FotolEdhar/Fotolia; kerdazz/Fotolia; p 59 Rido/Fotolia; desaif/Fotolia; p 60 michaeljung/Fotolia; p 61 auremar/Fotolia; p 62 wong yu liang/Fotolia; p 63 viappy/Fotolia; emde71/Fotolia; Monkey Business/Fotolia; Monkey Business/Fotolia; Andres Rodriguez/Fotolia; kerdazz/Fotolia; p 64 TessarTheTegu/Fotolia; p 65 Monkey Business/Fotolia; desaif/Fotolia; p 66 dmitrimaruta/Fotolia; p 67 aijohn784/Fotolia; p 68 ndoeljindoel/Fotolia; p 69 Ariwasabi/Fotolia; Tsiumpa/Fotolia; p 70 Semen Barkovskiy/Fotolia; p 71 CandyBox Images/Fotolia; p 72 paffy/Fotolia; p 73 FotolEdhar/Fotolia; kerdazz/Fotolia; desaif/Fotolia; p 74 ryflip/Fotolia; p 75 Andres Rodriguez/Fotolia; Web Buttons Inc/Fotolia

Chapter 5: p 77 Andres Rodriguez/Fotolia; p 78 Glenda Powers/Fotolia; kerdazz/Fotolia; p 79 Monkey Business/Fotolia; p 80 Alliance/Fotolia; p 81 Monkey Business/Fotolia; ra25 studio/Fotolia; p 82 rnl/Fotolia; p 83 rnl/Fotolia; michaeljung/Fotolia; p 84 Tsiumpa/Fotolia; p 85 Monkey Business/Fotolia; p 86 viappy/Fotolia; emde71/Fotolia; Monkey Business/Fotolia; Monkey Business/Fotolia; Andres Rodriguez/Fotolia; auremar/Fotolia; p 87 WavebreakMediaMicro/Fotolia; p 88 ryflip/Fotolia ; p 89 amanaimages/Fotolia; p. 92 ryflip/Fotolia; p. 93 Monkey Business/Fotolia; michaeljung/Fotolia; p 94 Andres Rodriguez/Fotolia;

Part II: p 95 DenisNata/Fotolia

Chapter 6: p 97 phoenix021/Fotolia; p 98 Kalim/Fotolia; p 99 Monkey Business/Fotolia; p 100 Gajus/Fotolia; desaif/Fotolia; p 101 Monkey Business/Fotolia; biglama/Fotolia; p 102 lucadp/Fotolia; chagin/Fotolia; p 104 viappy/Fotolia; emde71/Fotolia; Monkey Business/Fotolia; Monkey Business/Fotolia; Andres Rodriguez/Fotolia; nyul/Fotolia; Tsiumpa/Fotolia; p 105 Rido/Fotolia; p 106 michaeljung/Fotolia; p 107 Minerva Studio/Fotolia; p 108 Monkey Business/Fotolia; p 109 spaxiax/Fotolia; p 111 Syda Productions/Fotolia; ra2 studio/Fotolia; p 112 ryflip/Fotolia; p 113 Minerva Studio/Fotolia; michaeljung/Fotolia; p 114 Andres Rodriguez/Fotolia

Chapter 7: p 115 Hugo Félix/Fotolia; p 116 michaeljung/Fotolia; p 117 Monkey Business/Fotolia; Tsiumpa/Fotolia; p 118 Lisa F. Young/Fotolia; p 119 Arto/Fotolia; p 120 ra2 studio/Fotolia; p 122 apops/Fotolia: p 123 cfarmer/Fotolia; p 124 hartphotography/Fotolia; p 125 EdieLayland/Fotolia; p 126 Felix Mizioznikov/Fotolia; P 127 wong yu liang/Fotolia; desaif/Fotolia; p 128 SeanPavonePhoto/Fotolia; Sam Spiro/Fotolia;p. 130 desaif/Fotolia; p 13 viappy/Fotolia; emde71/Fotolia; Monkey Business/Fotolia; Monkey Business/Fotolia; Andres Rodriguez/Fotolia; George Wada/Fotolia; p 132 chunumunu/Fotolia; p 133 mangostock/Fotolia; p 136 ryflip/Fotolia; p 137 Bronwyn Photo/Fotolia; michaeljung/Fotolia; p 138 Andres Rodriguez/Fotolia

Chapter 8: p 139 eurobanks/Fotolia; p 140 Ljupco Smokovski/Fotolia; p 141 Tsiumpa/Fotolia; bst2012/Fotolia; p 142 kerdazz/Fotolia; p 143 michaeljung/Fotolia; p 144 fotosmile777/Fotolia; pressmaster/Fotolia; p 145 michaeljung/Fotolia; p 146 Andy Dean/Fotolia; p 147 xy/Fotolia; viappy/Fotolia; emde71/Fotolia; Monkey Business/Fotolia; Monkey Business/Fotolia; Andres Rodriguez/Fotolia; p 148 slasnyi/Fotolia; Wavebreakmedia-Micro/Fotolia; WavebreakmediaMicro/Fotolia; p 149 Monkey Business/Fotolia; p 150 Photographee.eu/Fotolia; p 151 Ljupco Smokovski/Fotolia; Ljupco Smokovski/Fotolia; p 152 desaif/Fotolia; Monkey Business/Fotolia; p 153 Glenda Powers/Fotolia; ra2 studio/Fotolia; p 154 DragonImages/Fotolia; p 155 michaeljung/Fotolia; p 156 ryflip/Fotolia; p 157 Andres Rodriguez/Fotolia; michaeljung/Fotolia; p 158 Andres Rodriguez/Fotolia

Chapter 9: p 159 Rido/Fotolia; p 160 Andres Rodriguez/Fotolia; p 161 desaif/Fotolia; auremar/Fotolia; p 163 Photographee.eu/Fotolia; kerdazz/Fotolia; p 164 Syda Productions/Fotolia; p 165 Minerva Studio/Fotolia; viappy/Fotolia; emde71/Fotolia; Monkey Business/Fotolia; Monkey Business/Fotolia; Andres Rodriguez/Fotolia; _robbie_/Fotolia; p 166 Tsiumpa/Fotolia; p 167 iceteastock/Fotolia; ra2studio/Fotolia; p 168 vladimirfloyd/Fotolia; Kablonk Micro/Fotolia; p 169 Gelpi/Fotolia; p 171 Minerva Studio/Fotolia; p 172 ryflip/Fotolia; p 173 JackF/Fotolia; michaeljung/Fotolia; p 174; Andres Rodriguez/Fotolia

Chapter 10: p 175 Leah-Anne Thompson/Fotolia; p 176 Nikolai Sorokin/Fotolia; p 177 xy/Fotolia; kerdazz/Fotolia; p 178 Imre Forgo/Fotolia; p 179 oneblink1/Fotolia; p 180 Tsiumpa/Fotolia; desaif/Fotolia; p 181 ra2 studio/Fotolia; Jonnystockphoto/Fotolia; p 182 alisonhancock/Fotolia; p 183 Olga Kovalenko/Fotolia; p 184 stockyimages/Fotolia; p 185 chalabala/Fotolia; p 186 viappy/Fotolia; emde71/Fotolia; Monkey Business/Fotolia; Monkey Business/Fotolia; Andres Rodriguez/Fotolia; emiliezhang/Fotolia; p 187 and.one/Fotolia; p 188 ryflip/Fotolia; p 189 Lorraine Swanson/Fotolia; michaeljung/Fotolia; Andres Rodriguez/Fotolia

Chapter 11: p 191 Andres Rodriguez/Fotolia; p 192 ryanking999/Fotolia; carlosseller/Fotolia; p 194 Tsiumpa/Fotolia; martin_matthews/Fotolia; p 195 Nataliya Hora/Fotolia; p 197 Brian Jackson/Fotolia; p 198 viappy/Fotolia; emde71/Fotolia; Monkey Business/Fotolia; Monkey Business/Fotolia; Andres Rodriguez/Fotolia; auremar/Fotolia; kerdazz/Fotolia; p 199 Johan Larson/Fotolia; ra2 studio/Fotolia; p 200 Intellistudies/Fotolia; desaif/Fotolia; p 202 ndoeljindoel/Fotolia; tororo reaction/Fotolia; p 203 Sertac Sakarya/Fotolia; p 204 ryflip/Fotolia; p 205 Coloures-Pic/Fotolia; michaeljung/Fotolia p 206 Andres Rodriguez/Fotolia

Part III: p 207 WONG SZE FEI/Fotolia

Chapter 12: p 209 auremar/Fotolia; p 210 lunamarina/Fotolia; Monkey Business/Fotolia; p 213 WavebreakmediaMicro/Fotolia; Rob/Fotolia; p214 desaif/Fotolia; p 216 Alexander Raths/Fotolia; p 218 rnl/Fotolia; p 219 nyul/Fotolia; Tsiumpa/Fotolia; p 221 Andres Rodriguez/Fotolia; viappy/Fotolia; emde71/Fotolia; Monkey Business/Fotolia; Monkey Business/Fotolia; Andres Rodriguez/Fotolia; p 222 thinglass/Fotolia; p 223 Elenathewise/Fotolia; ra2 studio/Fotolia; p 224 ryflip/Fotolia; p 225 OduaImages/Fotolia; michaeljung/Fotolia; p 226 Andres Rodriguez/Fotolia

Chapter 13: p 227 fuzzbones/Fotolia; p 228 iko/Fotolia; desaif/Fotolia; p 229 Monkey Business/Fotolia; p 230 Tsiumpa/Fotolia; p 231 Elenathewise/Fotolia; desaif/Fotolia; p 232 ra2 studio/Fotolia; p 233 vgstudio/Fotolia; p 235 Viorel Sima/Fotolia; p 236 Tanusha/Fotolia; desaif/Fotolia; p 237 WavebreakmediaMicro/Fotolia; p 238 Rob/Fotolia; p 239 Pius Lee/Fotolia; p 240 auremar/Fotolia; p 241 viappy/Fotolia; emde71/Fotolia; Monkey Business/Fotolia; Monkey Business/Fotolia; Andres Rodriguez/Fotolia; Minerva Studio/Fotolia; p 242 Monika Olszewska/Fotolia; p 243 kerdazz/Fotolia; Burlingham/Fotolia; p 244 Monkey Business/Fotolia; p 246 ryflip/Fotolia; p 247 Petr Malyshev/Fotolia; michaeljung/Fotolia; p 248 Andres Rodrigues/Fotolia

Photo Credits

Chapter 14: p 249 nyul/Fotolia; p 250 auremar/Fotolia; Scott Griessel/Fotolia; p 251 Halfpoint/Fotolia; pixs4u/Fotolia; p 252 Halfpoint/Fotolia; p 253 vgstudio/Fotolia; Monkey Business/Fotolia; p 254 kerdazz/Fotolia; mangostock/Fotolia; p 255 desaif/Fotolia; p 256 Ljupco Smokovski/Fotolia; duckman76/Fotolia; p 257 micromonkey/Fotolia; kerdazz/Fotolia; Tsiumpa/Fotolia; p 258 berc/Fotolia; p 259 Dario Lo Presti/Fotolia; desaif/Fotolia; Monkey Business/Fotolia; poco_bw/Fotolia; p 261 Monkey Business/Fotolia; p 262 Felix Mizioznikov/Fotolia; corepics/Fotolia; arekmalang/Fotolia; ra2 studio/Fotolia; p 263 vgstudio/Fotolia; Vibe Images/Fotolia; p 264 viappy/Fotolia; emde71/Fotolia; Monkey Business/Fotolia; Monkey Business/Fotolia; Andres Rodriguez/Fotolia; Ljupco Smokovski/Fotolia; p 265 doomu/Fotolia; pressmaster/Fotolia; p 267 desaif/Fotolia; p 268 Robert Kneschke/Fotolia; p 269 diego cervo/Fotolia; p 270 ryflip/Fotolia; p 271 pressmaster/Fotolia; Michaeljung/Fotolia; p 272 Andres Rodriguez/Fotolia

Chapter 15: p 273 dzimin/Fotolia; p 274 duckman76/Fotolia; p. 275 apops/Fotolia; Monkey Business/Fotolia; gwimages/Fotolia; ndoeljindoel/Fotolia; p 276 vedran79/Fotolia; auremar/Fotolia; p 277 Klaus Eppele/Fotolia; p 278 hyunsuss/Fotolia; arekmalang/Fotolia; ra2 studio/Fotolia; p 279 desaif/Fotolia; Monkey Business/Fotolia; 280 lefata/Fotolia; p 281 viappy/Fotolia; emde71/Fotolia; Monkey Business/Fotolia; Monkey Business/Fotolia; Andres Rodriguez/Fotolia; Robert Kneschke/Fotolia; p 282 David Gilder/Fotolia; p 283 WavebreakMediaMicro/Fotolia; p 285 Piotr Marcinski/Fotolia; ndoeljindoel/Fotolia; p 285 Coka/Fotolia; p 286 Andres Rodriguez/Fotolia; p 288 aleksey kashin/Fotolia; il-fede/Fotolia; Tsiumpa/Fotolia; p 289 RobertNyholm/Fotolia; desaif/Fotolia; p 290 ryflip/Fotolia; p 291 Africa Studio/Fotolia; michaeljung/Fotolia; p 292 Andres Rodriguez/Fotolia

Chapter 16: p 293 Grafvision/Fotolia; p 294 Jason Stitt/Fotolia; p 295 ra2 studio/Fotolia; p 296 Carlos Santa Maria/Fotolia; Tsiumpa/Fotolia; p 297 micromonkey/Fotolia; iofoto/Fotolia; ndoeljindoel/Fotolia; desaif/Fotolia; p 298 micromonkey/Fotolia; p 299 bst2012/Fotolia; p 300 iofoto/Fotolia; viappy/Fotolia; emde71/Fotolia; Monkey Business/Fotolia; Monkey Business/Fotolia; Andres Rodriguez/Fotolia; p 301 Warren Goldswain/Fotolia; p 302 gwimages/Fotolia; p 303 Siberia/Fotolia; p 304 opolja/Fotolia; p 305 Monkey Business/Fotolia; p 306 nyul/Fotolia; p 307 ra2 studio/Fotolia; apops/Fotolia; p 309 Monkey Business/Fotolia; p 310 kerdazz/Fotolia; Monkey Business/Fotolia; p 311 USDA; p 312 Jaren Wicklund/Fotolia; p 313 Monkey Business/Fotolia; p 314 ryflip/Fotolia; p 315 Siberia/Fotolia; michaeljung/Fotolia; Andres Rodriguez/Fotolia;

Part IV: p 317 iko/Fotolia

Chapter 17: p 319 WavebreakMediaMicro/Fotolia; p 320 slasnyi/Fotolia; selensergen/Fotolia; p 321 Konstantin Sutyagin/Fotolia; p 322 Vladimir Wrangel/Fotolia; p 323 vikarayu/Fotolia; Monkey Business/Fotolia; nyul/Fotolia; p 324 Maridav/Fotolia; p 325 Monkey Business/Fotolia; bst2012/Fotolia; p 326 gcpics/Fotolia; p 327 desaif/Fotolia; Monkey Business/Fotolia; p 328 kerdazz/Fotolia; p 329 bst2012/Fotolia; p 330 Tsiumpa/Fotolia; ra2 studio/Fotolia; p 331 Antonio Gravante/Fotolia; p 332 viappy/Fotolia; emde71/Fotolia; Monkey Business/Fotolia; Monkey Business Business/Fotolia: Andres Rodriguez/Fotolia; p 333 desaif/Fotolia; p 334 Dragonimages/Fotolia; p 335 joe/Fotolia; p 336 Odua Images/Fotolia; p 337 sanjagrujic/Fotolia; p338 diego cervo/Fotolia; p 339 Odua Images/Fotolia; ra2 studio/Fotolia; p 340 ryflip/Fotolia; p 341 bst2012/Fotolia; michaeljung/Fotolia; p 342 Andres Rodriguez/Fotolia

Chapter 18: p 343 vadymvdrobot/Fotolia; p 344 lucadp/Fotolia; p 345 luminastock/Fotolia; p 346 mast3r/Fotolia; kerdazz/Fotolia; p 347 Photographee.eu/Fotolia; p 349 iQoncept/Fotolia; p 349 vladimirfloyd/Fotolia; p 350 JohnKwan/Fotolia; Irina Brinza/Fotolia; p 351 desaif/Fotolia; HappyAlex/Fotolia; p 352 Tsiumpa/Fotolia;Tom Wang/Fotolia; p 353 desaif/Fotolia; p 354 diego cervo/Fotolia; p 355 Tomasz Zajda/Fotolia; p 356 gwimages/Fotolia; p. 357 snowwhiteimages/Fotolia; p 358 viappy/Fotolia; emde71/Fotolia; Monkey Business/Fotolia; Monkey Business/Fotolia; Andres Rodriguez/Fotolia; Minerva Studio/Fotolia; p 359 Monkey Business/Fotolia; p 360 Adrin Shamsudin/Fotolia; Tsiumpa/Fotolia; p 361 alexskopje/Fotolia; p 364 Creativa/Fotolia; Blue Moon/Fotolia; p 365 diego cervo/Fotolia; p 366 pkstock/Fotolia; ra2 studio/Fotolia; p 367 alexyndr/Fotolia; dyoma/Fotolia; p 368 Innovated Captures/Fotolia; p 369 apops/Fotolia; p 370 ryflip/Fotolia; p 371 Samir_/Fotolia; michaeljung/Fotolia; p 372 Andres Rodriguez/Fotolia;